MINERAL RECOGNITION

MINERAL RECOGNITION

IRIS VANDERS

PAUL F. KERR

NEWBERRY PROFESSOR OF MINERALOGY, COLUMBIA UNIVERSITY

John Wiley & Sons, Inc., New York | London | Sydney

67/   3812

Dedicated to
Helen Kerr
and to the Memory of
Florence Victor

Preface

This book is intended for the mineral collector, the nonprofessional reader with an interest in minerals, the beginning student of mineral science, the geologist, or the scientist in related fields. It is designed to provide an introductory background in mineralogy as well as a procedure for the rapid recognition of common minerals. The descriptions and illustrations are intended to establish a familiarity with the more important mineral species sufficient to enable any one not previously trained in the subject to become reasonably proficient in observational ability. Precision in identification ordinarily requires instruments such as a polarizing microscope, x-ray apparatus, or physical testing equipment. However, many mineral specimens exhibit distinctive features that facilitate reasonably accurate identification on the basis of carefully considered visual data.

The greater part of the book attempts to develop the criteria for rapid recognition. When possible, the explanations of phenomena are simplified, but in no place is there any intention to sacrifice accuracy for the mere sake of simplicity. The book may either stand by itself as an introductory text or serve as an introduction to well-known textbooks on mineralogy.

A supplementary discussion provides an outline of the history, nature, and utilization of minerals, as well as an introduction to the role of mineralogy as a science. Some aspects of crystal growth are discussed in the hope that such a discussion will stimulate further interest in this fascinating and challenging field. Data on crystallography which constitute an important part of standard texts are greatly abbreviated, and technical terms considered essential are explained either in the text or in footnotes.

The numerous photographs, diagrams, and other illustrations are intended to furnish a realistic representation of minerals and mineralogical features. A great effort has been made to provide color photographs adequately clear to enable an experienced mineralogist to recognize the specimens shown without reference to the accompanying captions. At the same time, the student, or any one with a newly awakened interest in minerals, will find the photographs an important reference library, one that facilitates comparison in the identification of unknown specimens. Photographs are intended to exhibit maximum color fidelity and clarity, and to this end over 1000 color photos of minerals were taken in order to obtain those included in the text. The mineral specimens photographed are from the Columbia University Research Collection, The American Museum of Natural History, and the U.S. National Museum. These collections contain some of the finest mineral specimens to be observed.

Identification tables are provided to show a systematic tabulation of data. They are based on such observational properties as crystal habit, color, luster, hardness, cleavage, and specific gravity.

The illustrations made available by individuals, museums, and companies are credited in the text. In addition, the authors wish to thank Dr. Clifford Frondel for reviewing the entire manuscript and for his suggested improvements.

IRIS VANDERS
PAUL F. KERR

Columbia University
New York

Contents

CHAPTER 3. CRYSTAL CHEMISTRY   37

CHAPTER 4. SYMMETRY OF MINERAL CRYSTALS   57

CHAPTER 5. PHYSICAL PROPERTIES OF CRYSTALS   75

CHAPTER 6. CHEMICAL TESTS   95

CHAPTER 7. MINERAL IDENTIFICATION TABLES   107

CHAPTER 8. MINERAL DESCRIPTIONS   154

1.

Minerals Past and Present

The first human who noticed a lump of bright yellow gold in a stream course, picked it out, balanced it in his hands, noticed its weight, and beat it between two rocks to mould it was engaged in mineral recognition. However, we have come a long way since. Modern science provides sophisticated equipment for making recognition more reliable. Yet even as in the earliest times a keen eye and careful observation are still important working prerequisites for anyone interested in minerals.

The utility, peculiarity, and even rare beauty of minerals have merited attention since the beginning of human activity. Before we undertake the problems of recognizing minerals, it is of interest to examine their nature, consider their distribution, and inquire briefly into the impact of minerals on both modern and ancient society.

NATURE OF A MINERAL

Minerals have been described as "Nature's Chemicals." More precisely, a mineral is A NATURALLY OCCURRING SOLID* SUBSTANCE—SUCH AS AN ELEMENT, A SOLID SOLUTION, OR A COMPOUND—OF INORGANIC ORIGIN.

*A few exceptions, such as mercury, may be liquid.

Minerals are homogeneous, and are distinct from mixed heterogeneous assemblages, such as rocks and ores.

Two fundamental features ordinarily characterize a mineral: (1) A chemical composition, expressed by a chemical formula; and (2) a distinctive crystallization (atomic structure). Many minerals belong to groups with a range in both chemical composition and internal structural dimensions, but chemical composition and arrangement of atoms still remain the two fundamental factors in precise mineral identification.

Substances synthesized in chemical laboratories may have the same atomic structure and chemical composition as minerals, but such products are artificial in origin and, hence, are not minerals. For example, many of the gemstones are produced synthetically and are thus identical in composition and structure with naturally occurring gem minerals. These man-made gems, however, are referred to as synthetic emerald, synthetic ruby, synthetic sapphire, synthetic diamond, and so on.

Minerals are formed by inorganic processes, and, strictly speaking, substances—such as coal, amber, pearls, and asphaltite—that originate in plant or animal life, are not minerals, although they do occur naturally and although some do conform structurally and chemically to minerals.

1

DISTRIBUTION OF MINERALS

Minerals are made up of chemical elements. These are derived from the earth, the oceans, or the atmosphere, or fall from outer space in meteorites. Among the 103 known elements, only a few are predominant. The 12 most abundant elements found in the surface portion of the earth are indicated graphically in Figure 1-1. Relative abundance, however, is an unreliable index of the availability of any element for industrial use, for most are widely disseminated in small amounts, or occur in economically unrecoverable mixtures.

Some minerals, such as quartz and feldspar, are widespread in occurrence. Others, such as platinum and diamond, are limited and are concentrated only in occasional locations. Most mineral deposits are distinctly local accumulations and do not occur uniformly over wide areas. The molybdenum deposit at Climax, Colorado, occupies an area of less than one square mile, yet for many years it has supplied about 85 percent of the world's molybdenum. A large part

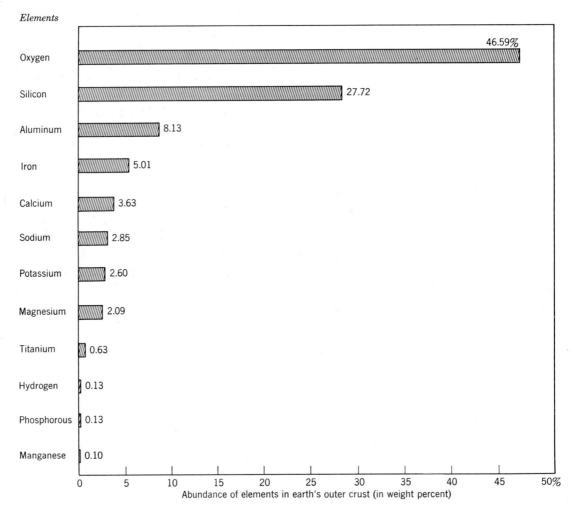

Fig. 1-1. The 12 most abundant elements in the earth's outer crust.

Plate 1. REPRESENTATIVE MINERAL AGGREGATES

A   x 0.75

COPPER

Dendritic Aggregate—branching, tree-like aggregate of subhedral crystals (see dendritic growth). Keweenaw Peninsula, Michigan.

B   x 1.00

RHODOCHROSITE

Concentric—circular, banded layers of microcrystalline crystals arranged about a common center. Capillitas—Catamarca, Argentina.

C   x 1.70

WULFENITE

Tabular—composed of flat, plate-like, well-developed crystals. Ahumada, Mexico.

D   x 1.60

HEMIMORPHITE

Botryoidal—spherical or hemispherical groups resembling a bunch of grapes. Cumberland, England.

Plate 2. REPRESENTATIVE MINERAL AGGREGATES

A   x 1.05

NATROLITE

Acicular (needle-like)—aggregate of radiating needle-like crystals. Bergen Hill, New Jersey.

B   x 0.40

STIBNITE

Bladed—flat elongated crystals. Ichinokaiva, Japan.

C   x 0.89

SERPENTINE (CHRYSOTILE ASBESTOS)

Fibrous—Long separable fibers of asbestos.

D   x 0.66

BARITE ("DESERT ROSES")

Rosette Shape—Plate-like barite crystals containing sand grouped like the petals of a flower. Cleveland County, Oklahoma.

(75 percent) of the world's production of nickel for a long time came from a comparatively small area in the vicinity of Sudbury, Ontario, Canada. The world's most important emerald locality is limited to a small district near Muzo in Colombia, South America.

Minerals containing specific groups of elements frequently occur together, and the knowledge of such associations is an important aid to understanding mineral occurrence. Examples of such groupings are copper, lead, zinc; gold and silver; silver and lead; copper and gold; tin, tungsten, and molybdenum; iron and manganese; iron and titanium; nickel and cobalt; nickel and copper; zinc and cadmium; and chromium and platinum. Some minerals also occur in metallogenic provinces; the copper-bearing area of the western United States, the tin-bearing region of Malaya, and the silver-producing belt of Mexico are some examples.

Impact of Mineral Distribution on Human Affairs

Mineral deposits often play a vital role in human affairs. Minerals are unevenly distributed geographically, and the best deposits may be found in widely separated parts of the world. No one nation is self-sufficient in all minerals; some countries have a surplus of some minerals, but are deficient in others.

The United States has large deposits of coal, iron, potash, molybdenum, copper, lead, zinc, phosphate, salt, sulfur, and magnesium, but it is strikingly deficient in tin, high-grade manganese ore, metallurgical chromium ore, nickel, platinum metals, tantalum, industrial diamonds, quartz crystals, and strategic mica. With the possible exception of industrial diamonds, quartz crystals, and mica, which are now in part produced synthetically, these minerals must be imported.

Minerals provide such great human benefits that control of mineral deposits has not only been a fundamental source of international jealousy but has also led to wars. For centuries, the mineral deposits in western Europe have been an underlying cause of strife, and much of the conflict between Germany and France has been influenced by the desire of each to control the iron deposits of Alsace-Lorraine.

Minerals which at one time or another have been the motivating force behind political action of one form or another are: the tungsten ores of China; the quartz of Brazil; the diamonds of Africa; the gold of the Transvaal; the tin of the Malays; the chrome of Turkey; the copper and cobalt of the Congo; and the manganese of the Ukraine.

THE EARLY USE OF MINERALS

Stone Age Mineralogy—The Dawn of Mineral Consciousness

The earliest knowledge of minerals is hidden in antiquity, but even at the beginning of human existence man employed rocks and minerals as tools. The oldest known stone tools, from Olduvai Gorge, in northern Tanganyika, are believed to date back some 1,750,000 years. The earliest tools (pebble tools) were fashioned from split pebbles, but later on man also utilized several types of chalcedony, including jasper, flint, and petrified wood.

Much later, but still at an early date, minerals were used as pigments. Mineral paint has been found in the caves of Neanderthal man (who lived from about 85,000 to 40,000 years ago). It is not known how the pigments were used, but it is believed they may have been utilized for body decoration.

Around 20,000 B.C., mineral pigments were employed by Cro-Magnon man to paint remarkably realistic animal pictures. These paintings are now preserved on the walls of caves in southern France and northern Spain.[2] Even in earliest times man recognized that ferric oxide (hematite) produced red pigment; hydrous ferric oxide (limonite) formed shades of yellow; and manganese oxide (pyrolusite) yielded a blue-black. These mineral pigments were finely ground and mixed with water or animal fats to produce liquid or paste-like paints.

In the Old Stone Age, clay was employed for sculpture and pottery, whereas somewhat later it was utilized for the manufacture of bricks. Thus, the first mineral industry to emerge from antiquity may have been clay digging.

Metals Used by Early Man

■ **GOLD**

Gold—shining in stream beds—was probably one of the first metals to be discovered. It is believed that gold was used before copper which may have been discovered as early as 18,000 B.C.

According to Greek mythology, Jason in his ship, the *Argo,* went in search of the Golden Fleece to Colchis, an ancient country on the eastern shore of the Black Sea. Strabo, an ancient Greek historian and geographer, wrote that in Colchis gold was washed down from the mountains by winter torrents and collected in troughs pierced with holes and lined with fleeces. Perhaps this legend of the Golden Fleece is an early recognition of placer mining.

Gold was used by the ancient Babylonians, Assyrians, and Egyptians for weapons and utensils

Fig. 1-2. Gold strainer from the XIX Dynasty of Egypt (1350–1200 B.C.). (Courtesy of the Metropolitan Museum of Art, New York; bequest of Theodore M. Davis, 1915.)

(Figure 1-2), but from the first, gold has been largely limited in use to decorative and monetary application.

■ **COPPER**

Probably copper was the first metal to be extracted from chemical compounds found in nature. Perhaps, some primitive man once built a campfire on a copper-bearing outcrop where metallic copper was reduced by the burning charcoal and heat of the fire. Several of the great copper mines of Africa are located near sites where early African tribes melted copper-bearing minerals in order to produce copper cross pieces that served as coins.

Early copper workings, discovered around the ancient Middle Eastern city of Ur in southern Mesopotamia, were operated about 6000 B.C., and it has been suggested that primitive metallurgy, which consisted of hardening copper by hammering, originated here. The ornaments and weapons found during the excavations at Ur imply that the casting of copper began about 3500 B.C. Thus, an interval of more than 2000 years may lie between the first crudely hammered copper and cast metal[4] (Figure 1-3).

■ **COPPER ALLOYS**

Alloys of copper and tin, nickel, or zinc followed the early production of the single metal. *Bronze,* the common alloy of copper and tin, was probably first produced by accident. Copper and tin minerals are so intermixed in ores worked long ago in England (Cornwall), western China, Bohemia, and Spain that the ancients could hardly have avoided producing a crude bronze by their primitive methods of smelting these mineral mixtures.

Certain copper ores also contain nickel and zinc minerals. Thus, it is not surprising that some of the first *brass* was smelted in India from ores containing mixtures of copper and zinc. In a similar manner, copper and nickel alloys, the forerunners of *Monel metal,* were produced in Germany; and alloys of copper-nickel–zinc, which many centuries later were produced in Germany (*German silver*), were first accidentally produced in China from complex ores.

It was not until about 2500 B.C. that bronze became widely used in the eastern Mediterranean coun-

Fig. 1-3. Copper head of a bull with inlaid eyes of shell and lapis lazuli. Mesopotamian (Sumerian), about 2600 B.C. (Courtesy of the Metropolitan Museum of Art, New York; Fletcher Fund, 1947.)

tries—Crete, Greece, and Turkey. Apparently, the Bronze Age did not immediately follow the smelting of the complex ores. Man first learned to extract one metal at a time, as, for example, the reduction of tin from cassiterite (SnO_2), followed later by alloying tin and copper to produce bronze.

■ IRON

Beads of meteoric iron were worn by the Egyptians as ornaments as early as 4000 B.C., and the earliest Egyptian word for iron was *benipe*, believed to mean "any metal from the sky." Around 1500 B.C. the art of extracting iron was developed by men living on the slopes of the Caucasus Mountains facing the Black Sea; thence it spread to Iran and Iraq.[5] The ancient Egyptians prized iron highly, and about 1350 B.C., Egypt's ruler, Tut-ankh-amen, had an iron dagger and an iron bracelet buried with him. The dagger (Figure

1-4) is believed to be of Hittite workmanship and not to have been made in Egypt. At that time only the rich and powerful could own objects made of iron.

The Early Search for Precious Stones

Turquoise was mined over 5000 years ago by the Egyptians, and was probably the first semiprecious stone to be extracted from the earth on a large scale. At times the Egyptian Pharaohs employed from 2000 to 3000 miners in the turquoise mines on the Sinai Peninsula,[1] where mining was conducted with the use of stone tools and other primitive implements.

Fig. 1-4. Iron dagger from the tomb of Tut-ankh-amen of Egypt fitted with gold hilt and knob of rock crystal, about 1350 B.C. (Courtesy of the Metropolitan Museum of Art, New York.) Photograph by Harry Burton.

Gems and decorative stones were among the earliest compact minerals to be mined. Prior to 3400 B.C., the Egyptians made ornaments of rock crystal, amethyst, lapis lazuli, malachite, turquoise, and carnelian. They were mined mostly in Egypt, but it is believed that even in early times the Egyptians also obtained lapis lazuli from Afghanistan, about 2400 miles away.[3] So great was the Egyptian desire for lapis lazuli that deep-blue imitations were created—one type by coloring an alkali silicate with copper carbonate, and another by forming a brilliant blue glass.

MODERN USES OF MINERALS AND MINERAL PRODUCTS

Modern science is actively searching for new uses for minerals; thus, many minerals which in the past were of little importance have become useful. In the search for new applications, the properties of minerals and mineral products are constantly under study. Quartz, for example, has the ability to produce an electric charge when pressure is applied to the ends of certain crystallographic axes (see the piezoelectric effect, p. 92), and, conversely, the ability to change shape when an electric field is applied. A quartz plate placed in a rapidly alternating field expands and contracts alternately, and can thus be used to control radio frequencies.

Selenium, obtained chiefly as a byproduct in the refining of copper, silver, lead, and iron sulfide ores, is light-sensitive, and has the unusual property of conducting electricity differently in darkness from the way it does in light. When illumination increases, the electrical conductivity of the "metallic" selenium increases; as a result, it is used in a variety of modern devices that are highly sensitive to light. Some of these devices are electric eyes and electrostatic copying equipment (such as xerographic process).

Germanium and silicon semiconductors, which control the flow of electrons in the solid state, are being used in transistors.

In recent years the uses of fluorine—chiefly derived from the mineral fluorite (CaF_2)—have been greatly expanded. The fluorocarbon plastics, for example, are not only extremely slippery but also heat and acid resistant. Thus, they are used as automobile bearings that never need greasing, and as the plastic in nonstick frying pans. Many refrigerators, air conditioners, and pressurized spray cans use fluorocarbon gases because they are nontoxic and nonexplosive.

Modern science is creating many new alloys to supply the demands of industry. It has long been known that certain metal wires, when cooled to $-270°C$, conduct electricity without resistance. Ordinarily, however, when these wires are wound into a coil to make an electromagnet, the superconductivity mysteriously disappears. However, in 1961, Bell Telephone Laboratories reported that wires made of a tin-niobium alloy could be wound into powerful magnets and still retain their superconducting ability. A new vanadium-gallium alloy may prove to be a suitable material for superconductors, even more powerful than the tin-niobium magnets.

Science and industry are searching for new high-temperature alloys suitable for spacecraft. Of late, much attention has been directed toward heat-resistant ceramics and metals. Boron is used in the manufacture of high-temperature glass and in high-energy fuels for rockets and jet aircraft. An alloy of pyrolytic graphite and boron is of interest as rocket nozzle material for a variety of solid-fueled missiles. Other temperature-resistant metals under study include molybdenum, tantalum, tungsten, niobium, titanium, and zirconium.

The construction of nuclear reactors required metals such as zirconium because of its excellent corrosion resistance at high temperatures and permeability to slow neutrons. Hafnium, on the other hand, has a high capacity for neutron absorption and is used for control rods in reactors. Cadmium and boron are also used to control the rate of nuclear fission. Beryllium slows down neutrons without absorbing them and is thus used as a moderator in reactors.

Modern science has learned to synthesize numerous minerals to improve upon nature's product or supply. For years man has dreamed of synthesizing diamonds, and in 1955 General Electric Company developed a high-temperature–high-pressure process to produce synthetic diamonds for industrial use (see

p. 26). Under an extreme pressure of about a million pounds per square inch and the intense heat that accompanies it, the carbon atoms in graphite rearrange themselves to form the diamond structure. Although diamonds are usually synthesized from graphite, many materials containing carbon can be transformed into diamonds. To illustrate this, a scientist at the General Electric Company Research Laboratory put a dab of peanut butter into the high-pressure press and a short time later took out a tiny diamond.

The synthetic material Borazon (cubic boron nitride) created under high pressures (45–75 kilobars) and high temperatures (1200–2000°C) compares with a diamond in hardness, but is far more heat resistant. Diamonds burn at about 1800°F, but Borazon will not burn at twice the temperature. Borazon has a potential use as an abrasive and also as a semiconductor in equipment that generates great heat. The possible use of hexagonal boron nitride (which has a layer structure) as a high-temperature lubricant is being investigated.

Synthetic mica has been created that can withstand higher temperatures than natural mica. This is possible through the substitution of fluorine ions for hydrogen in the synthetic fluoro-phlogopite mica ($KMg_3AlSi_3O_{10}F_2$).

Synthetic rubies and sapphires have long been widely used in industry for watch jewels, and bearings in scientific instruments. Because of its ability to withstand high temperatures, synthetic sapphire is also used in the nose-cone tips of guided missiles (see synthetic corundum).

Quartz crystals are produced artificially to supplement the supply of natural quartz, most of which is imported from Brazil. Synthetic minerals at times cost more than natural minerals, but constancy of supply and uniformity in quality cause an ever-increasing demand.

REFERENCES

1. S. H. Ball, "Historic Notes on Gem Mining," *Economic Geology,* Vol. 26, 1931, pp. 681–738.
2. G. Bataille, *Prehistoric Painting, Lascaux, Or The Birth Of Art,* Skira, 1955.
3. A. M. Bateman, *Economic Mineral Deposits,* John Wiley and Sons, New York, 2nd edition, 1954 (Chapter 2).
4. J. G. Parr, *Man, Metals, and Modern Magic,* American Society of Metals, Cleveland, Ohio, and the Iowa State College Press, 1958, pp. 12–13.
5. J. W. Sullivan, *The Story of Metals,* American Society of Metals, Cleveland, Ohio, and the Iowa State College Press, 1951, pp. 54–55.

2.

Crystal Growth

All minerals are created by growth. With a few exceptions that may be considered separately (large or small, singly or in groups), minerals grow as crystals. Ordinarily, crystals start as small nuclei in solution, and layers of atoms from the solution are added to these minute centers until eventually a completed edifice emerges.

This completed edifice may assume many forms: some of them are as simple as a cube; many are more complex and require careful observation in order to be recognized; some are large single crystals; and others are peculiar aggregates of minute individuals. Behind each mineral produced lies a significant history of origin.

THE LIFE OF A CRYSTAL IN NATURE

Some crystals appear formed by one type of crystal growth, but for most no one simple individual growth theory furnishes the whole story. Environment or environmental changes play a vital role. Thus, a hypothetical crystal growing from solution might start as a small branching growth with fragile arms slightly twisted. Then, as growth becomes slower, the spaces between branches are filled in, but the slight tilting of the branches produces dislocations. Impuri-

ties or vacant atomic sites in the crystal structure could also produce these dislocations. Spiral growth proceeds from the dislocations until an accident occurs and the dislocations are healed over. At this point the crystal stops growing, but a change in environment that produces more ideal growth conditions may cause our hypothetical crystal to build up its crystal faces by addition of noncontinuous layers of atoms.

The introduction of impurities may also greatly influence the crystal's growth. If, for example, a particular crystal face adsorbs traces of a foreign substance that slows down its growth, the shape of the entire crystal may be greatly modified. Thus, many changes can occur in the "life" of a growing crystal.

In the nineteenth century, the English writer John Ruskin fancifully described the "lives" of crystals in this way:

You will see crowds of unfortunate little crystals, who have been forced to constitute themselves in a hurry, their dissolving element being fiercely scorched away; you will see them doing their best, bright and numberless, but tiny. Then you will find indulged crystals, who have had centuries to form themselves in, and have changed their mind and ways continually; and have been tired, and then taken heart again; and have been sick, and got well again; and thought they would try a different diet, and then thought better of it; and made but a poor use of their advantages, after all.

And sometimes you may see hypocritical crystals taking the shape of others, though they are nothing like in their minds; and vampire crystals eating out the hearts of others; and hermit-crab crystals living on the shells of others; and parasite crystals living on the means of others; and courtier crystals glittering in the attendance upon others; and all these, besides the two great companies of war and peace, who ally themselves, resolutely to attack, or resolutely to defend. And for the close, you will see the broad shadow and deadly force of inevitable fate, above all this: you see the multitudes of crystals whose time has come; not a set time, as with us, but yet a time, sooner or later, when they all must give up their crystal ghosts—when the strength by which they grew, and the breath given them to breathe, pass away from them; and they fail, and are consumed, and vanish away; and another generation is brought to life, framed out of their ashes.

Emergence of Crystals

Formless gases and liquids can be solidified into crystals by decreasing the motion of their atoms.

In GASES the atoms or molecules move about rapidly and in complete disorder, and are essentially independent of each other except when they collide. Around volcanic vents and related phenomena, rock surfaces may exhibit small crystals formed from gases.

In LIQUIDS the atoms move about irregularly, but their motions are considerably restricted, as they are almost as close together as the atoms in a solid. The forces of attraction that hold atoms of a liquid together, combined with the disordered motion that tends to dissociate them, cause groups of atoms to be constantly shifting. Hence, although there are many small areas of orderly atomic arrangement in liquids, each area of orderliness is limited to a SHORT RANGE of a few hundred atoms. As liquids are cooled, the motions of their atoms are decreased and they may form

either solids (crystalline substances) or "rigid liquids" (glasses). When a liquid is cooled slowly its atoms usually have time to assume the orderly arrangement of a crystal (Figure 2-3), but if cooled rapidly a glass with a mixed-up structure may result (Figure 2-2).

In SOLIDS (crystalline substances with LONG-RANGE order) the atoms or identical groups of atoms are repeated in space at regular intervals to form an orderly geometric structure. In the structure, individual atoms vibrate about fixed positions. THIS PERIODIC ARRAY OF ATOMIC UNITS DISTINGUISHES A CRYSTALLINE FROM A NONCRYSTALLINE SUBSTANCE where such uniformity is lacking. Further, each crystalline mineral has a characteristic internal arrangement. It has been shown mathematically that 230 different internal atomic arrangements are possible for crystals. These are called SPACE GROUPS. The unique arrangement of atoms in such a group provides a characteristic mineral structure.

A few minerals lack the definite regularity of solids and approach the atomic arrangements of liquids that were originally viscous but finally became rigid. Rigid substances with a semiregular structure (short-range order) are called AMORPHOUS. Amorphous substances may be formed by the rapid cooling of a liquid to form a glass, or by the slow hardening of a gel-like substance.

Minerals with a semiregular structure, which may be formed from a gel, are sometimes composed of extremely small, more or less crystalline patches (crystallites). The total number of atoms is so small in each minute crystallite, that the relatively large number of atoms in surface positions tends to distort the crystallite's internal order; thus, most do not have the regularity of a true solid. Minerals with such a semiregular structure are unstable; hence with time, and under suitable conditions, crystallites will join together to form a true solid aggregate.

The mineral *lechatelierite,* formed in quartz sand when lightning strikes, is a natural silica glass and may be thought of as a liquid that has cooled so rapidly that its atoms have not had a chance to arrange themselves in a symmetrical manner before their motion was stopped. The atomic arrangement of such a "rigid liquid" as a silica glass lacks continuous order

Fig. 2-1. A single SiO₄ tetrahedron with silicon surrounded by four oxygens.

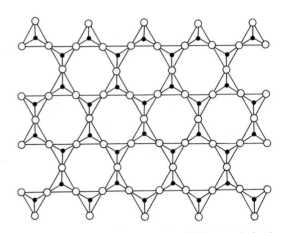

Fig. 2-3. Regularly repeating network of SiO₄ tetrahedra in a crystal.

throughout the mass, but exhibits local structural regularity, as shown by the grouping of individual silicon and oxygen atoms. These are arranged in a symmetrical tetrahedral pattern where each silicon is surrounded by four oxygens (Figure 2-1). However, these tetrahedral SiO₄ units are bound together in a random manner to form an overall disordered pattern (Figure 2-2). On the other hand, silica in quartz is arranged to form a continuous crystalline substance, all units of which are bound together in an orderly manner to form a geometric pattern. Figure 2-3 illustrates the regularly repeating network of SiO₄ tetrahedra in a crystal. The structure of a crystal, however, often contains local distortion caused by imperfections, such as missing or dislocated atoms or impurities.

According to Zachariasen,[19] a noncrystalline substance, such as a glass, does not melt at a definite temperature because the atoms in the random network have structural bonds differing in strength. Thus, the amount of energy required to detach an atom from its irregular structure differs for each atom. As the temperature is raised, an increasing number of atoms are able to break their unequal structural bonds. Therefore, the breakdown of a glass network is a gradual rather than an abrupt phenomenon; and glasses soften gradually when heated rather than melt abruptly as crystals do.

The melting point of a crystalline substance occurs at a definite temperature. At this temperature the energy of thermal agitation of atoms or groups of atoms is sufficient to break their interatomic bonds. Since the atoms* of a crystalline substance are arranged in a three-dimensional geometric pattern, the atomic bonds of uniform strength will abruptly break at the melting point.

The random organization of a noncrystalline substance is somewhat unstable and, after sufficient time, noncrystalline materials may become crystalline as their atoms rearrange themselves to form a more orderly pattern; this pattern has less energy than the disordered state and is thus more stable. Glasses can also be made to crystallize by prolonged heating below the melting temperature.

*With reference to crystal structure in general, where there is no danger of confusion, the term "atom" is used in describing ions, molecules, and ionic groups.

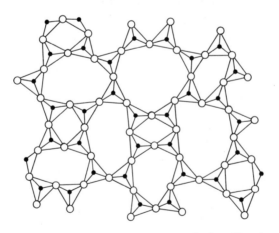

Fig. 2-2. A random network of SiO₄ tetrahedra in a silica glass.

Fig. 2-4. "Snowflake" obsidian (volcanic glass).

Volcanic glass sometimes contains white, gray, or reddish rounded bodies (spherulites) ranging in size from microscopic to several feet in diameter (Figure 2-4). Such volcanic glass is commonly called *"snowflake" obsidian*. The spherulites are often composed of radiating fiber-like crystals of feldspar and cristobalite, believed to have grown from scattered included nuclei (embryo crystals) after the glass became rigid.

Volcanic glass, like man-made glass, is a metastable material, and in time is likely to crystallize, especially if subjected to elevated temperatures resulting from being buried by overlying layers of rock or by the action of heated waters. Volcanic glass is rare in older geologic rocks. In their place, one finds fine-grained rocks containing numerous curved cracks (perlite shrinkage cracks) that perhaps survive as evidence of their former glassy state.

Nucleation and Growth

Crystallization may be described as a result of the process of NUCLEATION and GROWTH. Although formless solutions constitute the source from which crystals grow, the structure developed is an amazing internal and external geometric product.

The atoms in a liquid, for example, have a random motion, but at any given instant there is a chance that

small clusters will correspond to the ordered configuration of atoms in the solid state. These groups are called EMBRYOS or NUCLEI, and, at temperatures above the melting point, have only fleeting existence. At lower temperatures they may become stable. However, very small nuclei are relatively unstable and tend to dissolve because they have a large surface area in comparison to their volume. Since the surface ions are not bound on all sides by neighboring ions, they have much free energy, which serves to lower the stability of the solid. When the temperature is lowered, small nuclei may become stable.

As the freely moving atoms in the liquid attach themselves to the nuclei in an ordered geometrical arrangement, the nuclei "grow" into crystals. Theoretically, a crystal grows because the solid state is more stable than the liquid or vapor state. The atoms usually occupy structural sites whose occupancy provides the greatest DECREASE in free energy and, thus, the most stable structure.

When a liquid is cooled rapidly the rate of nucleation is greater than the rate of growth; hence, many small crystals or grains result. If a liquid is cooled slowly, the rate of growth will be faster than the rate of nucleation, and larger crystals will form (Figure 2-5). Foreign particles at times serve as nucleating centers for minerals if the particle and the mineral have a sufficiently similar crystal structure.

It has also been shown that a few large crystals tend to form when the supersaturation of a liquid is

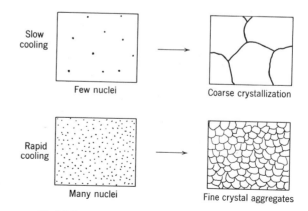

Fig. 2-5. Effect of the cooling rate on crystallization.

low, but that at higher degrees of supersaturation nuclei form more easily, resulting in the formation of many small crystals.

Magnitude of Crystals

Mineral crystals have internal crystallinity and may also form excellent visible crystals. Crystals are, in general, solid chemical elements, solid solutions, or compounds that grow symmetrically according to 3-dimensional atomic patterns. THE ATOMIC UNITS OF CRYSTALS ARE PACKED IN STRAIGHT LINES AND ARE RE-PEATED WITHOUT VARIATION IN SPACE (except for accidental imperfections, such as missing or dislocated atoms). Thus, WELL-FORMED CRYSTALS HAVE FLAT FACES, STRAIGHT EDGES, AND SHARP CORNERS, ALL OF WHICH ARE THE OUTWARD EXPRESSION OF AN ORDERED INTERNAL ATOMIC ARRANGEMENT.

Crystals of minerals range widely in size, shape, and perfection. Some are so minute that regularity can only be detected with a microscope, by x-rays, or with the electron microscope; others are found in exceptional size. The same mineral species may occur as either minute or large crystals, depending upon the conditions of crystallization.

Minerals such as *feldspar, quartz, beryl, mica, topaz, tourmaline, amblygonite, columbite-tantalite,* and *spodumene* frequently occur as large crystals in pegmatites, and may at times reach giant proportions. A few examples of giant crystals reported are:

BERYL: 200-ton crystal; Brazil. Crystal 19 feet long and 5 feet in diameter; Black Hills, South Dakota. Transparent aquamarine crystal; weight 243 pounds; sold uncut for $25,000; Brazil.

SPODUMENE: Crystal over 40 feet long; weight estimated at 90 tons; Etta Mine, Black Hills, South Dakota.

PHLOGOPITE MICA: Crystal 33 feet long, 14 feet wide; weight approximately 90 tons; Lacy Mine, Ontario, Canada.

QUARTZ: Milky white crystal; close to $11\frac{1}{2}$ feet long and $5\frac{1}{4}$ feet in diameter; weight about 13 tons; Siberia, U.S.S.R.

Fig. 2-6. Quartz crystal faces express the ordered internal atomic arrangement of the silicon and oxygen atoms.

COLUMBITE: One-ton crystal; Bob Ingersoll Mine, Keystone, South Dakota.

TOPAZ: Transparent crystals; weigh up to 600 pounds each; Minas Gerais, Brazil.

APATITE: 550-pound crystal; Ottawa County, Quebec, Canada.

Crystal Faces

The regular, smooth growth planes that form the surfaces of crystals are called FACES (Figure 2-6). During growth, a crystal is not always free to develop an external pattern of faces; or, after growth, the faces may have been broken, leaving behind an irregular surface. Such irregularly bounded masses, although lacking crystal faces, still have a regular internal atomic arrangement, and are CRYSTALLINE. For example, a piece of massive quartz (Figure 2-7) is crystalline, although it shows no outward sign of regular internal arrangement.

At the same temperature (and pressure), THE ANGLES BETWEEN CORRESPONDING FACES ON ALL CRYSTALS OF THE SAME SUBSTANCE ARE THE SAME. This fundamen-

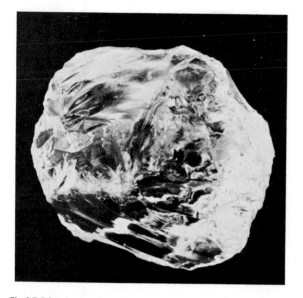

Fig. 2-7. Massive transparent quartz may resemble glass, but it is crystalline with an orderly internal structure, and the atoms lie in symmetrical arrangement.

nary point at the center of the crystal. According to this concept, similar pairs of faces on different crystals of the same substance range widely in size, but the angles between the face normals remain the same (Figure 2-8).

Crystals such as garnet (Figure 2-9) or magnetite (Figure 2-10) exhibit constant angular relationships, although individuals differ so much in size and facial development that they may appear different.

Crystals differ greatly in excellence of development. If crystal faces are well developed, crystals are said to be EUHEDRAL (well formed). Crystals of intermediate development with rounded outlines may be called SUBHEDRAL. Crystals that lack external faces but have internal regularity are ANHEDRAL (without form). In Plates 1 and 2 (facing pp. 2, 3), the crystals of wulfenite and stibnite are euhedral; those of copper and barite are subhedral; and the aggregates of rhodochrosite and hemimorphite are composed primarily of microscopic subhedral crystals.

tal behavior, known as the law of the CONSTANCY OF INTERFACIAL ANGLES, was first recorded by the Danish scientist Nicolaus Steno in 1669.

Later, in 1809, Bernhardi pointed out that the true interfacial angles are the angles between the normals to the crystal faces as projected from an imagi-

Striated Faces

■ **OSCILLATORY STRIATIONS**

Oscillation in growth between differently oriented faces may produce a single crystal face composed of narrow alternating furrows and ridges. These

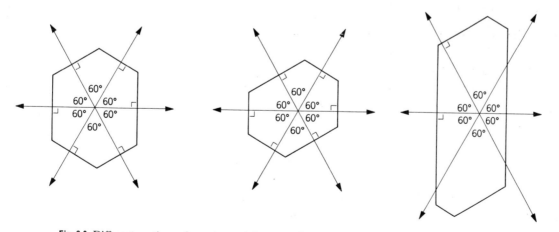

Fig. 2-8. Different sections of quartz crystals unequally developed, but all with the same interfacial angles.

Fig. 2-9. A comparison of two dodecahedral garnet crystals. The corresponding angles are the same but the faces differ in size.

Fig. 2-10. A comparison of two octahedral magnetite crystals. The corresponding angles are the same, but the face dimensions differ considerably.

narrow grooves appear as parallel lines or striations.

On PYRITE crystals, striations are caused by oscillation between the faces of the cube and the pyritohedron, as shown in Figure 2-11 and Plate 12 (facing p. 83). Parts of the cube faces form the tops of the ridges, whereas the furrows tend to be parallel to the faces of the pyritohedron.

The horizontal striations on prismatic crystals of QUARTZ are the result of the oscillation between the faces of the prism and the rhombohedron (Figure 2-12).

Oscillatory growth tends to produce rounded faces such as those on TOURMALINE crystals; they are commonly curved and vertically striated, as a result of the oscillatory combination of the prismatic faces (Figure 2-13).

■ TWINNING STRIATIONS

Striations may also be caused by repeated twinning lamellae, such as are commonly observed on plagioclase feldspars (see oligoclase, Plate 8, facing p. 51). Calcite and dolomite frequently exhibit fine striations caused by mechanical deformation (see glide twins, p. 73).

Modes of Aggregation

Crystal aggregates are more common than single crystals. Crystallization of a mineral from many different centers often produces a mass of interlocking

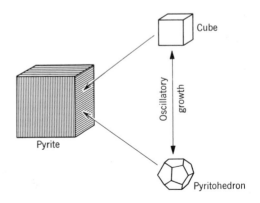

Fig. 2-11. Striated pyrite crystal.

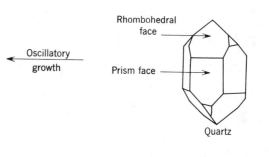

Fig. 2-12. Striated quartz crystal.

anhedral crystals (grains). Individually all the crystals possess the same ordered internal arrangement, but in the aggregate each is usually randomly oriented in relation to another (Figure 2-14).

Although the individuals of a crystalline aggregate are usually randomly oriented, in some cases they are related to each other in a significant fashion, such as the parallel grouping of galena crystals (Figure 2-48, p. 34). These "modes of aggregation" often yield information on the conditions under which a particu-

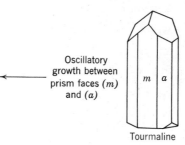

Fig. 2-13. Striated tourmaline crystals.

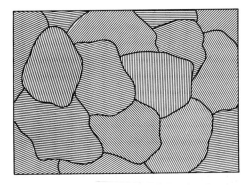

Fig. 2-14. Schematic diagram of the anhedral crystals of a mineral aggregate with atomic planes independently oriented.

lar mineral will form, and also aid in mineral recognition. Special illustrative terms have long been used to describe mineral aggregates, some of which suggest mineral identity. Several common aggregates are illustrated in Plates 1 and 2, while Table 7-2 (p. 130) lists a considerable number.

If the individual crystals of an aggregate may be resolved with a microscope but not with the unaided eye, the aggregate is called MICROCRYSTALLINE (Figure 2-15). If the crystallites of an aggregate are submicroscopic, the aggregate is called CRYPTOCRYSTALLINE.

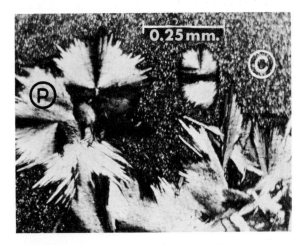

Fig. 2-15. Microcrystalline aggregates of radiating prochlorite crystals (P) surrounded by a finer almost submicroscopic aggregate of clinochlore (C). (x 90; x-Polaroids.)

Oriented Overgrowths

Crystals frequently grow in oriented crystallographic directions upon other crystals, such as hematite on quartz, rutile on hematite, or chalcopyrite on galena and sphalerite. These parallel deposits of one mineral on the surface of another are known as ORIENTED OVERGROWTHS, and frequently occur between mineral species which are not closely related chemically. The atoms in the surface layer of the host crystal extend their influence to orient the nuclei of the overcoating crystals. Apparently, in the majority of cases, the most important requirement for the formation of oriented overgrowths is that the structural planes of the two minerals in contact with each other must have similar atomic spacings.

One of the best known examples is the parallel growth of sodium nitrate crystals, $NaNO_3$ (soda-niter), on a cleavage surface of calcite. The atoms in soda-niter are arranged in the same way as calcite, with similar distances between atoms. If a drop or two of a warm concentrated solution of sodium nitrate is placed on a rhombohedral cleavage surface of calcite and covered with a glass plate, small rhombohedral crystals of soda-niter will grow on the calcite in perfect parallel alignment.

Cavernous Crystals

If atoms are added more rapidly to the edges of a crystal, rather than building up complete layers across the faces, a crystal with deep depressions in the center

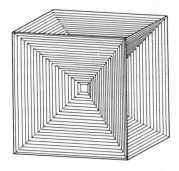

Fig. 2-16. Hopper-shaped crystal.

(such as the HOPPER crystal in Figure 2-16) may result. A higher degree of supersaturation near the edges of a growing crystal may cause the edges to build up faster than the center of the face. When crystals grow in a well-stirred medium, the supersaturation tends to be more uniform over the crystal faces and cavernous crystals are less likely to develop.

The mineral halite commonly forms hopper-shaped crystals. Other minerals, such as pyromorphite (Plate 31, facing p. 187) or vanadinite, frequently form hollow 6-sided prisms; they do this because during growth new growth layers form at the periphery before earlier ones complete their travel to the face center (Figure 2-17). Peripheral growth may be partly caused by defects developing along the edges of a growing crystal, as well as a higher degree of supersaturation near the edges, and the interaction (bunching) of growth layers.

Some crystals are bounded by well-formed faces but have internal cavities beneath some or all of the faces. These cavities may be caused by a lag or a halt in growth of the face centers while the rest of the face continues to grow. Possibly, this occurs because the supersaturation of the growth solution happens to sink to zero in the region of the face center, thus making growth there impossible; or perhaps the edges of the crystal receive more growth material and build up faster than the face centers. With a change in the growth environment, subsequent growth produces complete faces and the cavities are "healed over."

Certain quartz crystals, for example, contain such large cavities beneath the faces that the internal crys-

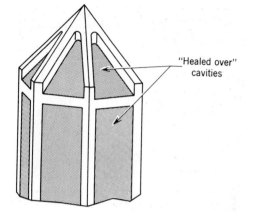

Fig. 2-18. "Skeletal" quartz.

tal appears to be skeletal. Such skeletal quartz crystals are common in Minas Gerais, Brazil (Figure 2-18). The cavities, which are sometimes filled with liquid, are more commonly parallel to the quartz rhombohedral faces than to the prism faces.

Phantom Crystals

Minerals such as quartz, calcite, or fluorite frequently exhibit growth outlines of crystals within crystals. The earlier-formed inner crystals are called PHANTOMS because of their ghostly appearance (Figure 2-19). During the growth of a crystal, impurities from the growth solution may at various intervals cover the crystal faces and become included. These thin layers of included foreign material delineate the phantom surface of an earlier crystal from the subsequent growth. Numerous quartz phantoms often occur within a single crystal and are frequently coated with chlorite, clay, or hematite. Changes in the color of a growing crystal may also produce phantoms.

The shape and color of a phantom may correspond to the outer crystal, but may differ if a change occurs in the growth environment. Striking amethyst phantoms are sometimes enclosed in colorless quartz crystals.

If the layer of foreign material covering an inner crystal is dissolved, the inner and overcoating crystals can sometimes be separated (Figure 2-20).

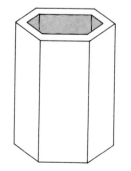

Fig. 2-17. Hollow prismatic crystal.

Fig. 2-19. Phantom quartz crystal within larger quartz crystal (x 1.3).

Fig. 2-20. Capped quartz crystal (x 0.6).

CRYSTAL HABIT (SHAPE)

The possible faces for a given crystal are governed by the atomic structure, but the environment exerts a selective influence on the relative growth rates, and thus tends to determine the crystal faces that will be most prominent. Frequently, during crystal growth, a consistent set of conditions exists; they include temperature, pressure, and the chemical conditions of precipitation. The balance of these determines whether a crystal will be long or short, flat or stubby, or whether they will assume some other characteristic development. The resulting uniformity of shape for a particular mineral crystal has led to the term CRYSTAL HABIT. The term describes the composite size and shape customarily assumed by the crystal faces, and represents THE RESULTANT OF THE INTERACTION OF THE ATOMIC STRUCTURE OF A MINERAL WITH THE ENVIRONMENT IN WHICH IT GROWS.

Many crystals occur with habits so characteristic that the identity of the mineral is suggested by observation alone. Large numbers of crystals with a similar habit formed in a mineral deposit may also suggest uniform conditions of mineral deposition.

Crystals of the same mineral may occur with different habits, representing different conditions of crystallization. Although internal structures are the same, the unit cells are stacked in different ways to produce different external shapes. For example, it was long ago shown by Haüy that identical cubic unit cells may be used to construct the octahedron or other forms by varying the rate of growth in different directions (Figure 2-21).

Relative Growth Rates of Crystal Faces

The shape of a crystal is influenced by the relative growth rates of its faces (Figure 2-22). Faces that grow SLOWLY in relation to others on the same crystal are generally the LARGEST, and rapid growth may entirely eliminate otherwise appropriately oriented faces. Thus, the crystal is bounded by its slowest-growing faces. Should crystals grow at the same rate in all directions they would be spherical.

Plate 3. ISOMETRIC CRYSTALS AND CORRESPONDING WOODEN MODELS

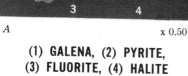
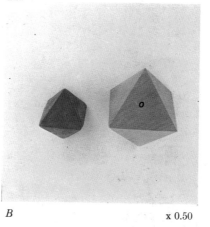

A   x 0.50

(1) GALENA, (2) PYRITE, (3) FLUORITE, (4) HALITE

Form: cube (6 square faces).

B   x 0.50

HAUERITE
MnS_2

Form: (*o*) octahedron (8 equilateral triangular faces). △

C   x 0.64

GARNET

With surface alteration of chlorite. Form: (*d*) dodecahedron (12 rhomb-shaped faces). ◇

D   x 0.53

KAOLINITE

Retains the form of leucite. Form: (*n*) trapezohedron (24 trapezium-shaped faces). ☐

E   x 0.46

PYRITE

Form: (*e*) pyritohedron (12 pentagonal-shaped faces). ⬠

Plate 4. ISOMETRIC CRYSTALS AND CORRESPONDING WOODEN MODELS

A   x 0.50

GALENA

Forms: cube (*a*) and octahedron (*o*).

B   x 0.63

MAGNETITE

Forms: octahedron (*o*) and dodeca-hedron (*d*).

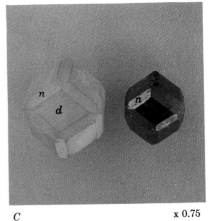

C   x 0.75

GARNET

Forms: dodecahedron (*d*) and trapezo-hedron (*n*).

D   x 0.54

HAUERITE

Forms: octahedron (*o*) and cube (*a*).

E   x 0.43

GARNET

Forms: trapezohedron (*n*) and dodeca-hedron (*d*).

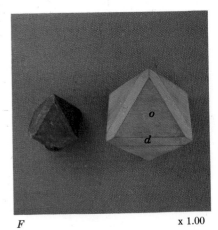

F   x 1.00

CUPRITE

Cuprite crystals altered to green mala-chite. Forms: octahedron (*o*) and dodeca-hedron (*d*).

Fig. 2-21. Octahedron built up of small cubes (after Haüy).

Although the growth velocities of crystal faces tend to be inversely proportional to their atomic density, other factors (such as the distribution of electrical charges of atoms in different directions, and environmental factors of temperature, pressure, im-

purities, and the concentration and movement of solutions) also influence the rate of growth.

Crystals of some minerals are always elongated. The end faces, which differ in atomic display from the side faces, grow more rapidly because the forces of attraction are stronger in this direction as a result of the electrical and the structural nature of their atoms. The fast-growing end faces are smaller than the relatively slow-growing side faces, resulting in the formation of an elongated crystal (Figure 2-23).

Changes in crystal shape may result when the supply of material for growth is nonuniform around the growing crystal. Some of the faces may receive more growth material than other faces less favorably situated. For example, when common salt crystals, NaCl, are grown on the bottom of a container, they assume square plate-like shapes rather than the usual cube shape. The simple explanation for this change in crystal habit is that the crystals resting on a surface can grow upward and sideways but not significantly downward, and thus the crystals are only about half as high as they are wide.

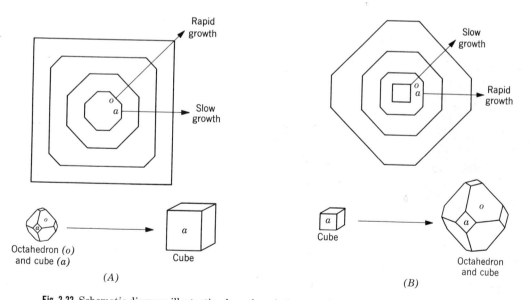

Fig. 2-22. Schematic diagram illustrating how the relative growth rates of crystal faces influence crystal habit. In (*A*) the crystal starts out as a cube-octahedron, but rapid growth on the octahedral faces (*o*) results in progressively smaller faces until the octahedral faces finally disappear, and the relatively slow-growing cube faces (*a*) are prominent. (*B*) shows the reverse process, where the crystal starts out as a cube but grows into a cube-octahedron.

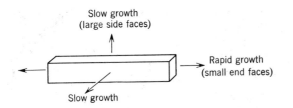

Fig. 2-23. Elongated crystal formed by rapid growth on the end faces.

Effects of Impurities on Crystal Habit

The presence of small amounts of certain impurities in the growth solutions often causes marked changes in crystal shape. This results because the adsorption of traces of a foreign substance on a crystal face affects the rate of growth. Usually, growth is retarded and previously fast-growing faces may become relatively slow growing, giving rise to an entirely new crystal habit.

Crystals grown from solutions containing small amounts of organic dye molecules often show marked changes in habit. For example, the addition of minute amounts of methylene blue to solutions in which lead nitrate crystals form causes a change in habit (Figure 2-24). As the concentration of methylene blue rises from 0.0003 to 0.003 weight percent, the cube faces gradually become larger and more deeply colored. Relatively rapid growth eliminates the octahedral faces, and the final result is a dark blue cubic crystal.[15]

In some cases, a dye is deposited on the disappearing faces. For example, the dye No. 26 Crocein Orange changes the habit of sodium fluoride crystals, NaF, to octahedral, but the deposition occurs on the cube faces.[3]

Experiments also indicate that impurities may alter the habits of certain crystals even when they are not appreciably incorporated into the growing crystal. Many crystals of potassium alum, for example, have been grown in the presence of strong concentrations of the dye, Bismark Brown, and suffered marked habit change without absorbing appreciable amounts of color. Thus, it is believed that the adsorption process on a growing crystal surface can be more or less independent of the habit-modification process, since some crystals that undergo strong habit changes absorb almost all the dye from solution and become deeply colored, whereas others are scarcely tinted at all (H. E. Buckley).[3]

Figure 2-41 (p. 31) illustrates an example of habit modification in which the adsorption of a "poison" causes part of a spiral growth step to stop, resulting in the formation of a **WHISKER** (a filamentary single crystal).

■ HOURGLASS AND MALTESE CROSS STRUCTURES

In some cases, impurities are selectively adsorbed only on certain crystal planes. An **HOURGLASS** or **MALTESE CROSS** structure may result if at least two types of crystal faces of different atomic arrangement are present from the initial to final stages of growth—one type covered by impurities, and the other avoided (Figures 2-25 and 2-26). Figure 2-25 shows an hourglass structure on a tetrahedral mineral formed by the selective deposition of impurities on the basal planes (c) during successive stages of growth.

Figure 2-26 illustrates the formation of a maltese cross type of hourglass on an isometric crystal. The crystal starts out as an octahedron (o) but the cube faces (a) are progressively developed during growth.

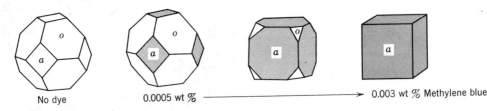

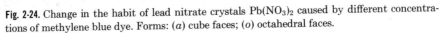

Fig. 2-24. Change in the habit of lead nitrate crystals $Pb(NO_3)_2$ caused by different concentrations of methylene blue dye. Forms: (a) cube faces; (o) octahedral faces.

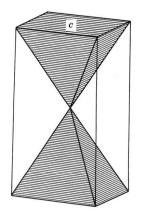

Fig. 2-25. Hourglass structure formed by the deposition of impurities on the basal planes (*c*) during the initial to final stages of growth.

Fig. 2-27. Sand gypsum crystals; Oklahoma (x 0.75).

The dye deposits on the cube faces and avoids the disappearing octahedral faces.

Plate 49 (facing p. 307) illustrates examples of *chiastolite* crystals, which show in cross section a black maltese cross; this cross is formed as a result of selectively distributed carbonaceous material that was pushed aside into definite areas as the crystal grew in metamorphosed shales.

■ **SAND CRYSTALS**

The forces of crystallization are strong, as illustrated by crystals of sand calcite, gypsum, or barite,

where grains of quartz are included during crystallization. Large single crystals several inches long may result, yet more than half of the volume may consist of included sand grains (Figures 2-27 and 2-28). In gypsum, the sand inclusions are sometimes distributed to

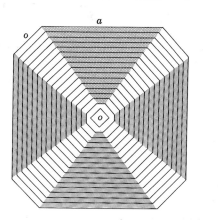

Fig. 2-26. Maltese cross type of hourglass structure on an isometric crystal (after H. E. Buckley, *Crystal Growth,* John Wiley and Sons, New York, 1951).

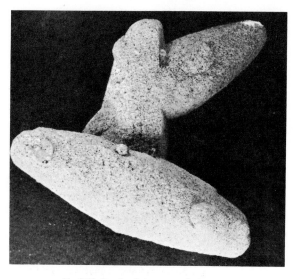

Fig. 2-28. Sand calcite crystals (x 0.65).

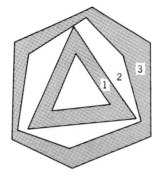

Fig. 2-29. Cross section of a zoned tourmaline crystal showing changes in habit from a trigonal prism (1) to a ditrigonal prism (2) to a hexagonal prism (3).

form an hourglass structure. Sand barite crystals are illustrated in Plate 2, facing p. 3.

■ **COLOR ZONING**

Color zoning in minerals can be caused by changes in the composition of the host solution during crystallization as well as by distortions in the crystal structure that form color centers. The presence of certain "impurities" may influence both the color of the crystal and its shape.

Figure 2-29 shows the zoned structure of a tourmaline crystal. The chemistry of tourmaline is complex. The composition ranges widely, forming the various colored varieties (see Plate 45, facing p. 275). The change in color and shape of the color bands is caused by changes in composition that differ from band to band.

Minerals such as corundum, tourmaline, cassiterite, and quartz frequently form remarkable zoned structures.

Color zoning in some cases is caused by selectively distributed impurities that impart no color to a nonirradiated crystal, but produce color centers (structural defects that absorb light) when the colorless crystal is irradiated with x-rays or exposed to radioactive material. An example of this type of zoning is illustrated in Figure 2-30.

The crystal in (*A*) is colorless but has certain impurities selectively distributed in zones 2 and 4. The outer zone 4 contains iron impurities. An amethystine color is produced when the crystal is irradiated (*B*). Experiments show that synthetic colorless quartz containing ferric iron (Fe^{3+}) can be colored amethyst by irradiation with x-rays, and that natural amethyst quartz, which has been decolorized by heat treatment, can be colored purple again by irradiation

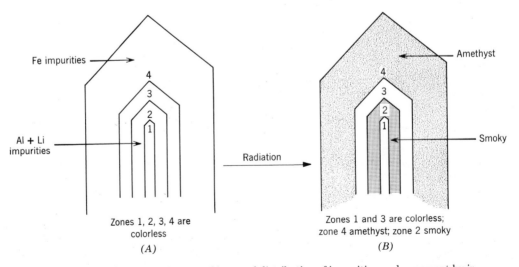

Fig. 2-30. Color zoning in quartz caused by zonal distribution of impurities made apparent by irradiation.

with x-rays or radium. Irradiation with x-rays has increased the intensity of color of natural amethyst up to five times. It is believed that the EPR spectrum of some of the color centers corresponds to a positive hole trapped on a substitutional Fe^{3+} ion in the quartz structure.[12]

Zone 2 contains aluminum + interstitial lithium impurities in substitution for silicon, which impart no color to the crystal until after irradiation (see smoky quartz).

Zoning in certain minerals may be caused by variations in temperature and pressure, since these environmental changes influence the ability of chemical compounds to crystallize. For example, if the temperature is raised, certain chemical compounds with low freezing points cannot crystallize. The zonal growth of some feldspar crystals, as shown under the microscope, is believed to be caused by oscillations in temperature and pressure of a magmatic melt.

Temperature-Pressure Relations

■ THE HABITS OF ICE CRYSTALS

Experiments in the growth of ice crystals indicate that the habit can be governed by the temperature of the air in which the crystals grow. B. J. Mason[9,13] grew ice crystals on a thin fiber running vertically through a diffusion cloud chamber in which the vertical gradients of temperature and vapor density were accurately controlled and measured. The experiments covered temperature ranges from 0°C to −50°C and vapor supersaturations ranging from a few percent to about 400 percent.

The results showed that the *basic shapes* of the crystals (such as prisms and plates) were dependent upon the TEMPERATURE. Figure 2-31 and Table 2-1 illustrate the range in shape of crystals with temperature along the length of the fiber. The most striking feature is the remarkable sequence of habit change. The crystals change from plate-like to column-like (prismatic) three times over a narrow temperature range from 0° to −25°C. These habit changes reflect changes in the relative growth rates of the prism and basal faces. The transitions between hexagonal plates

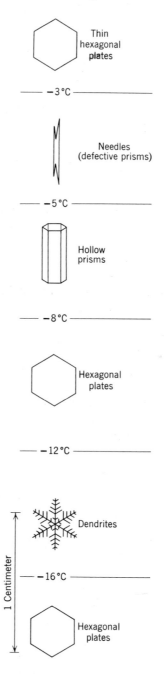

Fig. 2-31. Schematic diagram showing the range in the shape of ice crystals with temperature along a filament. The crystals were grown on a thin fiber suspended in a diffusion cloud chamber with controlled temperature gradient (after B. J. Mason[13]).

Table 2-1

Variation of Ice Crystal Habit with Temperature
(Hallett and Mason, 1958)

0°C to −3°C	Thin hexagonal plates
−3°C to −5°C	Needles
−5°C to −8°C	Hollow prisms
−8°C to −12°C	Hexagonal plates
−12°C to −16°C	Dendrites
−16°C to −25°C	Hexagonal plates
−25°C to −50°C	Hollow prisms

and needles occurred at −3°C and hollow prisms gave way to hexagonal plates at −8°C. These changes took place within temperature intervals of less than 1°C and the boundaries between one shape and another were sharp.

A growing ice crystal, when transferred to a new temperature environment by the raising or lowering of the fiber in the diffusion chamber, assumed a growth characteristic of the new environment. Thus, it was possible to produce hybrid combinations. For example, when needles grown at temperatures between −3° and −5°C were suddenly moved up in the chamber to about −2°C, hexagonal plates developed on their ends. Such radical changes were produced by a change in temperature of a degree or two at a constant supersaturation, but could not be produced by varying the supersaturation at a constant temperature.

Although variations in the degree of supersaturation of the surrounding vapor do not change the *basic crystal habit,* as between prisms and plates, the growth rates are greatly affected; that is, the greater the supersaturation the faster the growth. High supersaturation also appears to govern the development of secondary features such as needle-like extensions of hollow prisms, spikes and sectors at the corners of six-sided plates, and the fern-like development of star-shaped crystals.

It has also been observed that traces of organic vapors profoundly influence the shapes of ice crystals grown in a diffusion chamber.

SNOW CRYSTALS (ATMOSPHERIC ICE CRYSTALS) occur in a great variety of intricate patterns. During their formation from water vapor in the atmosphere, as the snow crystals fall to earth they frequently pass through regions of different temperatures and vapor pressures. Hence, as they go through various stages of development they produce an almost endless variety of complex hybrid combinations (Figure 2-32). However, in the atmosphere the delicate snow crystals often collide with each other, and the snowflakes reaching earth are, thus, usually damaged.

■ **MECHANISM OF ICE HABIT VARIATION**

It is believed that what constitutes the critical growth process of many crystals is the migration of material across the surface of a growing crystal (surface diffusion) to growth sites where the material becomes built into the crystal structure. Mason[9] is of the opinion that variations in the relative rates of surface diffusion on the basal and prism faces of ice crystals are responsible for determining their habits in the early stages of development. The surface properties of the base and prism faces differ, since unlike faces differ in atomic arrangement.

Water molecules arriving on the crystal surface have a surface migration length of several microns and therefore may reach an adjacent face before being built into the crystal structure. Thus, the INITIAL HABIT DEVELOPMENT may be determined by the relative distances that the molecules travel. For example, if the migration distance is greater for the basal face, there will be a net transport of material to the prism faces, and the crystal will start to develop a plate-like habit. The reverse will be true if the surface diffusion is greater for the prism faces. *The surface diffusion varies markedly with the temperature.*

Some ice crystals grown at low supersaturations were found to develop considerably in diameter with no discernible change in thickness, a fact which suggests that water molecules arriving on the upper basal surface were not being built into the basal face, but were migrating over the surface to the prism faces.

Once a habit is established in the early stages of growth, it is maintained by the arrival of water vapor on, and the latent heat of fusion away from, the crystal surface. At moderate supersaturations, crystals continue to develop as polyhedra, suggesting that excess water molecules arriving at the corners and edges are redistributed over the crystal surface by surface migration. However, at high supersaturations the surface diffusion is unable to handle the nonuniform deposition, and thus the corners and edges of the ice crystals begin to sprout to form sector plates, hopper crystals, dendrites (branching crystals), and other skeletal forms.

■ **DIAMOND CRYSTALS— CHANGES IN HABIT AND COLOR**

Synthetic Diamonds. In 1955, the General Electric Company engineers succeeded in building a high-pressure–high-temperature apparatus that could withstand and maintain the high temperatures and

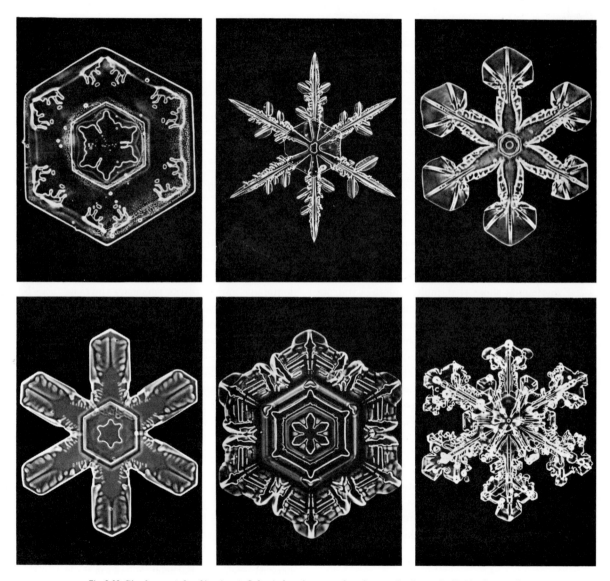

Fig. 2-32. Single crystals of ice (snowflakes) showing complex changes in shape. Individual crystals exhibit hybrid combinations of six-sided plates and branching dendrites (photomicrographs, courtesy of Carl Zeiss, New York).

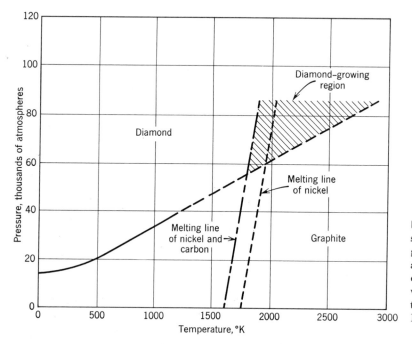

Fig. 2-33. Pressure-temperature diagram showing a portion of the diamond-graphite equilibrium line. Diamonds are formed when the pressure is high enough for diamond to be stable and when the temperature is high enough to melt the nickel-carbon alloy (from R. H. Wentorf, Jr.[9]).

pressures required to allow the carbon atoms in graphite to rearrange themselves to form the diamond structure (see Figure 3-6, p. 42).

Experiments show that molten metal catalysts such as nickel, platinum, iron, or manganese greatly facilitate the diamond synthesis, but small amounts of a "poison" such as lithium or iodine halt the formation of diamond (Wentorf, 1963).

In the diamond synthesis, mixtures of graphite and catalyst metal are subjected to pressures of 45 kilobars* or more and temperatures hot enough to melt the catalyst-carbon alloy (Figure 2-33). The transformation from graphite to diamond usually occurs in a few minutes and proceeds via a thin film of molten catalyst alloy.

The rate of growth and nucleation of diamond is sensitive to pressure but much less dependent upon temperature. The more the pressure is increased beyond that needed for equilibrium, the greater the rate of nucleation and growth.

*A bar is slightly less than 1 atmosphere. A kilobar is 1000 bars or 14,500 pounds per square inch.

The CRYSTAL HABITS of synthetic diamonds are affected by pressure.[9] At HIGH SYNTHESIS PRESSURES of 60 kilobars and above, the OCTAHEDRAL habit is predominant; and at LOWER PRESSURES the CUBE faces become more prevalent (Figure 2-34). The shape of a crystal is influenced by the relative growth rates of its faces, and growth on different faces apparently requires slightly different energy conditions.

Rhombic dodecahedra are common in natural diamonds but are rarely found in laboratory-grown

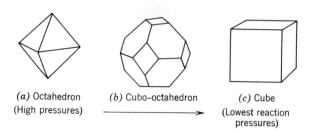

(a) Octahedron (High pressures)   *(b)* Cubo-octahedron   *(c)* Cube (Lowest reaction pressures)

Fig. 2-34. Influence of pressure on some habits of synthetic diamond. The octahedral habit *(a)* forms in the highest pressure range and the cubic habit *(c)* forms in the lowest pressure-reaction range.

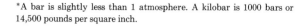

diamonds. It is probable that preferred solution can take place to produce dodecahedral planes, and that the rhombic dodecahedron is not a true growth form for the synthetic diamond.[2] The rounded diamonds found in nature may also have been subjected to solution following growth.

The COLOR of man-made diamond crystals is governed by temperature, pressure, and chemical environment. The color is usually affected by the type and concentration of included impurities, and can be changed by varying the temperature at which growth takes place. At the lowest possible growth temperatures, VERY DARK GREEN or BLACK diamonds tend to form with a distorted structure caused by included material. As the temperature is raised, blackish green or black is replaced by shades of LIGHT GREEN and YELLOW until an almost COLORLESS diamond is formed at the highest temperatures. Diamonds formed at higher temperatures and pressures generally have a greater chemical purity and fewer light-absorbing structural defects (color centers).

DARK BLUE DIAMONDS[17] are produced when small amounts of boron (0.1 percent) are added to the synthesis mixture. Lower boron concentrations produce lighter blue crystals. The presence of boron also greatly accelerates the formation of diamonds. Blue semiconducting diamonds can also be made from a mixture of nickel and carbon containing small amounts of borax.

Natural Phantom Diamonds. In nature there are many cases of a diamond crystallizing around another diamond to enclose a phantom crystal. The inner diamond often differs in color and habit from the outside crystal. Some natural diamond crystals are coated with graphite and then enclosed in a second diamond. After the first diamond forms, a change in environmental conditions occurs, which probably causes the diamond to be partially converted to graphite. When conditions under which carbon can crystallize as diamond are restored, a second diamond forms to enclose the graphite-coated diamond.

The growth of a synthetic diamond can be stopped and the diamond converted to graphite if the pressure is not maintained at the required high temperatures.

■ HIGH SILICA (STISHOVITE)

The crystal habit of *stishovite* (a dense high-pressure polymorph of silica (SiO_2) with a rutile structure) changes with temperature of crystallization. Experiments by Sclar, Carrison, and Cocks (1964) show that when synthetic stishovite is grown in a purely siliceous environment at a pressure of about 120 kilobars and a temperature of 550°C it assumes a BIPYRAMIDAL HABIT; between about 600° and 900°C, the habit is dominantly GRANULAR with some tendency to form prismatic or tabular crystals; and, crystallized above 900°C at 120 kilobars, stishovite has a NEEDLE-LIKE (acicular) habit.

When grown in a magnesium-rich environment between 500° and 650°C at 115- to 125-kilobar pressure, stishovite crystals have a stubby prismatic to tabular habit, indicating that the chemical environment of growth also influences the crystal habit.

The formation of *natural stishovite* is attributed to the transient high shock, pressure, and temperature accompanying the impact of a huge meteorite. Most of the natural stishovite from the highly siliceous Coconino Sandstone of Meteor Crater in Arizona has a NEEDLE-LIKE habit, which suggests that it crystallized at temperatures above 900°C. Additional data are needed, but it is probable that the effects of pressure on the boundary between acicular and non-acicular stishovite is small. The thermal dependence of the crystal habit of stishovite may thus constitute a high temperature geologic gauge, which could indicate the limiting values for the peak temperatures that prevailed at impact craters in highly siliceous rocks.[14]

TYPES OF GROWTH

Growth in Separate Layers (2-dimensional nucleation)

It has been proposed that under IDEAL GROWTH CONDITIONS crystal faces are slowly built up by the accretion of groups of atoms in separate layers, as

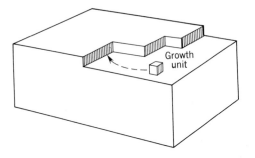
Growth
unit

Fig. 2-35. Two-dimensional nucleation. Growth units (atoms, ions, molecules) seek positions where the greatest decrease in energy will be provided by their presence; thus, they stick to re-entrant corners in partially formed atomic layers. Continued growth requires the nucleation of new layers as old layers are completed.

bricks are stacked to construct a wall. The atomic units find sites where their presence will produce the greatest decrease in free energy. Thus, it is easier for growth units to add to partially formed atomic layers than to start new layers (Figure 2-35). However, when a growth layer is completed, a new layer must be nucleated. A high supersaturation is necessary for this type of growth.

Spiral Growth

Many crystals form under conditions that differ from the ideal, and grow quite rapidly at low supersaturations. For example, calculations indicate that in the crystallization from a vapor the formation of a nucleus large enough to start a new layer requires a vapor pressure many times greater than saturation; yet crystals can be grown from vapor with a supersaturation of less than one percent. Thus, the theory of growth in separate layers cannot account for the formation of many crystals.

The explanation of the growth of crystals under conditions of low supersaturations was baffling up until 1949 when an English physicist, F. C. Frank, proposed that crystals can develop by means of SPIRAL GROWTH on screw dislocations (Figure 2-36). For example, if a screw dislocation is present it is not necessary to nucleate a new layer, since growth proceeds by a CONTINUOUS GROWTH OF THE SAME LAYER IN THE FORM OF

A SPIRAL. The screw dislocation provides a continuous open step where atoms easily attach themselves and build up the crystal without having to start a new layer. The dislocation theory of crystal growth is supported by a large number of observations of spiral growth on crystals, such as shown by Dawson and Vand's electron micrograph of a paraffin crystal (Figure 2-37).

The dislocation of adjacent atoms in crystals may have such causes as curvature of the crystal structure caused by impurities or vacant atomic sites; twinning in crystals; or mechanical twisting of sections of the crystal structure during growth.

Branching Growth—Dendrites

Under certain conditions, frequently rapid growth in limited directions, some minerals develop a branching tree-like structure, called DENDRITIC (Figure 2-38). Gold, silver, and copper often produce such forms (Plate 1, facing p. 2, and Plate 10, facing p. 67). Some snow crystals with their branching lacelike forms are symmetrical dendrites.

Dendritic growth results from different rates of growth in different crystallographic directions. One theory explains this type of growth in a liquid metal. The metal nucleus starts to grow by accretion of atoms parallel to crystal planes. The planes selected are most densely packed with atoms, and thus have the lowest surface energy. Growth may also occur in preferred directions influenced by impurities and dislocations. As the crystal nucleus starts to grow in the most favorable direction, heat is given off. The heat is evolved as the atoms in the liquid state give up their random motion for a more stable position in the crystal structure. When the latent heat of fusion reaches the melting point of the metal, growth "momentarily" stops in the most favorable direction and starts in a less favorable direction. The resulting structure, with its branching arms, forms a DENDRITE THAT HAS GROWN FROM A SINGLE NUCLEUS AND HAS A SINGLE CRYSTALLOGRAPHIC ORIENTATION, UNLESS DEFORMED.

Since the interface between the growing solid and the liquid is the place where latent heat is liberated, it should be slightly warmer than neighboring regions.

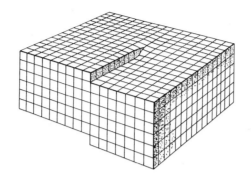

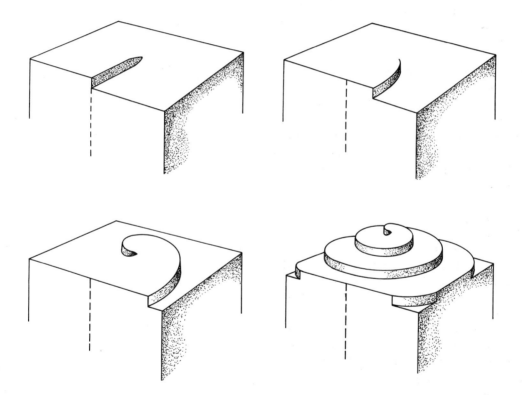

Fig. 2-36. *Screw dislocation,* illustrated at left, produces *spiral growth steps* as indicated by the succeeding diagrams. A screw dislocation is an imperfection that would result if a cut were made part way through a crystal and the two sides slipped over each other. The result is a permanent step extending across a portion of the crystal face and anchored at the boundary of the cut. Molecules depositing on the surface from a vapor lodge against the step, causing it to advance. Since one end is fixed, the step pivots around that point, the outer points falling behind the inner, and producing a never-ending spiral layer on the surface (from "The Growth of Crystals," by R. L. Fullman, *Scientific American,* March 1955).

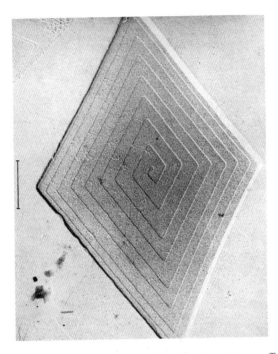

Fig. 2-37. Electron micrograph of growth steps on a paraffin crystal produced by a single dislocation. The crystal differs in rate of growth with direction, which yields straight rather than circular sides. Sides are normal to the direction of slow growth (after I. M. Dawson and V. Vand, 1951, *Proceedings, Royal Society,* London, A206, 555).

However, growth will only occur if the interface temperature is lower than, or equal to, the freezing point of the solid. Thus, the rate of growth is controlled by the rate of removal of latent heat. If, for example, heat is lost more rapidly at the corners of a crystal, the crystal will grow most rapidly from corners (Figure 2-39). However, as heat accumulates around the many dendrite arms, growth becomes slower, and if there is sufficient liquid between the arms these areas will be filled in until a crystal that has a single orientation is formed (Figure 2-40). If solidification is completed, all evidence of dendritic growth may be obliterated, and the dendrite may be visualized as a SKELETON crystal.

It has been suggested that metals grow by a dendrite process, regardless of their ultimate shape. Although dendrites have been observed in many miner-

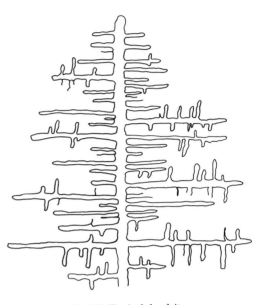

Fig. 2-38. Typical dendrite.

als, there is not sufficient evidence to suggest that mineral crystals in general can be considered as filled-in dendrites.

■ **TEMPERATURE VARIATION WITH COMPOSITION**

There is a tendency for dendrites to form in a liquid if the temperature of the liquid varies with the composition, as, for example, when solute atoms lower the freezing point of a solvent. As the liquid next to the solid-liquid interface is continually enriched with solute atoms not accepted by the solidifying material,

Fig. 2-39. Development of a dendrite arm.

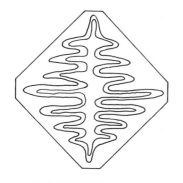

Fig. 2-40. Filled-in dendrite.

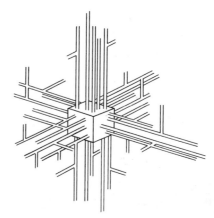

Fig. 2-42. Schematic diagram of alkali halide dendrites grown in a solution containing polyvinyl alcohol as a poison (from S. Amelinckx[1]).

a concentration gradient develops into the liquid from the growth surface. Since the temperature varies with the composition in this system, it will also, through the liquid, vary with distance from the interface. The liquid for some distance ahead of the interface will be

in a supercooled state, especially during rapid growth when the diffusion and convection currents in the liquid are not apt to be effective in smoothing out the uneven distribution of solute atoms.

■ **INFLUENCE OF IMPURITIES**

Experiments in crystal growth show that impurities greatly influence the formation of dendrites by

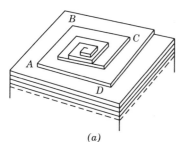

(a)

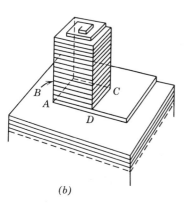

(b)

Fig. 2-41. The development of a dendrite branch (whisker) as a consequence of poisoning of the growth fronts *ABCDA* ... (*a*) The poisoned step is held up and a closed terrace is formed. (*b*) A whisker develops (from S. Amelinckx[1]).

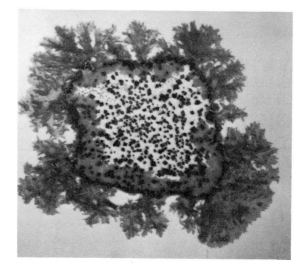

Fig. 2-43. Dendrites of potassium ferricyanide around a droplet. Small crystals form within the liquid (x 0.7).

Fig. 2-44. Dendritic freezing patterns on the surface of synthetic diamond (courtesy of R. H. Wentorf, Jr.[9]).

ber of dislocations from which spiral growth takes place. Then, as shown in Figure 2-41a traces of a poison are adsorbed preferentially along spiral steps *ABCDA* and growth is held up. In (b), successive steps emitted by the spiral-growth center form behind the poisoned step, creating a small terrace at its top face from which further spiral growth produces a dendrite arm or whisker. More complicated situations may arise if, instead of a simple spiral, a complicated spiral system exists.

Figure 2-42 shows an alkali halide crystal (such as halite (NaCl) or sylvite (KCl)) grown in dendritic form from solutions containing a suitable poison. The arms of the dendrites, many of which have side branches, are oriented in the six cube directions and are bounded by cube faces. Some of the atoms in the dendrite arms are slightly dislocated. The dendrite arms (whiskers) usually contain several to a large number of dislocations.

Dislocations may also be responsible for cubic minerals such as cuprite (Cu_2O) forming hair-like whiskers in the variety *chalcotrichite* (Plate 17, facing p. 114); or for the growth of wire-like silver (Plate 10, facing p. 67, and Figure 7-18, p. 135).

altering the rates of growth in certain crystallographic directions. It has been suggested that certain impurities adsorbed on a growing crystal face will hinder growth that is normal to the face by forming a mixed crystal that is unstable and tends to dissolve. On the other hand, experiments indicate that some impurities influence the crystal habit not so much by hindering face growth as by establishing a preference in dendritic limbs.[3] In the growth of synthetic diamonds, certain catalysts tend to produce dendritic structures under **NONRAPID** growth conditions.

A "poison" (an impurity that inhibits growth) combines with spiral growth in inducing dendrites in alkali halide crystals as explained by Amelinckx.[1] At first a small cubic crystal is formed, containing a num-

Fig. 2-45. Aggregates of dendrites of pyrolusite on limestone.

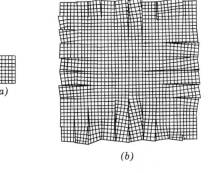

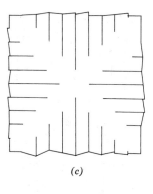

Fig. 2-46. Two-dimensional diagram of lineage structure. (*a*) Cells of a small perfect crystal. (*b*) Crystal structure branching into lineages. (*c*) Discontinuities in the structure of *b* that produce offsets at the surface of the crystal (after M. J. Buerger[4]).

(*a*)

(*b*)   (*c*)

■ HIGH SUPERSATURATION—RAPID GROWTH

The following simple experiment indicates that HIGH SUPERSATURATION, which induces RAPID GROWTH, can be a decisive factor in the formation of dendrites. That is to say, if a few drops of a saturated solution of potassium ferricyanide, $K_3Fe(CN)_6$, are placed on a glass surface and allowed to evaporate at room temperature, orange dendrites of potassium ferricyanide will form around the rim of the liquid where the evaporation and subsequent growth are rapid. Small reddish-orange crystals form in the center of the liquid where evaporation and growth are slower (Figure 2-43).

■ DENDRITIC IMPRINTS ON SYNTHETIC DIAMOND

Dendritic markings on the surface of synthetic diamond represent dendritic imprints of a metal catalyst such as nickel (Figure 2-44). The crystal structure of diamond has almost the same unit-cell size as nickel; thus, it can influence the crystallization of nickel,

which freezes as oriented dendrites on the diamond surface. Diamond growth ceases where the nickel dendrites have frozen. However, where the nickel is still molten, ridges of diamond continue to form until the remaining nickel also freezes. The resulting dendritic freezing patterns, consisting of ridges and valleys, are formed in the last stages of the diamond growth process.

■ DENDRITIC AGGREGATES

Dendritic mineral specimens may represent aggregates of dendrites (Figure 2-45). Dendritic crystals that grow from several centers often meet, and the resulting aggregate is composed of intergrown dendritic crystals.

If the arms of a dendritic crystal are mechanically twisted during growth, subsequent growth may at times produce a branching aggregate of randomly oriented crystals. The dendritic copper specimen illustrated in Plate 1(*A*), facing p. 2, may have been formed in this manner. (See also lineage structure—composite crystals.)

■ SUMMARY

Factors that tend to favor the growth of dendrites are rapid growth (which facilitates the accumulation of heat at certain points); a high value of latent heat; a positive gradient of supercooling from the interface into the liquid; impurities; dislocations; surface diffusion; and anisotropy of crystal shape. (In a cubic crystal, screw dislocations and impurities may introduce the anisotropy necessary to give one growth direction preference over another.)

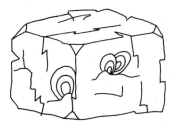

Fig. 2-47. Composite apophyllite crystal showing spiral growth hillocks and numerous offsets.

Fig. 2-48. "Parallel" grouping of galena.

SURFACE IMPERFECTIONS

Lineage Structure—Composite Crystals

M. J. Buerger[4] has proposed a theory of lineage structure in minerals, similar to dendritic growth. The crystal starts to grow from a small nucleus or seed crystal. As growth continues, separate portions of the crystal structure advance independently and produce LINEAGES or BRANCH-LIKE EXTENSIONS on the original nucleus. In growth, these lineages may twist slightly in relation to each other. When the areas between the lineages are filled in, the displaced lineages do not meet in perfect alignment; thus, they form screw dislocations that produce OFFSETS* or steps on the crystal surface (Figure 2-46). The resulting structures are commonly called COMPOSITE or OFFSET CRYSTALS.

Offset crystals of galena and some apophyllite are common (Figure 2-47). Overdeveloped offsets may result in a parallel grouping (Figure 2-48) with corresponding faces in parallel or almost parallel position.

*Offsets are also sometimes produced by twinning, such as Dauphiné twinning on quartz.

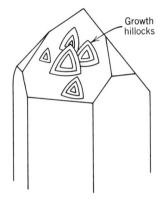

Fig. 2-49. Spiral growth hillocks on the rhombohedral faces of a quartz crystal.

Growth Hillocks

Spiral growth steps moving over the surface may interact with each other to form prominent pyramidal terraces or GROWTH HILLOCKS on the surface of the crystal. Impurities may also alter rates of growth in certain areas of a growing face. Growth hillocks are frequently observed on the faces of crystals, examples being the rhombohedral faces of quartz (Figure 2-49).

Growth Depressions

Processes of growth may at times produce geometrical depressions on the surface of a crystal. For

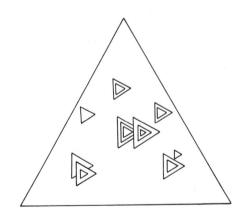

Fig. 2-50. Schematic diagram showing triangular growth pits ("trigons") on an octahedral face of diamond.

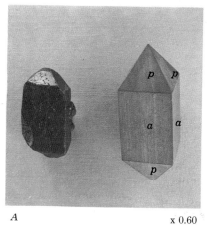

A   x 0.60

ZIRCON

Forms: (*a*) tetragonal prism; (*p*) tetragonal bipyramid.

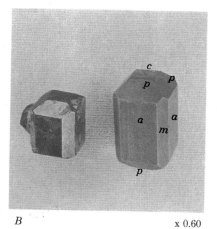

B   x 0.60

IDOCRASE

Forms: (*a*), (*m*) tetragonal prisms; (*p*) tetragonal bipyramid; (*c*) basal pinacoid.

C   x 0.50

SCAPOLITE

Forms: (*a*) tetragonal prisms; (*p*) tetragonal bipyramid.

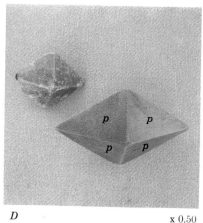

D   x 0.50

THENARDITE

Form: (*p*) rhombic bipyramid.

E   x 0.50

SULFUR

Forms: (*p*), (*s*) rhombic bipyramids; (*n*) rhombic prism; (*c*) basal pinacoid.

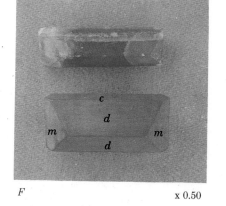

F   x 0.50

BARITE

Forms: (*d*), (*m*) rhombic prisms; (*c*) basal pinacoid.

Plate 6. HEXAGONAL CRYSTALS AND CORRESPONDING WOODEN MODELS

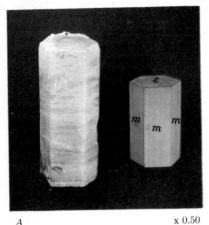

A   x 0.50

BERYL

Forms: (*m*) hexagonal prism; (*c*) pinacoid.

B   x 0.58

APATITE

Forms: (*m*) hexagonal prism; (*p*) hexagonal bipyramid.

C   x 0.55

APATITE

Forms: (*m*) hexagonal prism; (*p*) hexagonal bipyramid; (*c*) pinacoid.

D   x 0.63

TOURMALINE

Forms: (*a*) trigonal prism; (*m*) hexagonal prism; (*r*) rhombohedron.

E   x 0.50

CALCITE

Form: scalenohedron.

F   x 0.50

CALCITE

Form: rhombohedron showing polysynthetic twinning with striations parallel to short diagonal of rhombic cleavage plane.

Fig. 2-51. Etched topaz crystal illustrating that etch patterns differ on unlike crystal faces (after A. P. Honess[11]).

example, certain hollow triangular depressions frequently observed on the octahedral faces of diamond crystals are believed to be products of the **GROWTH PROCESS** (Figure 2-50). The corners of the triangular growth pits point toward the edges of the octahedral faces, and thus have a nonparallel orientation to the face edges. (The sides of triangular **SOLUTION PITS** on diamond are parallel to the octahedron edges.)

Solution Effects (Etch Figures)

When a crystal begins to dissolve it develops either shallow symmetrical pits or elevations on the surface. The dissolution does not take place uniformly over the whole surface but begins at certain points which often mark the location of emergent dislocations. Reagents frequently attack growth spirals on the surface by dissolving holes in the crystal at the centers of the growth spirals, which indicate that the core of a screw dislocation is chemically more reactive than the other more perfect parts of the crystal.[6]

As the etch pits are further developed by dissolution their boundaries extend. Then, if dissolution is allowed to continue, the etch pits merge, leaving raised residual islands (**ETCH HILLOCKS**) between the pits where the rest of the crystal face has been eaten away. It is sometimes difficult to determine whether a feature is a growth or an etch figure.

Since the rate of growth of a crystal varies with direction, it follows that the rate of solution should also vary with direction. The faces of different crystal forms generally dissolve at different rates, and **ETCH PATTERNS ARE SIMILAR FOR LIKE FACES ON A CRYSTAL AND DISSIMILAR FOR UNLIKE FACES** (Figure 2-51). The shapes of the pits and their angles relative to the faces reveal the true symmetry of the crystal.

REFERENCES

1. S. Amelinckx, "Dislocations in Alkali Halide Whiskers, Growth and Perfection of Crystals," *Proceedings of the International Conference on Crystal Growth* held at Cooperstown, New York, 1958; edited by R. H. Doremus, B. W. Roberts, and D. Turnbull, John Wiley and Sons, New York, 1958, p. 139.
2. H. P. Bovenkerk, "Some Observations on the Morphology and Physical Characteristics of Synthetic Diamond," *American Mineralogist*, Vol. 46, 1961, p. 952.
3. H. E. Buckley, *Crystal Growth*, John Wiley and Sons, New York, 1951.
4. M. J. Buerger, "Lineage Structure of Crystals," *Z. Krist.*, Vol. 89, 1934, p. 195.
5. C. Bunn, *Crystals: Their Role in Nature and Science*, Academic Press, New York, 1964.
6. W. C. Dash and A. G. Tweet, "Observing Dislocations in Crystals," *Scientific American*, October, 1961, p. 110.
7. I. M. Dawson and V. Vand, *Proc. Roy. Soc.*, London, Vol. A206, 1951, p. 555.
8. R. L. Fullman, "The Growth of Crystals," *Scientific American*, March, 1955.
9. J. J. Gilman, *The Art and Science of Growing Crystals*, John Wiley and Sons, New York, 1963. Chapter 7, "Ice," by B. J. Mason, pp. 119–150; Chapter 10, "Elements," by R. H. Wentorf, Jr., pp. 187–193.
10. A. Holden and P. Singer, *Crystals and Crystal Growing*, Doubleday and Company, New York, 1960.
11. A. P. Honess, *The Nature, Origin and Interpretation of Etch Figures on Crystals*, John Wiley and Sons, New York, 1927.
12. G. Lehmann and W. J. Moore, "Color Center in Ame-

thyst Quartz," *Science,* Vol. 152, No. 3725, May 20, 1966, p. 1061.

13. B. J. Mason, "The Growth of Snow Crystals," *Scientific American,* January, 1961, p. 120.

14. C. B. Sclar, L. C. Carrison, and G. G. Cocks, "Stishovite: Thermal Dependence of Crystal Habit," *Science,* Vol. 144, May, 1964, p. 833.

15. E. N. Slavnova, "New Data on the Interaction of Organic Impurities with Inorganic Crystals, Growth of Crystals," *Reports at the First Conference on Crystal Growth,* March 5–10, 1956 (original Russian text published by Academy of Sciences, USSR Press, Moscow, 1957; Eng-

lish translation, Consultants Bureau, New York, 1958, p. 117).

16. Ajit Ram Verma, "Crystal Growth and Dislocations," Butterworth's Scientific Publications, London, 1953.

17. R. H. Wentorf, Jr. and H. P. Bovenkerk, "Preparation of Semiconducting Diamonds," *J. Chem. Phys.* Vol. 36, 1962, p. 1987.

18. A. F. Williams, *Genesis of the Diamond,* Vol. 2, Ernest Benn, London, 1932.

19. W. H. Zachariasen, *J. Am. Chem. Soc.* Vol. 54, 1932, p. 3841.

3.

Crystal Chemistry

Crystal chemistry deals with the relationship between the chemical composition of a mineral, the geometrical arrangement of its atoms (structure), and the binding forces that hold the atoms together. This relationship determines the chemical and physical properties of minerals.

Since sight identification of minerals is based largely on physical properties such as specific gravity, hardness, cleavage, and magnetism, it is of interest to understand how atomic bonding and structure can affect such properties.

ELECTRONIC CONFIGURATION OF ATOMS

An atom may be pictured as a positive nucleus surrounded by electrons in rapid orbital motion. The number of electrons around the nucleus of a neutral atom is equal to the number of protons in the nucleus, which equals the ATOMIC NUMBER. For example, a sodium atom has an atomic number of 11; therefore, it has eleven protons in the nucleus surrounded by eleven orbiting electrons.

The electrons are not distributed at random but are arranged in energy levels. However, it is impossible (according to Heisenberg's uncertainty principle) to determine at the same time both the position and the velocity of an electron. Thus, the electron is pictured as an electron cloud whose density at any given point represents the probability of finding the electron at that point. The electron cloud is called an ORBITAL, and may be visualized as the electron's region of influence.

An orbital has a certain shape and direction (Figure 3-1), and may be either "SINGLY" OCCUPIED BY ONE ELECTRON OR "FILLED" BY TWO ELECTRONS THAT HAVE OPPOSITE SPINS. The orbitals are determined by quantum numbers that represent certain shells and subshells of electrons. No two electrons in an atom, or in atoms sharing electrons, can have the same four quantum numbers (Pauli's exclusion principle). Consequently, when an orbital contains two electrons, each must have an opposite spin; that is, the spin quantum numbers differ.

The electron SHELLS or ENERGY LEVELS are expressed by the quantum numbers (1, 2, 3, 4, 5, 6, and 7), and indicate the orbiting electron's average distance from the nucleus.

Subshells. The SUBSHELL number, which is expressed by the letters s, p, d, or f, determines the electron's angular momentum and the shape of its orbit. The orbital is the distribution of electrons within the subshell, and the number of orbitals within each subshell is limited. For example, s subshells have only

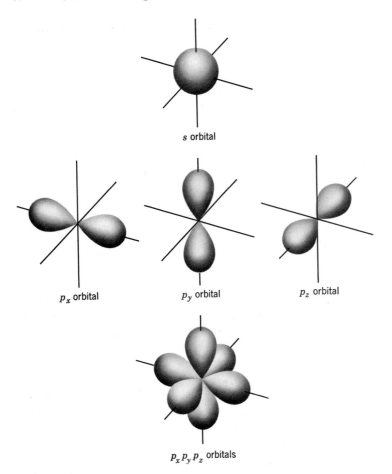

s orbital

p_x orbital p_y orbital p_z orbital

$p_x p_y p_z$ orbitals

Fig. 3-1. Models showing arrangements in space for the s and p_x, p_y, and p_z orbitals with respect to the nucleus of an atom. The s orbital is spherical and the p orbitals are dumbbell-shaped (after *Chemical Systems* by the Chemical Bond Approach Project, McGraw-Hill, New York, 1964).

one orbital and can therefore be singly occupied by one electron or can be "filled" with two electrons. A p subshell contains three orbitals, so that it can contain as many as six electrons. The three p orbitals are arranged in directions perpendicular to one another, since the exclusion principle prohibits any two orbitals from having the same direction. Subshell (d) can contain a maximum of five orbitals; and subshell (f), up to seven orbitals.

In general, electrons first occupy energy levels nearest the nucleus. When an atom contains only a few shells, one electron goes into each possible orbital in a given subshell. These single electrons have parallel spins. Then another electron of opposite spin is added to each of the subshell's orbitals. When atoms contain many shells the situation becomes more com-

plicated, since there is considerable overlapping of sublevels.

Shells. Shell 1, closest to the atomic nucleus, has only one s subshell with one orbital. In the case of the element hydrogen, with atomic number 1, the subshell s contains one electron ($1s^1$). In all higher elements the s subshell of shell 1 contains two electrons of opposite spins ($1s^2$).

Shell 2 contains an s subshell with one orbital limited to two electrons and a p subshell with three orbitals to yield a maximum of six electrons. Hence, shell 2 when filled can contain a total of eight electrons ($2s^2 2p^6$). The element neon, for example, has an atomic number of 10 with ten electrons, two of which are in shell 1 and eight in shell 2. The electronic notation for neon is

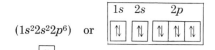

$(1s^2 2s^2 2p^6)$ or

The symbol ⬆⬇ represents two paired electrons of opposite spins.

Shell 3 contains s, p, and d subshells. When filled, these contain a total of two s electrons, six p electrons, and ten d electrons. The s and p electrons of shell 3 are filled in sequence, but the d electrons are added after the s electrons of shell 4; this is done because the s electrons of shell 4 ($4s$) actually have a slightly lower energy state than the $3d$ electrons. In succeeding shells, 4, 5, and 6, the d subshells are also filled after the next higher s subshell. The $4f$ subshell is not filled until after the $5s$, $5p$, and $6s$ subshells. Figure 3-2 shows the sequence in which the shells are filled.

The atomic orbital model is convenient for explaining bonding in atoms. However, recent calculations indicate that because of electron-electron repulsion, electrons will form two-electron orbitals only if these result in a considerable increase in charge density close to the nuclei.

ATOMIC BONDS

Many factors influence the way in which atoms unite and the type of bond formed. Among these are the atomic number, the electron distribution in various subshells, the size of the combining atoms, and environmental conditions such as temperature and pressure.

The bonding abilities of atoms depend largely upon the extent to which their OUTER electron shells are completed. The electronic configuration $s^2 p^6$ in an atom is called a COMPLETE or FILLED SHELL. An atom with an outer shell nearly filled tends to accept electrons, whereas an atom with only one or two electrons in its outermost shell tends to give up electrons. The loss of an electron by an atom is known as IONIZATION, and the energy required to remove an electron from an atom is called its IONIZATION POTENTIAL. The atom that gives up electrons becomes a POSITIVE ION or CATION, whereas an atom that receives an electron to complete its orbital becomes a NEGATIVE ION or ANION. Atoms with outer electron shells that are about half

filled form complete or closed shells by sharing electrons. The VALENCE of an atom is usually equal to the number of electrons that must be gained, lost, or shared by the atom in order to obtain a stable electronic configuration; and the outer electrons that form the bonds are called VALENCE ELECTRONS.

The atoms of the noble gases (helium, neon, argon, krypton, xenon, and radon) have, with the exception of helium, a "complete" outer shell of eight electrons ($s^2 p^6$), and for many years it was believed that these atoms could not form chemical bonds because they would neither accept nor donate electrons. In 1962, however, compounds of noble gases were produced with relative ease. Beautiful colorless crystals of *xenon tetrafluoride* (XeF_4) were made at the Argonne National Laboratory by heating a mixture of xenon and fluorine in a nickel container for one hour at 400°C.[8] Since xenon has a "complete" outer shell of eight electrons, its valence shell of shared electrons with fluoride must contain more than eight electrons.

Types of Bonds

In general, atomic bonds may be grouped into four types: COVALENT BONDS, METALLIC BONDS, IONIC BONDS, and VAN DER WAALS' BONDS. Two or more bond types may coexist in one mineral. All atomic bonds impart characteristic properties to the minerals in which they occur. It must be emphasized, however, that although the four bond types have well-defined characteristics, the bonding in minerals is usually intermediate in character. A transition between the various bond types occurs in most crystal structures. REAL BONDS IN MINERALS ARE RARELY IF EVER PURE BOND TYPES. For example, whether a bond will be ionic, or covalent, or intermediate in character depends upon where the two electrons involved spend their time. The bond is ionic if the electrons spend all their time near one atom, and covalent if the electrons divide their time equally between the two atoms.

■ COVALENT BONDING

COVALENT BONDS BETWEEN ATOMS, ACCORDING TO THE ATOMIC-ORBITAL MODEL, ARE FORMED BY THE OVERLAPPING OF ATOMIC ORBITALS, EACH OF WHICH IS SINGLY OCCUPIED BY ONE ELECTRON (THAT IS HALF FILLED). The outer or

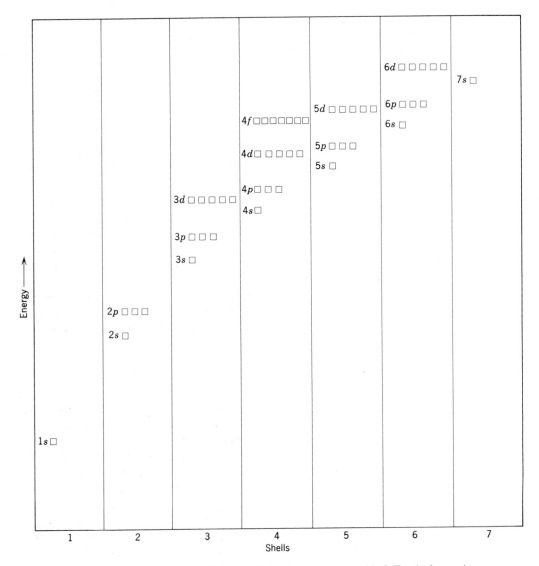

Fig. 3-2. Sequence in which energy shells are filled. □ represents an orbital. Two is the maximum number of electrons that can occupy any one orbital.

valence electrons normally participate in the formation of bonds. The covalent bonds, unlike the ionic and metallic bonds, are oriented in space at definite angles. The strength of the covalent bond depends upon the amount of overlap between the orbitals; the greater the overlap, the stronger the bond. On the other hand, when two "filled" orbitals (each contain-

ing two electrons) overlap, they repel each other and push the atoms apart. A filled orbital that cannot take part in bonding is called a LONE PAIR. Atoms with half-filled outer electron shells tend to form covalent bonds by the sharing of pairs of electrons.

The environmental conditions existing during crystallization may control the type of bonds formed and the resulting atomic structure. For example, although *diamond* and *graphite* are both composed of carbon atoms, diamond is formed under extremely high temperature-pressure conditions, whereas graphite is formed under lower temperatures and pressures. Although both are chemically identical, graphite is soft, black, and submetallic, whereas diamond is extremely hard, brilliant, and transparent. Thus, the physical properties depend not only upon the nature of the constituent atoms but also upon the way in which the atoms are bonded and arranged in the structure.

Carbon atoms in DIAMOND furnish an example of covalent bonding. A carbon atom has an atomic number of 6 and an electronic configuration of $1s^22s^22p^2$ or, better still, $1s^22s^22p_x^12p_y^1$ in the GROUND STATE, the lowest energy state (Figure 3-3a).

From the ground state of carbon, it may be reasonable to conclude that carbon will form only two covalent bonds with other atoms, since it has only two singly occupied (half-filled) *p* orbitals. However, in the diamond structure and many other carbon compounds, the carbon atom forms *four* covalent bonds. To account for this tetra-bonding (four atoms), the carbon atom must attain four unpaired electrons. This can be achieved by exciting one of the 2s electrons to the third 2p orbital, with the resulting electronic configuration of $1s^22s^12p_x^12p_y^12p_z^1$ (Figure 3-3b). However, the four new half-filled orbitals are not made up of one s orbital and three p orbitals, but are a new set of four mutually equivalent orbitals, each of which is called a sp^3 HYBRID ORBITAL. The redistribution of electron densities (orbitals) in the ground state of an isolated atom to form a new set of equivalent orbitals in

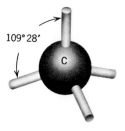

Fig. 3-4. Schematic diagram showing the directional nature of the covalent bonds in diamond that are directed toward the four corners of a tetrahedron.

the bonded atom is called HYBRIDIZATION. The carbon atom has slightly more energy in the excited state than in the ground state, but this increase in energy is then compensated by the formation of two extra bonds that decrease the atom's energy and thus form a more stable structure. The four sp^3 hybrid orbitals in the excited carbon atom point toward the four corners of a tetrahedron, making an angle of 109°28′ with each other (Figure 3-4).

In diamond, the bond between two adjacent carbon atoms is a sp^3–sp^3 bond formed by the overlap of two sp^3 orbitals (Figure 3-5). Each carbon atom has four nearest neighbors located at the four corners of a tetrahedron, and these tetrahedra are repeated throughout the diamond structure to form a GIANT MOLECULE that is as large as the crystal itself. In Figure 3-6, the relative sizes of the carbon atoms are reduced to show more clearly the structural arrangement of diamond.

Such covalent crystals as diamond, which are held together by strong covalent bonds in *three dimensions,* are STRONG and HARD, and have HIGH MELTING POINTS. The melting point of diamond is 3500°C. Pure diamond is a NONCONDUCTOR OF ELECTRICITY and a GOOD INSULATOR because the valence electrons of its carbon atoms are all LOCALIZED in strong covalent bonds between adjacent atoms (sp^3–sp^3 bonds), and are therefore not free to migrate through the crystal.

GRAPHITE is also composed of carbon atoms, but in the graphite structure each carbon atom is linked by strong covalent bonds to *three* others in the same plane, forming sheet-like layers composed of six-membered rings of atoms (Figure 3-7). The covalent

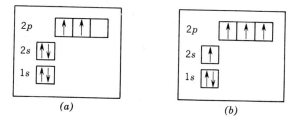

Fig. 3-3. (a) Ground state of a carbon atom ($1s^22s^22p^2$). (b) Excited state of carbon ($1s^22s^12p_x^12p_y^12p_z^1$).

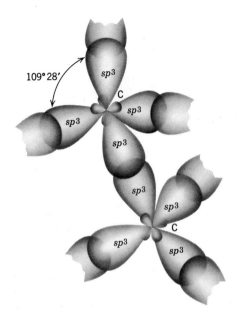

Fig. 3-5. sp^3–sp^3 bonding in diamond (after *Chemical Systems,* by the Chemical Bond Approach Project, McGraw-Hill, New York, 1964).

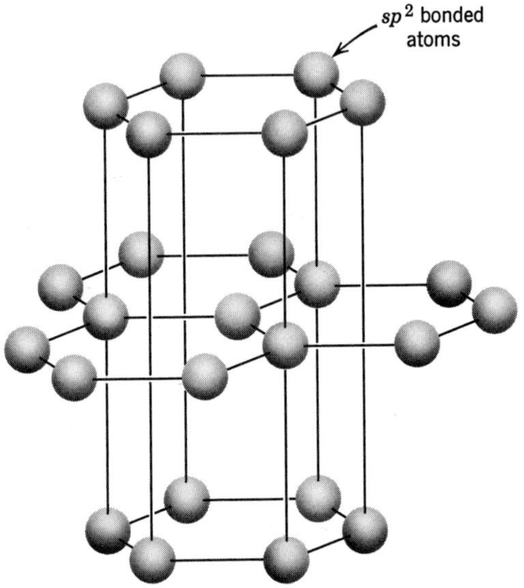

bonds in the graphite sheets involve three sp^2 hybrid orbitals that originate from each carbon atom (sp^2–sp^2 bond). (See Figure 3-8.) However, this leaves each carbon atom with one half-filled p orbital, whose

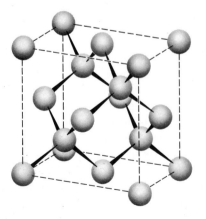

Fig. 3-6. Diamond structure—formed under extremely high-temperature-high-pressure conditions. Each carbon atom has four near neighbors located at the corners of a tetrahedron. Repetition of the tetrahedra produces a giant molecule.

Fig. 3-7. Graphite structure showing sheet-like layers of sp^2–sp^2 bonded atoms. Weak bonds exist between the widely separated layers, and cause graphite to split easily into sheets.

lobes extend at right angles above and below the plane of the sp^2 orbitals. The p orbitals overlap SIDE-WISE to form π bonds (pronounced $p\bar{\imath}$). The electrons in the π bonds do not belong to any specific atom and are said to be DELOCALIZED.* The delocalized π electrons are smeared out above and below the plane of the sp^2–sp^2 bonded layer of atoms.

The strong covalently bonded sheet-like layers in graphite are held apart by the repulsion of the carbon nuclei in adjacent layers and by the repulsion of the π electrons for the next layer.[2] The weak attractive forces between layers cause graphite to easily *split into thin sheets.* Graphite is a FAIRLY GOOD CONDUCTOR OF ELECTRICITY and has a SEMIMETALLIC appearance because of its DELOCALIZED π electrons. The SOFTNESS of graphite is attributed to the weak bonding between sheet-like layers, and this bonding permits adjacent layers to slide over each other. Graphite has a *high melting point* because of the strong covalent bonds within the sheets (sp^2–sp^2 bonds) that are difficult to disrupt.

* Electron waves that belong to more than two atoms are delocalized.

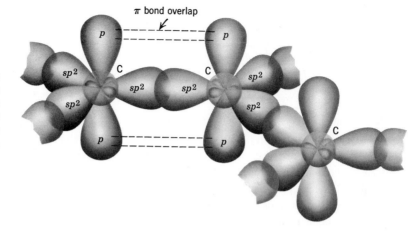

Fig. 3-8. Orbital model for graphite showing the sp^2–sp^2 bonds that form the sheet-like layers. The p orbitals overlap sidewise to form π bonds (after *Chemical Systems,* by the Chemical Bond Approach Project, McGraw-Hill, New York, 1964).

■ **METALLIC BONDING**

If one strikes a nonmetallic covalently bonded crystal (such as diamond or ice) or an ionic crystal (such as halite) with a hammer it will shatter. On the other hand, most metals DEFORM PLASTICALLY, and are described as MALLEABLE when they can be hammered into thin sheets without breaking, or as DUCTILE when they can be stretched out into a wire. The reason metals deform plastically and are good conductors of heat and electricity can be explained by the fact that the metallic bond involves DELOCALIZED MOBILE ELECTRONS.

Metals consist of an assemblage of closely packed positive ions that have more available orbitals of nearly the same energy level than they have valence electrons to occupy the orbitals. The valence electrons of each ion take turns forming electron-pair bonds with other ions and thus can move with ease through the solid metal. These mobile electrons do not belong to any one ion but to the crystal as a whole; hence, they are described as DELOCALIZED.

The ions in a metal can be moved past each other with considerable ease as long as the average distance between the ions is not greatly changed. Since the delocalized electrons surround each ion symmetrically, when neighboring ions are moved past each other new bonds readily form with the new nearest neighbors. Thus, the environment of the interior metal ions in the crystal is the same before and after displacement (Figure 3-9). In metal crystals, however, there are imperfections such as missing or extra ions, and these disturb the regularity of the crystal structure. The displacement of ions in the crystal proceeds at the points of these imperfections (dislocations). Thus, ions are moved row by row rather than by the

Fig. 3-9. (*a*) Metallic bonding—positive metallic ions in a sea of delocalized electrons. (*b*) Plastic deformation of a metallic crystal displaces groups of ions, but new bonds are readily formed with the new neighboring ions.

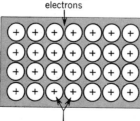

Delocalized electrons

Positive metal ions

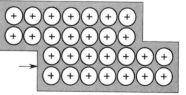

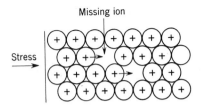

Fig. 3-10. Dislocations in crystals allow ions to migrate, row by row.

slippage of whole planes of ions past each other which would require much greater energy (Figure 3-10).

The mobility of the delocalized electrons causes metals to be GOOD CONDUCTORS OF ELECTRICITY, AND HEAT. The HIGH DENSITY of most metals may be attributed to the fact that the metallic bond permits the metal ions to be closely packed together. The majority of metals consist of atoms arranged so that each atom is surrounded by twelve nearest neighbors.

■ **IONIC BONDING**

The ions in an ionic crystal are held together by the strong electrical attractions between positive and negative ions. An ionic bond is produced when one or more electrons have been transferred from one atom to complete the orbitals of another atom. For example, the mineral halite (common salt) consists of a three-dimensional network of sodium (Na) and chlorine (Cl) ions. Sodium has an atomic number of 11 and an electronic configuration of $1s^2 2s^2 2p^6 3s^1$. The atomic number of chlorine is 17 and its electronic configuration is $1s^2 2s^2 2p^6 3s^2 3p^5$. The valence of sodium is $+1$; that is, it has one $3s$ electron that it can easily lose. On the other hand, chlorine has a valence of -1, which means it will accept an electron to complete its $3p$ shell.

Through electron transfer the sodium atoms become positive ions (Na^+) and the chlorine atoms become negative ions (Cl^-). Each ion tends to surround itself with as many oppositely charged ions as possible, but in order to maintain electrical neutrality the halite crystal must contain equal numbers of univalent sodium and chlorine ions. The sodium ions are smaller than the chlorine ions and are surrounded by six chlorine ions (Figures 3-11 and 3-12). The larger chlorine ions are similarly surrounded by six sodium ions.

No single chlorine ion is associated with a single sodium ion. Thus, the sodium chloride molecule (Na^+Cl^-) *does not exist* in the three-dimensional crystal. Such a molecule would require the electrostatic attraction between one sodium ion and one chlorine ion, as a result of the transfer of the valence electron of the sodium atom to the chlorine atom. The univalency of the sodium and chlorine ions expresses only the charge that the ions carry, and does not measure the number of other ions to which they can be linked.

The relative sizes of the ions and the energy requirements of the crystal play an important role in determining the number of ionic bonds originating from each ion. The arrangement of ions that will yield the lowest total energy is the most stable. Unlike the covalent bonds, the ionic bonds are NONDIRECTIONAL. In the sodium chloride crystal, the negative charge of the chlorine ion is neutralized by the sum of the partial positive charges of the six surrounding sodium ions.

Coordination of Ions. The way ions pack together in ionic crystals is, to a great extent, a function of their sizes and shapes. Each ion tends to surround itself with as many oppositely charged neighbors as possible. The number of nearest neighbors is the COORDINA-

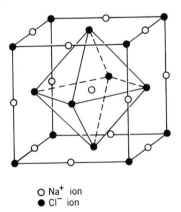

○ Na$^+$ ion
● Cl$^-$ ion

Fig. 3-11. Unit cell of halite (NaCl). Small black and white circles represent the positions of the centers of the ions, and do not represent relative sizes. Each atom is surrounded by six oppositely charged neighbors indicated by the sodium ion (white) surrounded by six chlorine ions (black) at the corners of an octahedron. The halite crystal is built up by a repetition of the unit cell in three dimensions. The unit cell is the atomic building block.

TION NUMBER of an ion. However, since ions are not rigid spheres with fixed surfaces, their pulsating electronic clouds may be deformed under the influences of surrounding ions. Thus, the sizes and shapes of ions are affected by contact with other charged bodies; and the same ion can have several different radii according to its coordination number and the forces existing between the ions. Therefore, although the coordination number is largely dependent upon the relative sizes of the ions in contact, the sizes of the ions are similarly influenced by the coordination number. The energy requirements of the crystal structure determine the type of coordination that will occur. For example, aluminum can occur in 4-fold or 6-fold coordination, depending upon the temperature and pressures that exist at the time of crystallization.

If it is assumed for convenience that ions are rigid spheres of fixed radii, then it is possible to calculate, from purely geometrical considerations, the radius of an ion around which the other ions of a given radius can be arranged. The RADIUS RATIO, which to a large extent determines the crystal structure, is the ratio of the radius of the cation to the radius of the anion (r_c/r_a). Table 3-1 shows the relationship of the radius ratio to the coordination number. The coordination numbers of cations most frequently found in crystal structures are 3, 4, 6, 8, and 12. Twelve-fold

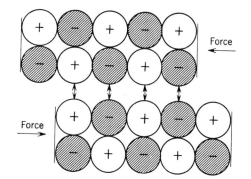

Fig. 3-13. Deformation of an ionic crystal in certain directions brings like ions into close contact, and destroys the electrostatic forces that hold the crystal together.

coordination occurs most frequently in metals that are not ionically bonded.

Physical Properties of Ionic Crystals. Ionic crystals are BRITTLE and tend to break along certain planes of ions when subjected to stress, rather than to deform like a metal in a malleable or ductile manner. The displacement of ions relative to each other in certain directions in an ionic crystal brings ions of LIKE CHARGE into close contact. The repulsion of the like ions causes the crystal to rupture (Figure 3-13).

Ionic crystals are POOR CONDUCTORS OF ELECTRICITY AND HEAT because their valence electrons are bound to individual ions (localized) and thus are not free to move through the crystal. However, the solutions of ionic compounds are good conductors of electricity, for they readily separate into positive and negative ions that are free to move about.

The HARDNESS of ionic crystals will vary over a wide range. Ionic bonds generally become progressively stronger with an increase in ionic charge (increase in valence) and a decrease in center-to-center distance between ions. Small ions have short interionic distances, since the distance between oppositely charged ions in contact with each other is determined by the sum of their radii. Since the hardness of an ionic crystal is a rough measure of its bond strength, SMALL IONS WITH MULTIPLE CHARGES WILL PRODUCE HARD CRYSTALS. For example, the compound BeO has an interionic distance of 1.65 Å and a hardness of 9.0 (according to the Mohs' hardness scale), whereas the

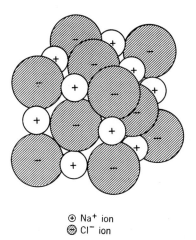

⊕ Na⁺ ion
⊘ Cl⁻ ion

Fig. 3-12. Packing of sodium and chlorine ions in the sodium chloride structure (halite), showing the relative sizes of the ions.

Table 3-1
Arrangements of ions

Coordination number (number of nearest neighbors)	Geometrical arrangement of ions	Radius ratio limits (r_c/r_a)

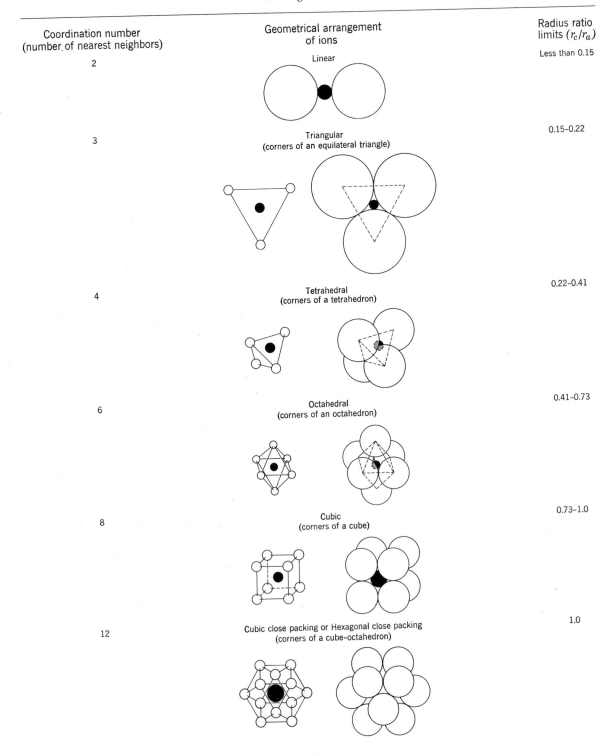

Coordination number (number of nearest neighbors)

Geometrical arrangement of ions

Radius ratio limits (r_c/r_a)

2 — Linear — Less than 0.15

3 — Triangular (corners of an equilateral triangle) — 0.15–0.22

4 — Tetrahedral (corners of a tetrahedron) — 0.22–0.41

6 — Octahedral (corners of an octahedron) — 0.41–0.73

8 — Cubic (corners of a cube) — 0.73–1.0

12 — Cubic close packing or Hexagonal close packing (corners of a cube-octahedron) — 1.0

compound NaF has an interionic distance of 2.31 Å and a hardness of only 3.2.

■ VAN DER WAALS BONDING

Neutral atoms and molecules that have no valence electrons available for ionic, covalent, or metallic bonding have a WEAK ATTRACTION for each other when brought into CLOSE PROXIMITY. This weak attractive force is called a RESIDUAL BOND or VAN DER WAALS BOND, and results from "momentary" shifts of electrons and nuclei toward opposite ends of atoms to produce temporary dipoles.

The electrons around the nucleus of an atom are in constant motion, and for a fraction of a microsecond their distribution may be greater at one end of the atom than at the other; this causes the atom to be temporarily more negative at one end. The positive end of the atom attracts and displaces the electrons of its neighboring atom; thus, each atom induces a temporary dipole in its neighbor. However, since the electrons are in constant motion, one portion of the atom or molecule does not remain positive or negative for any length of time, but fluctuates back and forth (Figure 3-14). These fluctuating dipoles attract each other and are referred to as Van der Waals forces.

Van der Waals forces contribute slightly to the bonding in ionic and covalent solids, but play an important role in the bonding of organic compounds. Van der Waals bonds are extremely weak. The weakness of these bonds is revealed by the *low melting points* and *softness* of residually bonded substances, and is a measure of the low energy needed to break the bonds.

INTERNAL STRUCTURE OF CRYSTALS

A crystal is built up by a repetition in three dimensions of IDENTICAL UNITS (atomic building blocks). Any repeating pattern, however, can be divided into identical units of pattern, all of which touch each other without leaving any spaces. For example, Figure 3-15 illustrates an appropriate unit of a two-dimensional pattern like the one common in wallpaper design. The three-dimensional atomic patterns of

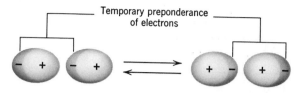

Temporary preponderance of electrons

Fig. 3-14. Fluctuation of electron distribution in two adjacent atoms.

crystals can also be divided into units. The units are usually chosen as the area with the shortest sides and angles nearest to being right angles. The box-like units of pattern in crystals are referred to as UNIT CELLS, and their repetition in straight lines in space constructs the crystal.

The unit cells of crystals are measured in Ångström units. An Ångström unit is 1/100,000,000th of a centimeter ($Å = 10^{-8}$ cm or 0.000,000,01 cm). For example, the length of the edge of a unit cell of copper is 3.6077 Ångström units (3.6077×10^{-8} cm or 0.000000014204 in.). One hundred million unit cells of copper side by side would measure slightly more than 1.4 inches.

The unit cell of copper is face-centered cubic (Figure 3-16), the 8 corners and 6 faces of a cubic lattice being occupied. Such cells, stacked together in three dimensions, form a copper crystal. Each corner atom is a common point for 8 unit cells, and each face-centered atom is shared by 2 cells. Thus, each unit cell of copper contains four atoms. It is not possible for

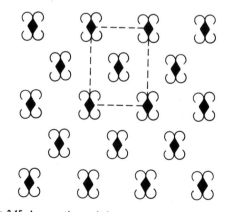

Fig. 3-15. A repeating unit in a two-dimensional pattern.

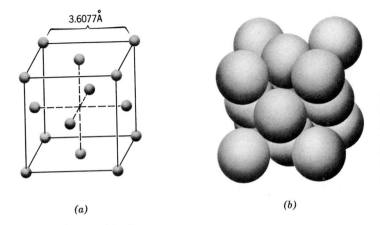

3.6077Å

(a)   (b)

Fig. 3-16. Diagrams of a face-centered cubic unit cell of copper. (*a*) The centers of the copper atoms are represented as small circles. (*b*) The "packing model" represents the copper atoms as spheres. Each face-centered atom touches its nearest corner atom, but the corner atoms do not touch each other.

a single unit cell to exist alone, since the cell boundaries of one unit are also the boundaries of neighboring units.

Each atom in the copper structure has 12 neighboring atoms, all at the same distance from each other. Gold and silver also occur in the "close-packed," face-centered cubic structure.

Space Lattice. A three-dimensional network of points that provides a geometrical framework to which the atoms of a crystal can be related is called a SPACE LATTICE (Figure 3-17). IDENTICAL atomic units are arranged around the points of the lattice; thus, the environment around each point is the same. As first demonstrated by Bravais, there are only fourteen fundamental arrangements that satisfy this condition (Figure 3-18).

A space lattice should not be confused with an established crystal structure of packed atoms, for the space lattice merely represents a three-dimensional periodic array of imaginary points. Because the points on a space lattice are positions that can be occupied either by single atoms or by groups of atoms, a large number of crystal structures may originate in the fourteen lattices (Figure 3-19). The atoms in the crystal VIBRATE about fixed positions in the space lattice, and the higher the temperature, the more violent the vibrations.

Although an ideal crystal would have all atoms symmetrically arranged around the points of a space lattice, real crystals, as ordinarily found, have imperfections in their structure that cause distortions in the crystal lattice array (see dislocations, Figure 2-36, p. 29). Imperfections are frequently caused by impurities or vacant atomic spaces in the crystal structure (Figure 3-20). Many interesting and useful properties of crystals (such as semiconductance, luminescence, or color) result from defects in the crystal structure.

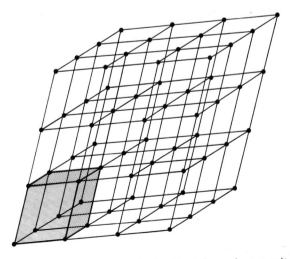

Fig. 3-17. A portion of a space lattice. Shaded area shows a unit of the lattice.

POLYMORPHISM

The ability of an element or compound to assume more than one atomic structure is referred to as POLY-

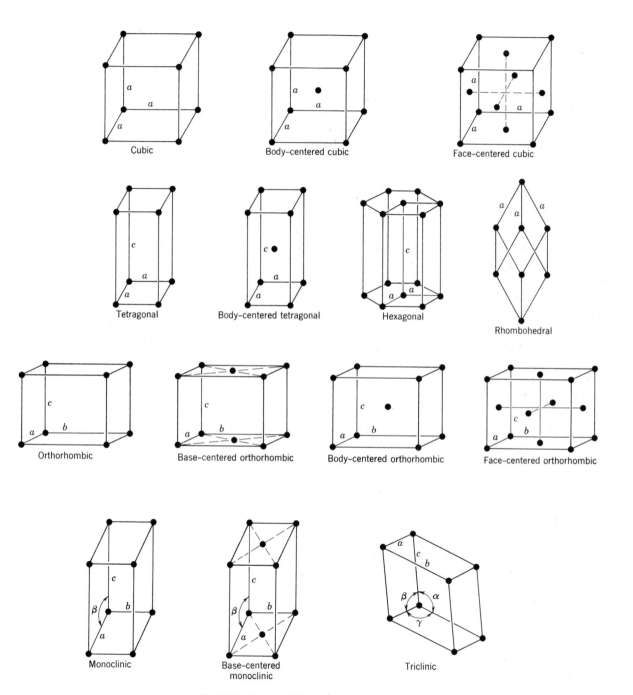

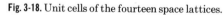

Fig. 3-18. Unit cells of the fourteen space lattices.

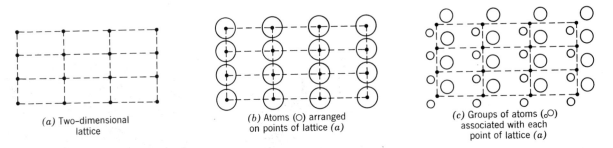

(a) Two-dimensional
lattice

(b) Atoms (O) arranged
on points of lattice *(a)*

(c) Groups of atoms (∘O)
associated with each
point of lattice *(a)*

Fig. 3-19. Schematic two-dimensional diagram showing two crystal structures originating from the same lattice. In structure *(c)* the atoms do not lie on the lattice points, but are symmetrically arranged about each point in identical groups.

MORPHISM or ALLOTROPY. Minerals of identical composition that occur in two distinct structural forms are called DIMORPHOUS; those with three different atomic structures are called TRIMORPHOUS. Table 3-2 compares the properties of some well-known polymorphous minerals.

Different polymorphs of the same mineral compound are stable under different conditions of pressure and temperature; and, frequently, as the environmental conditions change, the atoms or structural units rearrange themselves in the solid state to form the more stable structure. THE ATOMS OF A SUBSTANCE ALWAYS TEND TO ARRANGE THEMSELVES IN THE CONFIGURATION THAT WILL BE MOST STABLE UNDER GIVEN CONDITIONS OF PRESSURE AND TEMPERATURE.

The transition of one polymorph to another is RAPID when the transformation requires little energy and involves only minor rearrangements of atoms or displacement of structural units. A DISPLACIVE (OR HIGH-LOW) TRANSFORMATION proceeds rapidly because the structural units of the original mineral are only slightly displaced without breaking any of the atomic bonds (Figure 3-21, $(A \rightarrow B)$). The transition of high quartz to low quartz is an example of a high-low transformation. This reversible reaction is almost instantaneous at about 573°C and 1 atmosphere. High quartz is stable above 573°C and low quartz is stable below 573°C. That is,

$$\text{high quartz} \underset{\overrightarrow{}}{\overset{573°\text{C; 1 atm.}}{\rightleftharpoons}} \text{low quartz}$$

Other fairly rapid transformations occur when little energy is needed to break atomic bonds that immediately reform with other atoms.

The transformation of one polymorph to another is SLUGGISH when large amounts of energy are needed to break strong atomic bonds and reconstruct a new structure. The transition of graphite to diamond, or vice versa, is a RECONSTRUCTIVE transformation in which the original structure is torn apart and rebuilt with new bond types (see diamond and graphite structures, Figures 3-6 and 3-7, p. 42). Graphite is the

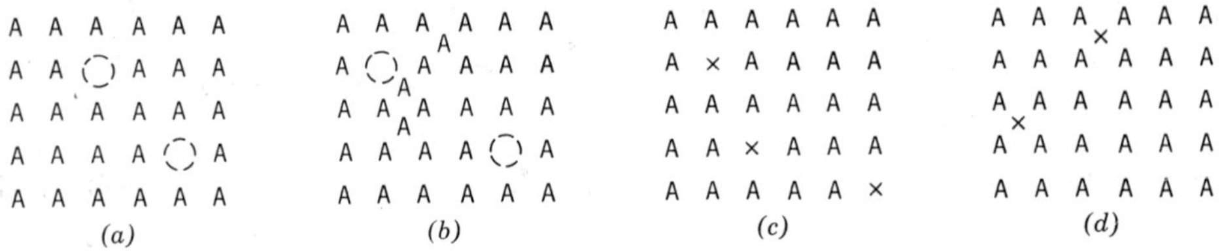

Fig. 3-20. Common defects in crystals. *(a)* Vacant lattice sites. *(b)* Atoms migrate from lattice sites to interstitial positions. *(c)* Foreign atoms in lattice positions. *(d)* Interstitial foreign atoms. Combinations of the various types of defects frequently occur in a single crystal.

Plate 7. MONOCLINIC (*A,B,C*) AND TRICLINIC (*D,E,F*) CRYSTALS

A   x 0.54

B   x 0.62

C   x 0.47

GYPSUM

Forms: (*b*) pinacoid; (*f, l*) monoclinic prisms.

SPHENE

Forms: (*c*) pinacoid; (*n, m*) prisms.

ORTHOCLASE

Forms: (*c, b, y*) pinacoids; (*m*) prism.

D   x 0.40

E   x 0.50

F   x 0.60

RHODONITE

Forms: pinacoids.

MICROCLINE

Forms: pinacoids.

CHALCANTHITE

Forms: pinacoids.

Plate 8. TWINNED CRYSTALS ■ CONTACT TWINS (A,B), PENETRATION TWINS (C,D), MULTIPLE TWINS (E,F)

A   **GYPSUM**   x 0.35

B   **CALCITE**   x 0.45

C   **ORTHOCLASE**   x 0.70

D   **STAUROLITE**   x 0.50

E   **RUTILE**   x 1.0

(cyclic twins)

F   **OLIGOCLASE**   x 0.70

(polysynthetic twins)

Table 3-2
Examples of Polymorphous Minerals

Chemical composition	Mineral	Crystal system	Hardness	Specific gravity
C	Diamond	Isometric	10	3.50–3.53
	Graphite	Hexagonal	1	2.09–2.23
$CaCO_3$	Calcite	Hexagonal	3	2.71
	Aragonite	Orthorhombic	$3\frac{1}{2}$	2.94
ZnS	Sphalerite	Isometric	$3\frac{1}{2}$–4	3.9–4.1
	Wurzite	Hexagonal	$3\frac{1}{2}$–4	3.98
FeS_2	Pyrite	Isometric	6	5.01
	Marcasite	Orthorhombic	6	4.88
TiO_2	Rutile	Tetragonal	6–$6\frac{1}{2}$	4.2
	Anatase (Octahedrite)	Tetragonal	$5\frac{1}{2}$–6	3.9
	Brookite	Orthorhombic	$5\frac{1}{2}$–6	4.14
Al_2SiO_5	Andalusite	Orthorhombic	$7\frac{1}{2}$	3.1–3.2
	Sillimanite	Orthorhombic	$6\frac{1}{2}$–$7\frac{1}{2}$	3.23
	Kyanite	Triclinic	5–7	3.53–3.65

stable polymorphous form of carbon at room temperature. At very high temperatures and pressures graphite transforms to diamond (Figure 2-33, p. 26). However, once formed, diamond is also a remarkably stable substance at room temperature and pressure, and does not readily change to graphite because of the high energy required to break down and rearrange its strongly bonded structure. Hence, the rate of change of diamond to graphite at room temperature is not capable of detection.

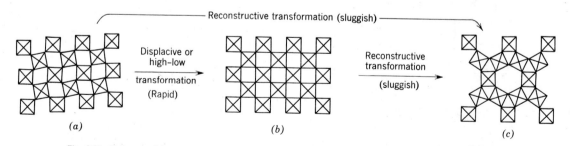

Fig. 3-21. Polymorphic transformations. Schematic crystal (*a*) consists of atomic polyhedra sharing corners. The displacive transformation (*a*) → (*b*) is rapid because none of the atomic bonds are broken; but the reconstructive transformations (*a*) → (*c*) or (*b*) → (*c*) are sluggish because high energy is needed to break the bonds between atoms and form them anew (after Azároff: *Introduction to Solids*, McGraw-Hill, New York, 1960).

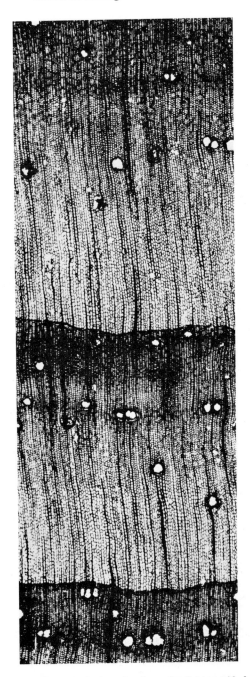

Fig. 3-22. Transverse section of a piece of extinct petrified pine wood showing an extremely well-preserved cellular structure. The annual growth rings are indicated by the dark horizontal bands (courtesy of Erling Dorf, "The Petrified Forests of Yellowstone Park, "*Scientific American,* April, 1964; photomircograph by Sol Mednick).

ORDER-DISORDER transformations (see p. 54) are believed to be responsible for the polymorphic transformations in the potassium feldspars (orthoclase, microcline, sanidine, and adularia).

PSEUDOMORPHS (FALSE FORMS)

At times a crystal may have the atomic structure and chemical composition of one mineral and the external crystal form of another. Such a material is said to be a FALSE FORM or PSEUDOMORPH. For example, octahedral crystals of the mineral magnetite (Fe_3O_4) frequently alter to hematite (Fe_2O_3) but still retain their original octahedral shape. Such crystals are pseudomorphs of HEMATITE AFTER MAGNETITE.

Pseudomorphs may be formed in several ways:

1. Molds or cavities previously occupied by one substance may later be filled by a mineral that preserved the original outline of the earlier substance. When wood is petrified, mineral matter (usually, silica, SiO_2) commonly fills the cavities inside empty wood cells and surrounds the tough residual cellular walls. This preserves even the microscopic structure of the original wood (Figure 3-22). The residual cellular walls can often be exposed by immersing a polished section of silicified wood in hydrofluoric acid. The hydrofluoric acid dissolves some of the silica and leaves a projecting residue of wood fibers.[3] A polished section of petrified wood from the Petrified Forest in Arizona is illustrated in Plate 35, facing p. 211.

2. A mineral may be deposited as a crust over crystals of another mineral, as in the case of quartz encrusting cubic fluorite (Figure 3-23), or marcasite covering calcite.

3. Pseudomorphs may be formed by alteration that involves a partial chemical change in a mineral. For example, *copper* may oxidize to *malachite*, $Cu_2CO_3(OH)_2$; *pyrite*, FeS_2, may alter to *goethite*, $HFeO_2$; *galena*, PbS, frequently alters to *anglesite*, $PbSO_4$ ("woody anglesite," Plate 28); and the *feldspars* and *feldspathoids* commonly alter to *clay* minerals. Plate 3, facing p. 18, shows trapezohedral leucite crystals which have altered to the clay mineral kaolinite and formed pseudomorphs of kaolinite after leucite.

Fig. 3-23. Pseudomorph (overgrowth) of quartz after fluorite formed by quartz encrusting cubic fluorite crystals: Derbyshire, England (x 0.78).

4. Minerals that are chemically unrelated may also form pseudomorphs, such as quartz replacing calcite. Frequently, one mineral may replace a number of unrelated minerals. On the other hand, a single mineral species may be replaced by several different minerals. The replacement of one mineral by a second mineral implies that a change in environmental conditions caused the original mineral to become unstable.

5. **PARAMORPHS** are formed when the atoms in a crystal are rearranged but the chemical composition and the external forms remain unchanged. Only polymorphous minerals can form paramorphs. The orthorhombic mineral aragonite, $CaCO_3$, can change to the more stable hexagonal mineral calcite, $CaCO_3$, but still retain its original external shape; this change is a paramorph of calcite after aragonite. Another example of paramorphism is the inversion of orthorhombic marcasite to isometric pyrite, which is the more stable form of FeS_2.

SOLID SOLUTIONS

Although most familiar examples of solutions are of the liquid type, many solutions of solids within solids occur. A **SOLID SOLUTION** is a homogeneous solid of two or more elements in solution with one another. Three types of solid solutions exist: **SUBSTITUTIONAL, INTERSTITIAL,** and **DEFECT.**

Substitutional Solid Solution

Substitutional solid solution occurs when solute atoms replace the solvent atoms on points of the lattice of the solvent mineral. However, if the atoms differ appreciably in size, the solute atoms will not fit into the parent crystal structure without causing excessive distortion, and the solid solubility will be limited or nonexistent. The type of atomic structure plays an important role in determining the amount of atomic substitution.

At elevated temperatures the amount of atomic substitution is usually increased; this occurs because the distances between atoms in a crystal structure are greater and therefore the similar size requirement for the atoms forming a solid solution is not as critical.

The usual type of substitutional solid solution is the **DISORDERED** type where the solute atoms are distributed at random on the points of the solvent lattice (Figure 3-24a). **ORDERED** substitutional solid solution (known as **SUPERSTRUCTURE**) occurs when the atoms are not randomly dispersed but occupy preferred lattice sites (Figure 3-24b). When the atoms in a disordered-type solid solution rearrange themselves to form a superstructure, the process is called **ORDER-DISORDER**

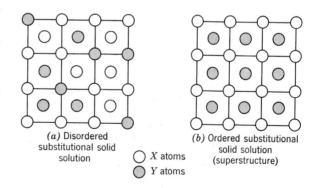

(a) Disordered substitutional solid solution

(b) Ordered substitutional solid solution (superstructure)

○ X atoms   ◉ Y atoms

Fig. 3-24. Order-disorder transitions. (a) disordered solid solution of X and Y atoms where Y atoms (black) are distributed at random on the solvent crystal lattice; and (b) ordered substitutional solid solution where Y atoms occupy preferred lattice sites.

TRANSFORMATION. The ordered arrangement has less energy than the disordered distribution, and is thus more stable. At elevated temperatures disorder is more common. When a random solid solution, formed at elevated temperatures, is slowly cooled, there may be a tendency for the atoms to order. An ordered structure may be disordered by heating it, and the disorder may be frozen in by rapid cooling.

The polymorphic forms of the potassium feldspars (orthoclase, microcline, sanidine, adularia) have the same chemical composition, $KAlSi_3O_8$, and are believed to result from order-disorder transformations. For example, the silicon and aluminum ions are randomly distributed over their lattice sites in the high-temperature monoclinic polymorph *sanidine,* but are ordered in triclinic *microcline.* Microcline can be transformed to sanidine by hydrothermal treatment.

COMPLETE SOLID SOLUBILITY occurs when atoms are soluble in all proportions. A SOLID SOLUTION SERIES EXHIBITING CONTINUOUS CHANGE IN COMPOSITION WITH VERY LTTLE OR NO CHANGE IN GEOMETRIC STRUCTURE IS CALLED ISOMORPHOUS (SAME FORM). For example, gold and silver atoms each with essentially the same atomic radius (1.44 Å) and similar chemical properties display complete solid solubility. All the members of the gold-silver series have a face-centered cubic atomic structure and thus are isomorphous. The plagioclase feldspars also form an isomorphous series. The end members of the series are *albite,* $Na(AlSi_3O_8)$, and *anorthite,* $Ca(Al_2Si_2O_8)$. Between the two extremes, calcium ions substitute for sodium ions that are almost identical in size. However, since the electrical charge

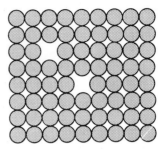

Fig. 3-26. Defect solid solution caused by the absence of atoms from their atomic sites in the crystal structure.

or valency of a Ca^{++} ion is $+2$ and the charge of Na^+ is $+1$, aluminum ions (Al^{3+}) substitute for silicon ions (Si^{4+}) to maintain the electrical neutrality of the mineral. Hence, $Ca^{2+} + Al^{3+}$ substitutes for $Na^+ + Si^{4+}$.

However, not all continuous solid solution series are isomorphous, since intermediate phases differing in crystal structure may form. For example, in the solid solution series of copper and zinc, three types of brass with different atomic structures form between certain percent ratios of copper and zinc.

LIMITED SOLID SOLUTION occurs when the atoms cannot combine in all proportions. Copper atoms, which have an atomic radius of 1.28 Å, can form only limited solid solutions with silver and gold atoms having an atomic radius of 1.44 Å; this is the case even though the atoms of all three metals are arranged in a face-centered cubic lattice.

Iron atoms can occupy positions normally occupied by zinc atoms in the mineral sphalerite, but only a limited amount of iron can replace the zinc. At elevated temperatures the amount of zinc replaced by iron is increased, but the solid solution is still limited.

Interstitial Solid Solution

The second major type of solid solution is INTERSTITIAL SOLID SOLUTION. Here the solute atoms do not replace the solvent atoms, but occupy positions within the interstices (open spaces between atoms) of the host structure (Figure 3-25). Since the atoms must fit into the interstices of the host structure, they are usually very small and will occupy spaces in the structure with the greatest linear dimensions. Hydrogen,

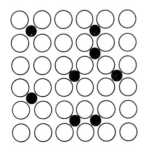

Fig. 3-25. Interstitial solid solution showing the random distribution of small atoms (black) in the spaces between the atoms in a crystal structure.

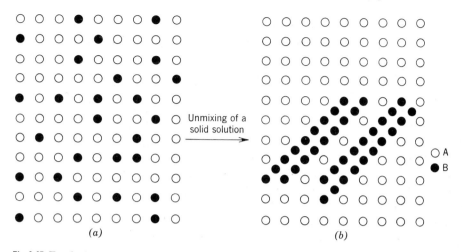

Fig. 3-27. Exsolution (unmixing of a solid solution). (*a*) Solid solution of solvent atoms A (○) and solute atoms B (●), which is stable only at elevated temperatures. (*b*) Precipitation in the solid state of solute atoms B (◉) in definite crystallographic directions of host lattice A (○).

carbon, boron, and nitrogen commonly form interstitial solid solutions in metals. Many interstitial combinations are very hard and have extremely high melting points, a fact which indicates that the interstitial elements may form strong covalent bonds with their neighboring metallic atoms.

Although small atoms usually occupy the interstitial spaces, if the interstices are very large, atoms of considerable size can occupy them. Thus, sodium atoms are able to occupy the large openings in the crystal structure of the minerals cristobalite (SiO_2) and tridymite (SiO_2).

Defect Solid Solution

Defect solid solution, otherwise called omission solid solution, occurs when some of the atomic sites in a crystal structure are vacant (Figure 3-26). For example, the mineral pyrrhotite ($Fe_{1-x}S$) has a deficiency of iron resulting from the absence of some of the atoms from their places in the crystal structure.

Unmixing of Solid Solutions (Exsolution)

Unmixing or exsolution of a solid solution involves the separation in the **SOLID STATE** of a single

homogeneous mineral into two minerals. Some solid solutions form at high temperatures and become unstable at lower temperatures. As the temperature is lowered the solubility of the solute in these solid solutions decreases, and the unstable solute atoms migrate through the parent crystal structure. These misfit atoms group themselves locally, and finally form their own crystal structure, which is oriented in a **DEFINITE CRYSTALLOGRAPHIC POSITION** with respect to the host structure in which it forms (Figure 3-27). The atoms of the two lattices form a partial registry at their mutual boundary. This process of segregation and growth in the solid state of rejected atoms or ions to locally form their own crystal structure in the host lattice is called **EXSOLUTION.** *

Corundum, rose quartz, and blue quartz commonly contain microscopic needle-like oriented inclusions of rutile; these have been precipitated from solid solution (see asterism, p. 85).

Figure A3 shows the Widmanstätten structure of an iron meteorite which has resulted from the unmixing of an original high-temperature nickel-iron alloy.

Various other mineral intergrowths are interpreted as exsolved crystals from solid solution. For example, unmixing of magnetite and ilmenite produces thin plates of ilmenite in magnetite. Perthite structure, consisting of thin plates of sodium feldspar in potassium feldspar, is sometimes attributed to exsolution.

REFERENCES

1. L. V. Azároff, *Introduction to Solids,* McGraw-Hill, New York, 1960, pp. 180–183.
2. Chemical Bond Approach Project, *Chemical Systems,* McGraw-Hill, New York, 1964.
3. E. Dorf, "The Petrified Forests of Yellowstone Park," *Scientific American,* April, 1964, pp. 106–114.
4. R. C. Evans, *An Introduction to Crystal Chemistry,* Cambridge University Press, Cambridge, 1952.
5. W. S. Fyfe, *Geochemistry of Solids: An Introduction,* McGraw-Hill, New York, 1964.
6. E. S. Gilreath, *Fundamental Concepts of Inorganic Chemistry,* McGraw-Hill, New York, 1958.
7. *Handbook of Chemistry and Physics,* 45th edition, the Chemical Rubber Company, Cleveland, Ohio, 1964.
8. H. Selig, J. G. Malm, and H. H. Classen, "The Chemistry of the Noble Gases," *Scientific American,* May, 1964.
9. M. J. Sienko and R. A. Plane, *Chemistry,* 2nd edition, McGraw-Hill, New York, 1961.

4.

Symmetry of Mineral Crystals

All crystals grow according to symmetrical patterns. These patterns extend in three directions and provide the basis for crystal classification. Crystal systems (6 in number) and crystal classes (numbering 32) are determined by such symmetry arrangements.

The geometrical forms exhibited by crystals frequently provide clues that enable an observer to recognize the system, or even the class, to which a crystal belongs. After starting this study with such simple forms as the cube or the octahedron, the observer may proceed until he becomes aware of the distribution of the 48 possible forms and the allowable combinations among the crystal classes. Since minerals often crystallize in simple combinations of forms, their recognition is frequently a useful aid in mineral identification.

EXTERNAL SYMMETRY OF CRYSTALS

Symmetry is so common that it is ordinarily accepted without comment. The pods on a tree, living creatures, works of art, architecture, and the plans of many cities exhibit symmetry. The absence of symmetry is always likely to be noticed—city streets without design; a man with one eye; an automobile with doors on only one side. Two or more units combined in a uniform pattern exhibit symmetry.

MINERAL CRYSTALS EXHIBIT EXTERNAL SYMMETRY BECAUSE THEY HAVE IDENTICAL AND REGULARLY REPEATED FACES. Knowledge of crystal symmetry is an important aid to crystal identification. The repetition of crystal faces may be described in terms of rotation about an axis, reflection across a plane (mirror reflection), and inversion through a center.

In terms of surficial symmetry, 32 patterns (and only 32) may be recognized on crystals. These are the 32 CRYSTAL CLASSES, each representing a UNIQUE GROUP OF SYMMETRY OPERATIONS. The growth of precise symmetrical crystals from unsymmetrical solutions is remarkable.

X-ray studies have shown that all crystals also grow internally according to a geometric pattern. The number of internal crystal patterns (space groups) is mathematically limited to 230.

When an element or chemical compound crystallizes it follows one of the known symmetrical arrangements. The symmetry arrangements of crystals may be grouped according to SYSTEMS, CLASSES, and SPACE GROUPS. The 6 systems and the 32 classes may be interpreted from a study of external features, but recognition of the 230 space groups depends upon interpretation of the relationship of atoms within the crystal ordinarily accomplished with the aid of x-ray reflections.

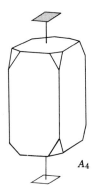

Fig. 4-1. A single axis of 4-fold symmetry (A_4). Rotation about the A_4 axis brings like faces into similar positions at 90° intervals.

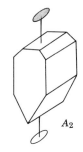

Fig. 4-3. An axis of 2-fold symmetry (A_2). Rotation about the axis brings like faces into similar positions at 180° intervals.

SYMMETRY OPERATIONS

A **symmetry axis** represents an imaginary line through the center of a crystal, and the crystal may be rotated about this axis so that **identical** groups of faces will fall into equivalent positions at regular intervals of rotation (Figures 4-1 to 4-5). Such an axis may be 2-fold, 3-fold, 4-fold, or 6-fold, depending upon the repetition of similar positions every 180°, 120°, 90°, or 60°, as the crystal is rotated. Convenient symbols may be used to designate the axes: A_2, A_3, A_4, A_6, as illustrated in Figures 4-1 to 4-5 (●A_2; ▲A_3; ■A_4; ⬤ A_6).

A **symmetry plane** is an imaginary plane through

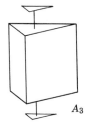

Fig. 4-4. An axis of 3-fold symmetry (A_3). Rotation about the A_3 axis brings like faces into similar positions at 120° intervals.

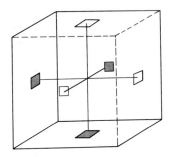

Fig. 4-2. A cube with three axes of 4-fold symmetry ($3A_4$).

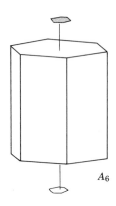

Fig. 4-5. An axis of 6-fold symmetry (A_6). Rotation about the A_6 axis brings like faces into similar positions at 60° intervals.

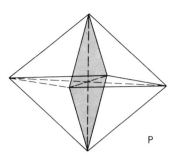

Fig. 4-6. Symmetry plane. The shaded symmetry plane divides the crystal into two halves—mirror images of each other.

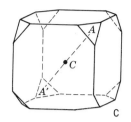

Fig. 4-8. Center of symmetry (cube cut by an octahedron).

the center of a crystal that divides the crystal into two halves, both being mirror images of each other (Figure 4-6).

There are **NINE** symmetry planes that divide a

cube into mirror images (Figure 4-7). Three planes are parallel to the cube faces, and six diagonal planes pass through opposite edges of the cube.

A crystal has a **CENTER OF SYMMETRY** if an imaginary straight line can be passed from a point on any face of the crystal, through a center, to a corresponding point on an opposite face (Figure 4-8). For each

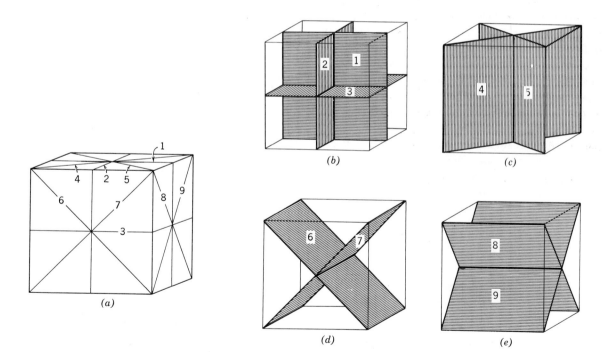

Fig. 4-7. The nine symmetry planes of a cube. (*a*) Traces of symmetry, planes 1 to 9. (*b*) Planes 1, 2, and 3. (*c*) Planes 4 and 5. (*d*) Planes 6 and 7. (*e*) Planes 8 and 9 (after Klug and Alexander, *X-ray Diffraction Procedures,* John Wiley and Sons, New York, 1954).

Fig. 4-9. The tetrahedron has no center of symmetry.

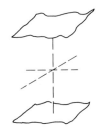

Fig. 4-11. Pinacoid—an open form consisting of two opposite parallel faces.

face on crystal with a symmetry center there is a corresponding parallel face on the opposite side.

A tetrahedron, which has four equilateral triangular faces, has no center of symmetry (Figure 4-9). A line drawn from the base of the tetrahedron through the center will not intersect a similar face on the opposite side.

The six crystal systems (ISOMETRIC, TETRAGONAL, HEXAGONAL, ORTHORHOMBIC, MONOCLINIC, and TRICLINIC) subdivide the 32 classes.

Crystals belonging to any one system can be referred to the same set of reference (CRYSTALLO-GRAPHIC) axes. These imaginary axes have definite unit and angular relations, and are oriented to conform to the symmetry elements of the crystal. The LENGTHS of the crystallographic axes are the RELATIVE lengths of the edges of a unit cell that is measured in the direction of the axes. An isometric unit cell, for example, is cube shaped, and therefore its edges are equal in length. The isometric system has 3 crystallographic axes of equal length at right angles to each other.

The crystallographic axes are distinct from the axes of symmetry, although, in a number of instances, an axis of reference and an axis of symmetry may coincide. Ordinarily, six sets* of crystallographic axes serve for the six crystal systems.

Each crystal system has distinctive symmetry features. For example, all members of the ISOMETRIC SYSTEM have 4 AXES OF 3-FOLD SYMMETRY. No other crystal system has more than 1 axis of 3-fold symmetry. Several classes of the TETRAGONAL SYSTEM have 1 AXIS OF 4-FOLD SYMMETRY; in others, the same position may be occupied by an axis of 2-fold symmetry. In the HEXAGONAL SYSTEM, 5 classes have a single axis of 6-fold symmetry, and in 7 classes a single 3-fold axis occupies the same position. The orthorhombic system is limited to axes of 2-fold symmetry. Many crystals of the ORTHORHOMBIC SYSTEM have 3 AXES OF 2-FOLD SYMMETRY.

* The rhombohedral division of the hexagonal system is described at times in terms of three equally inclined axes defined by the intersecting faces of a rhombohedron.

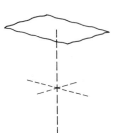

Fig. 4-10. Pedion—an open form consisting of a single face, with no symmetrically equivalent face.

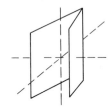

Fig. 4-12. Sphenoid or Dome—open form consisting of two nonparallel faces that intersect two of the crystallographic axes and are parallel to the third. The faces of a sphenoid are symmetrical to an axis of 2-fold symmetry. The dome is symmetrical with respect to a symmetry plane.

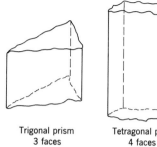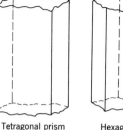

Trigonal prism
3 faces

Tetragonal prism
4 faces

Hexagonal prism
6 faces

Fig. 4-13. Prism—an open form composed of 3,4,6,8, or 12 faces, all parallel to, and symmetrical about, an axis.

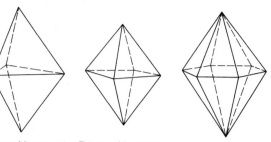

Trigonal bypyramid   Tetragonal bypyramid   Hexagonal bypyramid

Fig. 4-15. Bipyramid—a closed form composed of an upper and a lower pyramid.

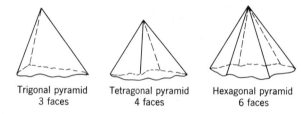

Trigonal pyramid
3 faces

Tetragonal pyramid
4 faces

Hexagonal pyramid
6 faces

Fig. 4-14. Pyramid—an open form composed of 3,4,6,8, or 12 nonparallel faces that intersect at a common point.

Fig. 4-16. Bisphenoid—a closed form consisting of two sphenoids.

CRYSTAL FORMS

The total number of like faces on a crystal constitutes a **FORM**. Symmetrical growth provides a wide variety of forms, a number of which correspond to the common forms of solid geometry. All faces in a form have the same atomic pattern, yield similar etch patterns, and occupy a similar position relative to the symmetry of the crystal. Certain forms are unique for one crystal class or system, whereas others may occur in several classes or systems. Some 48 forms are recognized on crystals.

In the isometric system, all crystal forms completely enclose space and are known as **CLOSED FORMS**. In other systems, forms may or may not completely enclose space. Hence, both closed and open forms exist. **OPEN FORMS** such as prisms and pinacoids cannot exist alone; they must combine with each other or with closed forms to complete the crystal.

The common forms and combinations of forms in each of the six crystal systems in (Figures 4-10 to 4-19)

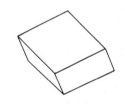

Fig. 4-17. Rhombohedron—a closed form consisting of six faces that are parallelograms.

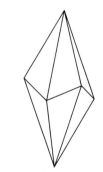

Fig. 4-18. Scalenohedron—a closed form whose faces are scalene triangles (sides and angles of scalene triangles are unequal).

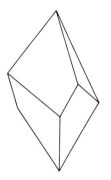

Fig. 4-19. Trapezohedron—a closed form made up of trapezium faces (which are plane figures consisting of four sides, no two of which are parallel).

illustrate important forms common to the crystal systems.

INDICES

The indices of crystal faces are designated by numbers in parentheses—such as (321)—and crystal forms are designated by numbers in brackets—such as {321}.

The index figures represent the relative intercepts of face intersections with axes, and are not coordinate distances as in geometry. Each group of indices is the reciprocal of a fractional value (Miller indices).

Where a crystal face is parallel to a crystal axis it meets the axis at infinity, and the reciprocal value is $\frac{1}{\infty}$, which equals 0. Thus, the index values that are parallel to axes are recorded as 0. Where the intercept is $\frac{1}{1}$ the reciprocal value is unchanged and the index is 1. Where values are $\frac{1}{2}$, $\frac{1}{3}$, $\frac{1}{4}$, etc., the indices become 2, 3, 4, etc.

In the isometric system, the axial ratios on the three axes are the same, and are equal to 1. Hence, a face with the indices (111) would be equally inclined to the three crystal axes. In the orthorhombic system, the axial ratio values on the three axes differ, and a face with the indices (111) is unequally inclined to the three axes. In the tetragonal system, a face with the indices (111) would be equally inclined to the two hori-

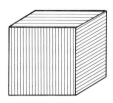

Fig. 4-20. A striated cube.

zontal axes, but would cut the vertical axis at a different angle.

Crystal faces may intersect axes on either negative or positive sides. A negative index figure will have a minus sign above it—such as $(1\bar{3}2)$.

The intercepts on the three axes are ordinarily designated (h,k,l), if unknown. In the hexagonal system they become $(h,k,\bar{\imath},l)$.

SYMMETRY INDICATED BY SURFACE FEATURES

The presence of such surface features on crystals as striations, luster, and etch marks (see p. 35) indicates symmetry.

Striations. Pyrite often occurs in cube-shaped crystals, but the atomic arrangement of pyrite has a lower symmetry than the cube. Cube faces of pyrite are striated and the striations on adjacent faces are perpendicular to each other (Figure 4-20 and Plate 12, facing p. 83). The striated cube has less symmetry than an unstriated cube. The faces of a striated cube show 2-fold symmetry and the faces of an unstriated cube, 4-fold symmetry.

Surface Luster. Luster represents the appear-

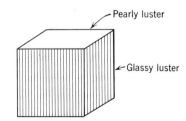

Fig. 4-21. Difference in luster on the faces of an apophyllite crystal. The faces belong to different forms.

ance of a mineral surface in reflected light. Crystal faces that differ in atomic arrangement, and therefore not symmetrically equivalent, often display a slightly different luster. For example, the mineral apophyllite is tetragonal but its crystals sometimes look like cubes. However, two opposite base faces have a pearly luster, and the other four prism faces, a glassy luster (Figure 4-21) with parallel striations. Thus, the six faces illustrate two forms; a pinacoid (two pearly faces) and a tetragonal prism (two glassy striated faces).

THE SIX CRYSTAL SYSTEMS

Isometric System

NAME OF CLASS	INT'L SYMBOL	FACES IN GENERAL FORM	SYMMETRY	EXAMPLE
Hexocta-hedral	$\frac{4}{m} \bar{3} \frac{2}{m}$	48	$3A_4, 4A_3, 6A_2,$ $9P, C$	galena
Gyroidal	4 3 2	24	$3A_4, 4A_3, 6A_2$	*cuprite
Hextetra-hedral	$\bar{4}$ 3 m	24	$4A_3, 3A_2, 6P$	tetrahedrite
Diploidal	$\frac{2}{m} \bar{3}$	24	$4A_3, 3A_2, 3P, C$	pyrite
Tetartoidal	2 3	12	$4A_3, 3A_2$	ullmanite

*Cuprite has been considered a member of the gyroidal class, but recent studies indicate that it may be hexoctahedral.

All the crystal forms of the isometric system are referred to three crystallographic axes of equal length at right angles to each other (Figure 4-22). The photographs of crystals with corresponding wooden models in Plates 3 and 4 (facing pp. 18; 19) illustrate some of the more common isometric forms and combinations of these forms. Each of the 15 forms in the isometric system has a specific name (Figure 4-24). Figure 4-33 shows the distribution of the 15 forms of the isometric system.

Tetragonal System

NAME OF CLASS	INT'L SYMBOL	FACES IN GENERAL FORM	SYMMETRY	EXAMPLE
Ditetragonal bipyramidal	$\frac{4}{m} \frac{2}{m} \frac{2}{m}$	16	$A_4, 4A_2, 5P, C$	zircon
Ditetragonal pyramidal	4 m m	8	$A_4, 4P$	diaboleite
Tetragonal bipyramidal	$\frac{4}{m}$	8	A_4, P, C	scheelite
Tetragonal trapezohedral	4 2 2	8	$A_4, 4A_2$	phosgenite
Tetragonal scalenohedral	$\bar{4}$ 2 m	8	$3A_2, 2P$	chalcopyrite
Tetragonal pyramidal	4	4	A_4	wulfenite
Tetragonal bisphenoidal	$\bar{4}$	4	A_2	cahnite, $CaB(OH)_4AsO_4$

The forms of the tetragonal system are referred to three mutually perpendicular crystallographic axes. The two horizontal a axes are equal in length, but the vertical c axis is longer or shorter than the horizontal axes (Figure 4-23).

The photographs of crystals with corresponding wooden models in Plate 5 (facing p. 34) illustrate common tetragonal crystal forms. The diagrams in Figure 4-25 illustrate some of the basic tetragonal forms.

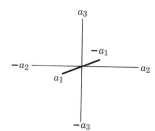

Fig. 4-22. Isometric crystal axes $a_1 = a_2 = a_3$.

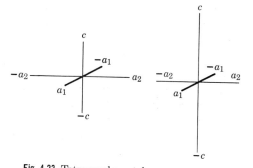

Fig. 4-23. Tetragonal crystal axes $a_1 = a_2 \neq c$.

64

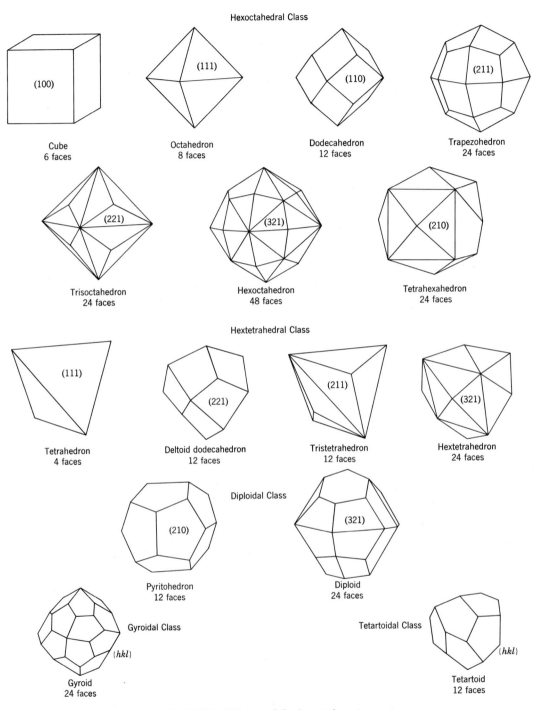

Fig. 4-24. The 15 forms of the isometric system.

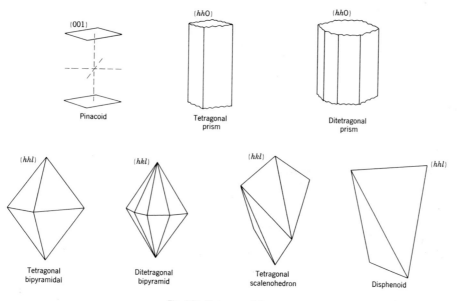

{001}

Pinacoid

{hhO}

Tetragonal
prism

{hhO}

Ditetragonal
prism

{hhl}

Tetragonal
bipyramidal

{hkl}

Ditetragonal
bipyramid

{hkl}

Tetragonal
scalenohedron

{hhl}

Disphenoid

Fig. 4-25. Tetragonal forms.

{100}

{010}

{001}

a, b and c pinacoids

c 001

a
100

b
010

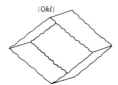

{Okl}

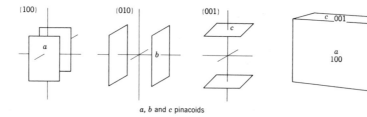

{hOl}

{hkO}

Orthorhombic prisms

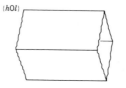

{hkl}

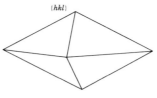

Orthorhombic bipyramids

Fig. 4-26. Orthorhombic forms.

Orthorhombic System

NAME OF CLASS	INT'L SYMBOL	FACES IN GENERAL FORM	SYMMETRY	EXAMPLE
Rhombic bipyramidal	$\frac{2}{m}\frac{2}{m}\frac{2}{m}$	8	$3A_2,3P,C$	barite
Rhombic pyramidal	$2\ m\ m$	4	$A_2,2P$	hemimorphite
Rhombic bisphenoidal	$2\ 2\ 2$	4	$3A_2$	epsomite

The forms of the orthorhombic system are referred to three mutually perpendicular crystallographic axes (a, b, and c) that are unequal in length (Figure 4-27).

The photographs in Plate 5 (facing p. 34) illustrate common orthorhombic crystals. The diagrams in Figure 4-26 illustrate basic orthorhombic forms.

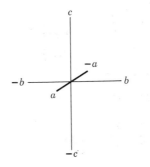

Fig. 4-27. Orthorhombic crystal axes $a \neq b \neq c$.

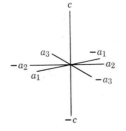

Fig. 4-28. Hexagonal crystal axes $a_1 = a_2 = a_3 \neq c$.

Hexagonal System

■ HEXAGONAL DIVISION

NAME OF CLASS	INT'L SYMBOL	FACES IN GENERAL FORM	SYMMETRY	EXAMPLE
Dihexagonal bipyramidal	$\frac{6}{m}\frac{2}{m}\frac{2}{m}$	24	$A_6,6A_2,7P,C$	beryl
Dihexagonal pyramidal	$6\ m\ m$	12	$A_6,6P$	zincite
Hexagonal trapezohedral	$6\ 2\ 2$	12	$A_6,6A_2$	β quartz (high temperature)
Hexagonal bipyramidal	$\frac{6}{m}$	12	A_6,P,C	apatite
Hexagonal pyramidal	6	6	A_6	nepheline
Ditrigonal bipyramidal	$\bar{6}\ m\ 2$	12	$A_3,3A_2,4P$	benitoite
Trigonal bipyramidal	$\bar{6}$	6	A_3,P	—

■ RHOMBOHEDRAL DIVISION

NAME OF CLASS	INT'L SYMBOL	FACES IN GENERAL FORM	SYMMETRY	EXAMPLE
Hexagonal scalenohedral	$\bar{3}\ \frac{2}{m}$	12	$A_3,3A_2,3P,C$	calcite
Ditrigonal pyramidal	$3\ m$	6	$A_3,3P$	tourmaline
Trigonal trapezohedral	$3\ 2$	6	$A_3,3A_2$	α quartz (low temperature)
Rhombohedral	$\bar{3}$	6	A_3,C	phenacite
Trigonal pyramidal	3	3	A_3	—

The forms of the hexagonal system are referred to four crystallographic axes. The three a axes are equal in length with angles of 120° between the positive ends. The fourth c axis is vertical (Figure 4-28).

The photographs in Plate 6 (facing p. 35) illustrate common hexagonal crystals. The following diagrams (Figure 4-29) illustrate some of the basic hexagonal forms.

Plate 9. LUMINESCENT MINERALS

WHITE LIGHT

SHORT-WAVE U.V. RADIATION (2537 Å)*

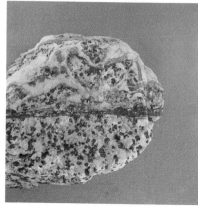

A   x 0.55

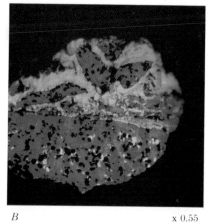

B   x 0.55

WILLEMITE, CALCITE, AND FRANKLINITE

Franklin, New Jersey. In ultraviolet light: calcite—red; willemite—green; non-luminescent franklinite—black.

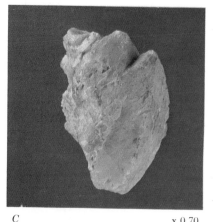

C   x 0.70

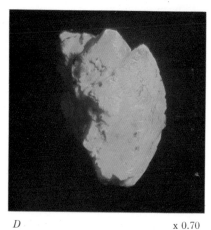

D   x 0.70

SCHEELITE

Kramat Pulai, Federated Malay States.

WHITE LIGHT

LONG-WAVE U.V. RADIATION (3660 Å)

E   x 0.32

F   x 0.32

SCAPOLITE

Grenville, Quebec Canada.

*(Å) is an Ångström unit equivalent to 1/100,000,000th of a centimeter (1 × 10⁻⁸ cm.).

A   x 0.50

GOLD
Au

A leaf with protruding crystals: Osceola, Nevada.

B   x 0.50

GOLD
nugget

Golden color and extreme malleability are highly distinctive.

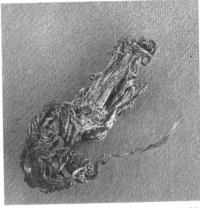

C   x 0.80

SILVER
Ag

Wire-like aggregate of silver: Batopilas, Mexico.

D   x 0.50

PLATINUM
Pt

Large platinum nugget: Ural Mountains, U.S.S.R.

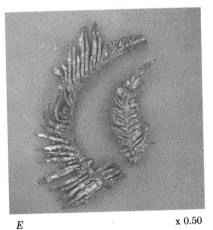

E   x 0.50

SILVER

Silver dendrites: Batopilas, Mexico.

F   x 0.50

COPPER
Cu

Aggregate of rounded distorted copper crystals: Keweenaw Peninsula, northern Michigan.

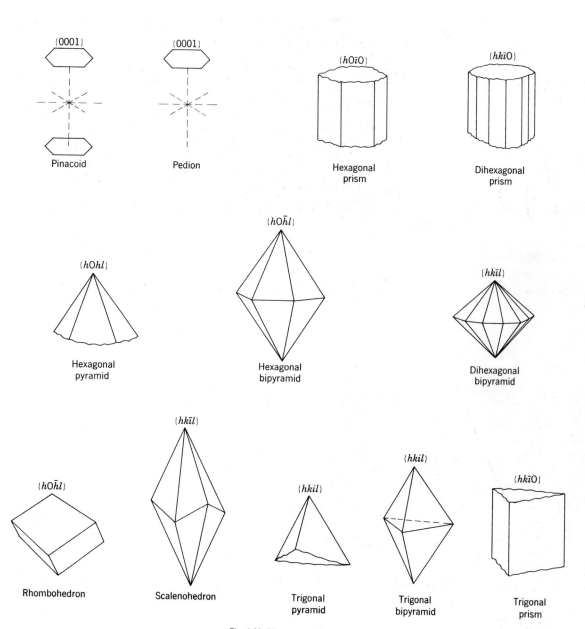

Fig. 4-29. Hexagonal forms.

Monoclinic System

NAME OF CLASS	INT'L SYMBOL	FACES IN GENERAL FORM	SYMMETRY	EXAMPLE
Prismatic	$\frac{2}{m}$	4	A_2, P, C	gypsum
Sphenoidal	2	2	A_2	(sucrose sugar) no common mineral
Domatic	m	2	P	clinohedrite

The forms of the monoclinic system are referred to three unequal axes (a, b, c) with one pair at right angles. The crystals are oriented so that the b axis is horizontal, the c axis is vertical, and the a axis is inclined downward (Figure 4-30).

The photographs in Plate 7 (facing p. 50) illustrate common monoclinic crystals. Figure 4-32 illustrates common monoclinic forms.

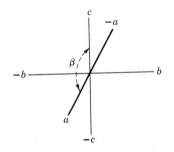

Fig. 4-30. Monoclinic crystal axes $a \neq b \neq c$.

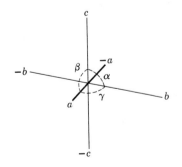

Fig. 4-31. Triclinic crystal axes $a \neq b \neq c$.

Triclinic System

NAME OF CLASS	INT'L SYMBOL	FACES IN GENERAL FORM	SYMMETRY	EXAMPLE
Pinacoidal	$\bar{1}$	2	C	albite
Pedial	1	1	none	

The forms of the triclinic system are referred to three unequal crystallographic axes (a, b, c) that make oblique angles with each other (Figure 4-31). The **PEDION** and the **PINACOID** are the only forms in the triclinic system. Plate 7 illustrates some common triclinic crystals.

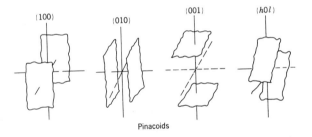

Pinacoids

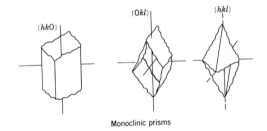

Monoclinic prisms

Fig. 4-32. Monoclinic crystal forms.

Form	$\frac{4}{m}\,\bar{3}\,\frac{2}{m}$ Hexoctahedral	$\bar{4}\,3\,m$ Hextetrahedral	$\frac{2}{m}\,\bar{3}$ Diploidal	$4\,3\,2$ Gyroidal	$2\,3$ Tetartoidal
Tetrahedron		×			×
Cube	×	×	×	×	×
Octahedron	×		×	×	
Deltohedron		×			×
Dodecahedron	×	×	×	×	×
Pyritohedron			×		×
Tetartoid					×
Tristetrahedron		×			×
Diploid			×		
Gyroid				×	
Hextetrahedron		×			
Tetrahexahedron	×	×		×	
Trapezohedron	×		×	×	
Trisoctahedron	×		×	×	
Hexoctahedron	×				

Class

Fig. 4-33. The distribution of the 15 forms in the isometric system.

Class

Form	$\frac{4}{m}\,\frac{2}{m}\,\frac{2}{m}$ Ditetragonal bipyramidal	$4\,m\,m$ Ditetragonal pyramidal	$\frac{4}{m}$ Tetragonal bipyramidal	$4\,2\,2$ Tetragonal trapezohedral	$\bar{4}\,2\,m$ Tetragonal scalenohedral	4 Tetragonal pyramidal	$\bar{4}$ Tetragonal bisphenoidal
Pedion		×				×	
Pinacoid	×		×	×	×		×
Tetragonal bisphenoid					×		×
Tetragonal prism	×	×	×	×	×	×	×
Tetragonal pyramid		×				×	
Tetragonal bipyramid	×		×	×	×		
Tetragonal scalenohedron					×		
Tetragonal trapezohedron				×			
Ditetragonal bipyramid	×						
Ditetragonal prism	×	×		×	×		
Ditetragonal pyramid		×					

Fig. 4-34. The distribution of forms in the tetragonal system.

Class

Form	$\frac{6}{m}\frac{2}{m}\frac{2}{m}$ Dihexagonal bipyramidal	$6mm$ Dihexagonal pyramidal	$\frac{6}{m}$ Hexagonal bipyramidal	622 Hexagonal trapezohedral	6 Hexagonal pyramidal	$\bar{6}m2$ Ditrigonal bipyramidal	$\bar{6}$ Trigonal bipyramidal	$\bar{3}\frac{2}{m}$ Hexagonal scalenohedral	$3m$ Ditrigonal pyramidal	32 Trigonal trapezohedral	$\bar{3}$ Rhombohedral	3 Trigonal pyramidal
Pedion		X			X				X			X
Pinacoid	X		X	X		X	X	X		X	X	
Trigonal prism						X	X		X	X		X
Trigonal pyramid									X			X
Ditrigonal prism						X			X	X		
Ditrigonal pyramid									X			
Hexagonal prism	X	X	X	X	X	X		X	X	X	X	
Hexagonal pyramid		X			X				X			
Rhombohedron								X		X	X	
Trigonal bipyramid						X	X					
Trigonal trapezohedron										X		
Dihexagonal prism	X	X		X				X				
Dihexagonal pyramid		X										
Ditrigonal bipyramid						X						
Hexagonal bipyramid	X		X	X								
Hexagonal scalenohedron								X				
Hexagonal trapezohedron				X								
Dihexagonal bipyramid	X											

Fig. 4-35. The distribution of forms in the hexagonal system.

Forms	Class		
	$\frac{2}{m}\frac{2}{m}\frac{2}{m}$	$2\ m\ m$	$2\ 2\ 2$
	Rhombic bipyramidal	Rhombic pyramidal	Rhombic bisphenoidal
Pedion		×	
Pinacoid	×	×	×
Rhombic bisphenoid			×
Rhombic prism	×	×	×
Rhombic pyramid		×	
Dome		×	
Rhombic bipyramid	×		

Fig. 4-36. The distribution of forms in the orthorhombic system.

Form	Class		
	$\frac{2}{m}$	m	2
	Prismatic	Domatic	Sphenoidal
Pedion		×	×
Pinacoid	×	×	×
Prism	×		
Dome		×	
Sphenoid			×

Fig. 4-37. The distribution of forms in the monoclinic system.

Form	Class	
	$\bar{1}$	1
	Pinacoidal	Pedial
Pedion		26 Pedions
Pinacoid	13 Pinacoids	

Fig. 4-38. The distribution of forms in the triclinic system.

TWINNED CRYSTALS

Twinned crystals are intergrowths of two or more individuals of the same substance united in a symmetrical manner. The individuals of a twin arrangement are related to each other by (1) rotation about a common axis (TWIN AXIS), which does not coincide with the 2-, 4-, or 6-fold axis of symmetry of the single crystal; (2) reflection over a common plane (TWIN PLANE), which is always parallel to a possible crystal face, but never parallels a plane of symmetry of the single crystal; or (3) by a combination of both reflection and rotation.

The plane along which the component parts of a twin crystal are united is called the *composition plane*. RE-ENTRANT ANGLES commonly occur on twinned crystals (Figure 4-39), but may also be found on "single" crystals that exhibit offsets or steps on their surfaces (lineage structure) and on related parallel growths.

In general, twinned crystals may be described as

CONTACT twins, PENETRATION twins, and REPEATER or MULTIPLE twins (Plate 8, facing p. 51). CONTACT TWINS occur when the components of the twin are united along the composition plane (Figure 4-39). Usually, if one unit of a contact twin is rotated 180° about the twinning axis, a simple crystal will result.

PENETRATION TWINS. The component part of two or more crystals appear to penetrate and cross through each other (Figure 4-40).

MULTIPLE TWINS are composed of three or more parts. If the component parts are PARALLEL to each other, the twin is called a multi-lamella or POLYSYN-THETIC TWIN (Figure 4-41). Polysynthetic twinning is commonly observed in the plagioclase feldspars, where it forms thin lines or striations. These striations are the edges of thin parallel plates. If the orientations of the multiple twinned units are NONPARALLEL, the twins are termed CYCLIC (Figure 4-42).

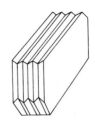

Fig. 4-41. Polysynthetic twinning in albite. Successive composition planes are parallel.

Causes of Twinning

Twinned crystals may be grouped into three types from the standpoint of origin: (1) growth twins; (2) deformation or glide twins, and (3) transformation twins.

GROWTH TWINS. Growth twins are caused by an accident in the growth of the crystal. The structural arrangement of the twinned crystal represents a slightly higher energy state than that of the single crystal, with the increased energy of the twin localized in a zone around the twin boundary. For example, as atoms are added to the external surface of a crystal in the early stages of crystal growth, the first atom to start a new atomic layer may have a choice of two structural sites THAT DIFFER ONLY SLIGHTLY IN ENERGY. The lower energy site is preferred since its occupation provides the greatest decrease in free energy and thus the most stable structure, but the atom may accidentally land on the slightly higher energy site. If crystallization is rapid, the misplaced atom may be unable

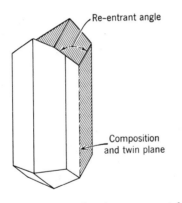

Fig. 4-39. Contact twinned gypsum crystal.

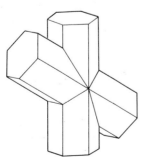

Fig. 4-40. Penetration twin of staurolite (see Plate 8, facing p. 51, and Plate 49, facing p. 307).

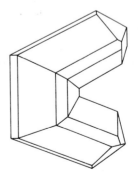

Fig. 4-42. Cyclic twin of rutile. Successive composition planes are nonparallel.

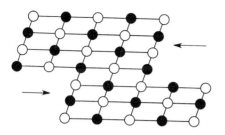

Fig. 4-43. Deformation by translation gliding (plastic deformation). Twins are not formed, because the atoms are moved to sites that are equivalent to positions in the undeformed crystal.

to adjust to the preferred site because it is locked into position by the arrival of new atoms. These atoms take similar positions and continue the atomic layer, because the forces between atoms that dictate the crystal structure do not extend their influence much beyond their nearest neighbors. The new layer of atoms becomes the twin boundary, and during rapid growth future layers of atoms continue the twin structure.

Growth twins commonly consist of two individuals, whereas glide and transformation twins are commonly composed of numerous components.

DEFORMATION OR GLIDE TWINS. When mechanical stress is applied, crystals may be deformed by the movement of atoms along certain planes. These planes of atoms do not move rigidly past each other because special structural defects known as dislocations enable the atoms to move row by row rather than whole planes slipping at one time. Slip occurs most easily on certain planes called slip or glide planes, and the direction along which the atoms move in these planes is the slip direction. Not all parallel planes slip with the same ease, for the dislocations that can lower resist-

ance to deformation vary from one parallel plane to another. Depending upon the amount of atomic displacement, slip may cause either a twin or a deformed crystal. The atoms in a plastically deformed crystal have been moved an integral number of atomic distances to sites that are EQUIVALENT to the positions in the undeformed crystal (Figure 4-43).

GLIDE TWINS are formed when the displaced atoms are moved to structural sites that DIFFER from the undeformed portion of the crystal. The atoms on one side of the slip plane are moved a distance of less than one lattice parameter past those on the other side. The boundary between the two regions is the twin plane. The displacement of atoms, which is proportional to the distance from the twin junction, occurs on successive atomic planes. The displacement increases with distance from the twin plane, so that the two parts of the crystal are so displaced that they form MIRROR IMAGES of each other across the twin plane (Figure 4-44). For slip and subsequent twinning to occur, adjacent atomic sites must be separated by low energy barriers, because if the energy barriers are high the amount of energy required for slip may equal that required for rupture, and the crystal will break rather than deform plastically.

The steps in the formation of glide twins are as follows:

1. Almost immediately following the application of the critical stress value to the crystal, movement occurs on parallel glide planes and many thin twin lamellae are almost instantaneously formed. Iron, for instance, forms numerous parallel twins (Neumann bands) on deformation by an impact blow. These are frequently observed in iron meteorites. After slip has occurred, the material becomes stronger and harder, so that more energy is needed for further deformation.

Fig. 4-44. Formation of glide twins. Displacement of atoms is proportional to distance from the twin plane.

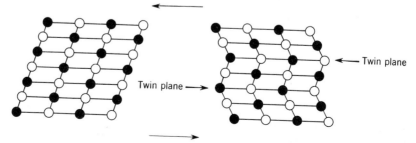

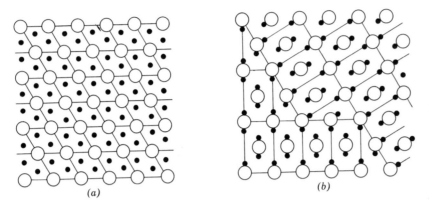

(a)   (b)

Fig. 4-45. Transformation twins formed by the inversion of high-temperature hexagonal structure (a) to twinned low-temperature orthorhombic structure (b) (from L. V. Azároff[1]).

2. As stress is continued, the thin lamellae slowly grow wider by atomic displacements on planes adjacent to the twin boundaries. Closely spaced twin lamellae that have the same crystallographic orientation frequently combine to form one twin. Under stress, the twin lamellae slowly continue to widen until they are blocked by some obstacle such as a foreign atom, a structural flaw, or a grain boundary. New twin lamellae will form in other parts of the crystal if the energy required to overcome the obstacle is greater than that required to start new twinning. Glide twins are normally parallel but they may be lenticular in shape, if during their formation obstacles such as grain boundaries are encountered.

Glide twins are common in metals and in metamorphic limestones where calcite or dolomite has been mechanically deformed.

TRANSFORMATION TWINS. The inversion of polymorphic* minerals is frequently accompanied by twinning (Figure 4-45). For example, when a high-temperature polymorph is cooled below the transition temperature, the low-temperature structure may form independently in different areas of the crystal; this occurs because crystal imperfections cause differences in thermal conductivity. As the transformation continues, the regions of low-temperature structure grow

*See p. 51.

in size. If two growing regions with the same crystallographic orientation meet, their boundaries will disappear, but if they have different orientations, their mutual boundary becomes a twin junction. The layer of atoms at the twin junction is common to both components of the twin. If many transformation centers develop with nonparallel orientations, numerous twins forming a mosaic pattern will result. The inversion from high-temperature to low-temperature quartz commonly forms a mosaic of low-temperature twins.

REFERENCES

1. L. V. Azároff, *Introduction to Solids,* McGraw-Hill, New York, 1960, pp. 154–161.

Mineralogy Textbooks Suggested for Further Reading

1. L. G. Berry and B. J. Mason, *Mineralogy,* W. H. Freeman and Company, San Francisco, 1959.
2. W. H. Dennen, *Principles of Mineralogy,* Ronald Press, New York, 1959.
3. C. S. Hurlbut, Jr., Edward S. Dana *Manual of Mineralogy* (17th edition), John Wiley and Sons, New York, 1959.
4. E. H. Kraus, W. F. Hunt, and L. S. Ramsdell, *Mineralogy* (5th edition), McGraw-Hill, New York.
5. F. C. Phillips, *An Introduction to Crystallography,* Longmans, Green and Company, New York, 1956.

5.

Physical Properties of Minerals

The physical properties of minerals provide convenient criteria that aid greatly in recognition. One mineral is hard, another soft; one mineral is heavy, another light; one is bright, another dull; and so on through a wide range of physical observations that describe the individualities of minerals.

Further, the physical properties are responsible for many of those features that make certain minerals objects of beauty; utility, and interest. The brilliance of the diamond, the magnetism of lodestone, the radio control embodied in quartz, the sectility of gold, the iridescence of opal, and the luminescence of willemite are examples selected at random from the wide range of physical properties inherent in minerals. Many are not simply explained, but even an introduction to such understanding is worthy of the effort.

THE ROLE OF PHYSICAL PROPERTIES

The physical properties of minerals range from such elementary characteristics as color, luster, hardness, and specific gravity to more involved optical properties, electrical characteristics, magnetism, and the relationship of structure to the physical nature of a material. In this book, those physical properties will be emphasized which are less complex and are particularly useful in rapid mineral identification.

KEYS TO MINERAL IDENTIFICATION

Color

COLOR IN MINERALS IS CAUSED NOT ONLY BY THE INTERACTION OF LIGHT WITH THE CHEMICAL CONSTITUENTS, BUT ALSO BY ATOMIC ARRANGEMENT, IMPURITIES, AND STRUCTURAL DEFECTS. Color in some minerals is a fundamental property because light-absorbing atoms are the principal constituents of the mineral structure; however, many minerals exhibit a range of colors caused by impurities and structural defects that absorb light (color centers). *Quartz,* for example, while at times clear, transparent, and colorless, may also be pink, red, blue, green, violet, purple, yellow, brown, or black. *Fluorite, calcite, beryl, apatite, corundum,* and *tourmaline* also occur in a range of colors.

A list of those minerals with characteristic colors is an important aid to identification. It should be remembered, however, that the mineral must always be examined on a fresh surface, since tarnish and surface alteration may change the normal color.

Some nonmetallic minerals with characteristic colors are:

BLUE	GREEN	YELLOW	ORANGE-RED	RED
Azurite	Malachite	Sulfur	Crocoite	Cuprite
Chalcanthite	Antlerite	Orpiment	Realgar	Cinnabar
Lazurite	Brochantite	Carnotite		
(lapis-lazuli)	Dioptase			
Lazulite				
Linarite				

The color of most metallic minerals—deep brass-yellow (chalcopyrite), pale brass-yellow (pyrite), bronze (pyrrhotite), and copper-red (copper and niccolite)—is characteristic.

Sensations of color are produced on the eye by light waves that range through the visible spectrum from violet 3900 Å to red 7600 Å (Figure 5-1). When white light (a combination of all wavelengths in the visible spectrum) falls on a colored mineral some of the wavelengths are absorbed. In general, the removal of absorbed wavelengths gives the mineral the color of the remaining wavelengths. Blue minerals, for example, absorb the long, red wavelengths, and red minerals absorb the short wavelengths.

Sensations of a single color can originate from different combinations of waves. For example, two minerals may appear to be of the same color, but the color in one may be produced by the action of a narrow band of wavelengths, and in the other, by a suitable combination of different wavelengths. The unaided eye is unable to detect whether a color is produced by one or a group of wavelengths.

Some transparent minerals selectively absorb light in different crystallographic directions, and thus the color may differ greatly when viewed in two directions at right angles. This property is called PLEOCHROISM. Such color changes are particularly noticeable in *gem andalusite,* which is green in one direction and red-brown in another. A number of precious stones like the *ruby, sapphire, topaz, cordierite, alexandrite,* and *emerald* are strongly pleochroic.

Changes in color may result from variations in chemical composition. For example, sphalerite, ZnS, may range from pale yellow through brown to black as ferrous iron partially substitutes for zinc in the structure.

Differences in structure rather than composition can also cause color variations. Gray-black opaque graphite or colorless transparent diamond results from different arrangements and bondings of carbon atoms (see p. 42).

Minerals are frequently colored by inclusions of other minerals in the host mineral. These inclusions can range from submicroscopic sizes to small crystals visible with a hand lens. For example, a variety of quartz called adventurine is often colored red-brown by minute scales of mica or hematite, or green by a chromium mica. Chalcedony is also commonly colored by admixed foreign material.

Color is sometimes produced by the selective scattering of light rays by minute included particles in minerals (TYNDALL EFFECT). Blue quartz in reflected light is commonly caused by the SCATTERING of blue wavelengths of ordinary light by minute needle-like rutile inclusions. In transmitted light, the color is reddish to yellowish-brown, apparently caused by the ABSORPTION of light by the included rutile needles. The scattering of light by tourmaline needles or other

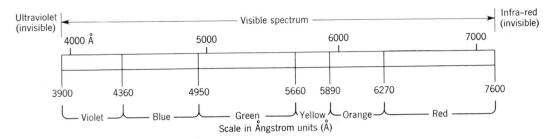

Fig. 5-1. Visible light spectrum.

microscopic included particles may also produce the blue color in quartz.

■ COLOR CENTERS

Color arising from imperfections in atomic structure is believed to be caused by the selective absorption of some component of visible light by certain imperfections in the crystal structure called COLOR CENTERS. Thus, A COLOR CENTER IS A STRUCTURAL DEFECT THAT ABSORBS LIGHT.

As a rule, minerals with characteristic colors that are composed of light-absorbing atoms are not apt to be greatly influenced by color centers such as impurity atoms or other light-absorbing structural defects. However, some colorless or light-colored minerals can be deeply colored by minute amounts—often only a few hundredths of a percent—of strong light-absorbing atoms, some examples being the transition elements (Sc, Ti, V, Cr, Mn, Fe, Co, Ni, Cu, Zn). It is interesting to note that the same element can cause different colors in different minerals. For example, colorless corundum (Al_2O_3) will become pink with the addition of about 0.05 percent of chromium. On the other hand, introduced chromium atoms in beryl, $Be_3Al_2(SiO_3)_6$, produce a deep-green emerald. The chromium atoms in synthetic pink corundum absorb ultraviolet radiation as well as a broad band of green and yellow light.

There are various types of color centers in addition to those produced by light-absorbing foreign atoms. The simplest is an F CENTER, derived from the German word for color (Farbe). THE F CENTER CONSISTS OF A NEGATIVE ION VACANCY COMBINED WITH AN ELECTRON (Figure 5-2).

F centers in sodium chloride result from heating a halite crystal (NaCl) in the presence of sodium vapor and then quickly cooling it. The crystal then becomes yellow. It is believed that excess sodium atoms deposit on the crystal surface, and that chlorine ions (negative ions) then diffuse to the surface where they combine with the sodium atoms that have become ionized by losing a valence electron. The electron can diffuse into the crystal until it encounters a negative ion vacancy and combines with it to form an F center. The energy value (quantum state) of the trapped electron depends upon the structural environment of the vacancy (crys-

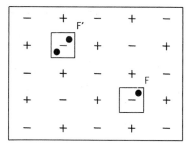

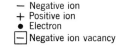

— Negative ion
+ Positive ion
● Electron
⊟ Negative ion vacancy

Fig. 5-2. Schematic diagram showing an F center and an F′ center. The F center consists of a negative ion vacancy and an associated electron. The F′ center is a negative ion vacancy with two bound electrons (adapted from C. Kittel[9]).

tal structure) and not upon the sodium atom from which the electron originally came. Impurity atoms that differ from the host atoms could also supply electrons to produce F centers and other types of color centers. Some halide crystal colors caused by the presence of F centers are:

> NaCl—yellow
> KCl—magenta
> KBr—blue
> KI—blue-green

F centers can also be produced by bombardment with x-rays or high-energy particles that remove electrons from lattice ions; these electrons are in turn captured by negative ion vacancies.

It is not completely understood how the F centers absorb light, but it is believed that some component of white light has the appropriate energy to excite the trapped electron from its ground state (lowest possible energy state) to a higher energy state; thus, in the process the appropriate wavelengths become absorbed by the crystal. A range of energies can excite the electron so that an absorption band called an F BAND is observed. It has been shown that the total absorption is proportional to the number of F centers produced.

It is of interest to consider an effect that occurs

in crystals with F centers when these are irradiated at elevated temperatures with light that is absorbed by the F centers. Here the absorption of the F band gradually decreases and a new band on the red side of the F band called an F′ band is formed. This process is called BLEACHING. It is believed that some of the electrons bound to F centers are excited by the incident light and are liberated by absorbing thermal energy. The "free" electrons combine with other F centers to form F′ centers. An F′ center consists of a negative ion vacancy with two bound electrons (Figure 5-2), and absorbs light on the long-wavelength side of the F band. It has been found that F′ centers can be destroyed by irradiating the crystal with light absorbed by the F′ band and F centers are formed again in the process.

Various types of color centers can be induced in minerals without a change in composition by exposure to high-energy radiation, such as x-rays, gamma rays, electrons, or neutrons. The color of smoky quartz is believed to be caused by radiation (see p. 22). Some crystals with color centers produced by high-energy radiation can be easily bleached by re-irradiating with visible light, or by heating.

The colors of many gem minerals may be altered by heat treatment. For example, yellow topaz assumes a pinkish hue when heated between 300° and 450°C. Heat-treated reddish-brown zircons become colorless or light blue. Greenish transparent beryl from Brazil heated to 450°C assumes a blue color that is prized in aquamarines. Bluish-violet and lilac transparent spodumene from Brazil when heated assumes a bright pink color (kunzite), and amethyst quartz turns yellow. Other minerals which may change color on heating include apatite, rutile, corundum, spinel, turquoise, crocoite, sphene, sodalite, cancrinite, lazulite, and rhodonite.

■ INTERFERENCE COLORS

As some minerals are turned they exhibit a brilliant patchwork of changing colors, similar to the interference colors of an oil film on water or on a soap bubble. Figure 5-3 illustrates the interference produced by the action of light on a thin film of oil. At times, the light-disturbing "films" in minerals are repeated narrow twinning plates (lamellae).

Incident rays of white light *A* and *B* strike a thin layer of oil and are immediately partially reflected at the surface. Part of the light rays, however, penetrate the oil film and are reflected at the interface between the oil and water. If the oil film is of the correct thickness, the portion of ray *B* that has traveled through the film emerges in the same place at the air-oil surface as the reflected portion of ray *A*. However, the portion of ray *B* that has traveled through the oil film is RETARDED because it has traveled a longer distance; thus, its vibrations are out of phase with the reflected portion of ray *A*. Since light waves can be added or subtracted, the resulting combination of wavelengths (*A* + *B*) differs from the original mixture of wavelengths in the incident white light, and thus is seen as color.

The color produced by the interference of light waves depends upon the angle of incidence of the original light rays and the thickness of the oil film, for these factors determine the amount of retardation of the light rays that travel the extra distance through the film. If the films are too thick or too thin, interference colors will not be observed.

In minerals, interference colors may be produced by the INTERFERENCE OF LIGHT from such factors as repeated twinning lamellae, minute air-filled cleavage cracks, inclusions, microscopic intergrowths of thin light-disturbing layers, or the regular stacking of minute spheres; this stacking behaves like a three-dimensional diffraction grating.

LABRADORITE, a calcium-rich plagioclase feldspar, frequently shows a brilliant play of iridescent blue and

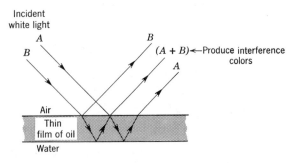

Fig. 5-3. Interference of light.

green and, sometimes, red and yellow colors (Plate 37, facing p. 219). The colors are caused by two combined factors: (1) the light interference from thin plagioclase twinning lamellae, and (2) the reflection of light from minute plate-like parallel inclusions of magnetite. The black magnetite inclusions also give labradorite that dark gray color which provides an excellent background for the brilliant display of iridescent patches.[5]

Interference light from repeated twinning lamellae also causes the blue "shiller" glow of PERISTERITE, a variety of plagioclase feldspar. MOONSTONE is composed of thin interlaminated plates of sodium-rich feldspar in potassium-rich feldspar. The blue glow of moonstone is produced by the interference of light from the thin plates.

■ CAUSE OF COLOR IN OPAL

Studies at high magnification (electron microscopy)[4] show that the brilliant play of colors of precious opal is caused by the DIFFRACTION OF LIGHT from the spaces between minute uniform silica spheres; these are stacked in an orderly three-dimensional pattern that behaves like a diffraction grating.

Light rays undergo certain changes when they strike the edge of an object. Part of the light ray goes straight ahead, while the other part that is closest to the edge of the object is bent from its expected path, and thus goes off in a different direction. This phenomenon (the bending of light) is called DIFFRACTION. The light waves in the bent or diffracted portion of the ray are out of phase with their neighboring waves, and are therefore in a position to cause interference colors. Diffraction occurs when light rays pass through extremely narrow slits or when they are reflected from a surface having many finely scratched lines (commonly, 30,000 or more to an inch), and results in the interference of light waves. Such devices are called diffraction gratings. The thickness and the spacing of the scratched lines or slits on a diffraction grating influence the interference colors produced. SIMILARLY, THE COLOR OF PRECIOUS OPAL IS RELATED TO THE SIZE AND SPACING OF THE REGULARLY STACKED SILICA SPHERES.

Electron microscope photographs of opal taken at magnifications of up to 40,000 times show that precious opal consists of layer upon layer of minute

spheres of silica orderly arranged with relatively uniform spaces between the spheres. The amorphous silica spheres are optically transparent but because there is a change in refractive index light is scattered at the interface between the spheres and the voids. The holes may even be filled as long as there is a change in refractive index at the interface. The spheres and the voids are arranged regularly in three dimensions (face-centered cubic pattern), an arrangement that produces a three-dimensional diffraction grating. A most important factor concerns the equal spacing of holes and spheres; when the spacing is about equal to the wavelengths of visible light, diffraction of light occurs. The process is similar to the diffraction of x-rays by orderly arranged atoms in crystals. The angle through which light is diffracted in precious opal varies continuously with the wavelength, so that a play of colors is produced by different colors appearing at different angles (Figure 5-4).

The diffraction colors of precious opal are seen in patches because the orderly grid-like arrangement of silica spheres occurs in patches that are oriented in different directions, thus producing different interference colors.

Common opal with a milky appearance (Plate 36, facing p. 218) also grows in tiny concentric spheres. However, the spheres lack the orderly arrangement of precious opal and therefore scatter light rather than diffract it. Hence, it is the lack of regularity in arrangement that prevents the interference which produces colors.

If interstitial silica completely fills the spaces

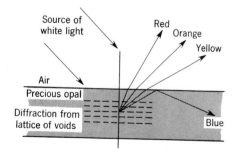

Fig. 5-4. Diffraction and refraction in precious opal (after P. J. Darragh and J. V. Sanders[4]).

between the silica spheres, the opal may be transparent; however, it will not act as a diffraction grating to produce interference colors.

Luster

Luster represents the appearance of a mineral surface in reflected light. Crystal faces that differ in atomic arrangement, and thus belong to different forms, often display slightly different lusters.

The two major types of luster are METALLIC and NONMETALLIC. Minerals with an intermediate luster are called SUBMETALLIC. The native metals and most of the sulfides have metallic lusters. The following terms are used to describe some of the varieties of nonmetallic luster:

Vitreous or Glassy: glass-like luster. Examples are quartz and beryl.

Adamantine or Brilliant: brilliant luster like that of a diamond. In general, the greater the difference between the velocity of light in air and in a mineral, the more brilliant the mineral will be. Thus, minerals with adamantine lusters have high indices of refraction. Examples are cassiterite, diamond, cerussite, and zircon.

Resinous: having a luster like that of rosin. Example, sulfur.

Silky: silk-like luster produced by light reflected from aggregates of parallel fibers. Examples are asbestos and fibrous gypsum.

Waxy: like wax. The microscopically rough surface scatters light to produce a waxy luster. Examples are chalcedony and cerargyrite.

Greasy: light is scattered by a microscopically rough surface. For example, rough diamond crystals may have a somewhat greasy luster.

Pearly: an iridescent pearl-like luster is often produced by reflections from layers of incipient cleavage cracks, and is usually observed on faces parallel to lamellar cleavage planes. Examples are talc, basal faces of apophyllite, and heulandite (see heulandite, Plate 39, facing p. 227).

Dull or Earthy: porous aggregates like chalk or clay, or minerals with weathered rough surfaces scatter incident light, and have a dull surface appearance.

Streak

Streak represents the color of the powder of a mineral, and is easily seen by rubbing the edge of a specimen across a surface of unglazed porcelain (streak plate). Though the hardness of the streak plate ($6\frac{1}{2}$) limits the range of minerals tested, it suffices for many common minerals. Dark metallic minerals often have streaks that differ from the color of the mineral. For example, the following metallic or submetallic minerals are often black, but the colors of their streaks are characteristic.

MINERAL	COMMON COLOR	STREAK
Goethite, $HFeO_2$	Black	Yellowish-brown
Hematite, Fe_2O_3	Black	Red-brown
Magnetite, Fe_3O_4	Black	Black
Alabandite, MnS	Black	Green

Hardness

Hardness concerns the extent to which a mineral resists abrasion. The time-honored mineral scale of hardness known as the Mohs scale (Table 5-1) lists ten common minerals, all arranged in sequence-of-hardness numbers from 1 to 10. Each entry on the list will scratch those with lower-hardness numbers and will

Table 5-1
Mohs Hardness Scale

1	Talc		
2	Gypsum	$2\frac{1}{4}$	Finger-nail
3	Calcite	3	Copper coin
4	Fluorite		
5	Apatite	$5\frac{1}{2}$	Knife-blade or
6	Orthoclase		window glass
7	Quartz	$6\frac{1}{2}$	Steel file
8	Topaz		
9	Corundum		
10	Diamond		

be scratched in turn by minerals with higher numbers. Diamond is the hardest mineral, having a hardness of 10 (H. = 10). The hardness numbers of the Mohs scale are only reference values, and intervals of hardness between numbers are not uniform. For example, the actual difference in hardness between diamond (H. = 10) and corundum (H. = 9) is greater than the difference between talc (H. = 1) and corundum.

If an unknown mineral can be scratched by a knife blade or window glass, but not by a copper coin, then its hardness lies somewhere between 5½ and 3. If the mineral will scratch fluorite, but not apatite, then its hardness is between 4 and 5.

Hardness is an extremely useful property but care should be taken to test hardness on a smooth fresh surface, for surface alterations are frequently much softer than the mineral. A simple and useful test consists in making a small scratch across a mineral surface with a knife blade to distinguish minerals with a hardness of above 5½ from those with a hardness below. The hardness test is apt to be deceptive on minerals that are granular or splintery. These minerals may appear to submit to a hardness test but the apparent scratch is only the displacement of granules or surface fragments.

The hardness of minerals is largely determined by the crystal structure and depends upon the bond strength holding the atoms together. These binding forces may differ in different crystallographic directions, causing some minerals to exhibit a change in hardness with direction. For example, kyanite has a hardness of 5, parallel to the length of the crystal (parallel to chains of atoms), but a hardness of 7 across the length of the crystal (perpendicular to chains of atoms) (Figure 5-5).

The same atoms in two different structural arrangements may differ greatly in hardness. For example, diamond (H. = 10) and graphite (H. = 1–2) are both composed of carbon atoms.

Tenacity

Tenacity describes the way in which a mineral breaks or deforms under stress. It depends upon the cohesive forces between atoms, and may vary with

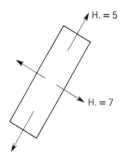

Fig. 5-5. Change in hardness with direction in a kyanite crystal.

direction in a mineral crystal. The following terms are used to describe tenacity:

Brittle. The mineral breaks or powders. Examples are quartz and diamond.

Malleable. The mineral can be hammered into thin sheets without breaking. Examples are gold, silver, and copper.

Ductile. The mineral can be drawn out into a thin wire. Examples are gold, silver, and copper.

Sectile. The mineral can be cut with a knife, like wax. An example is argentite. Some minerals, such as gypsum, are sectile but are not malleable, and will powder under the blow of a hammer. (Subsectile is the description applied to minerals less definitely sectile.)

Flexible. The mineral can be bent but does not return to its original shape after stress is released. Examples are talc and molybdenite.

Elastic. Thin layers, after being bent, return to their original shape. There is a temporary displacement of neighboring atoms, but when stress is released the atoms immediately return to their normal positions. Examples are muscovite, biotite, phlogopite, and lepidolite mica.

Cleavage

Cleavage represents the tendency of a mineral to break in certain directions parallel to atomic planes across which the binding forces are the weakest. Minerals like mica and graphite, for example, have perfect cleavage, parallel to one plane. They are composed of sheets of strongly bonded atoms that are separated by

Muscovite mica—single cleavage (1 plane)

Dolomite—rhombohedral cleavage (3 planes)

Fluorite—octahedral cleavage (4 planes)

Galena—cubic cleavage (3 planes)

Fig. 5-6. Cleavage fragments of mica, dolomite, fluorite, and galena.

weaker bonds between the sheets (see graphite structure Figure 3-7, p. 42). When a mineral breaks along such weak planes, the resulting surface is called a cleavage surface (Figure 5-6).

Cleavage is an important property. Not all minerals exhibit cleavage, but if it is present in a given mineral, ALL crystals of that mineral species exhibit the same cleavage. However, in some fine-grained massive specimens, the individual crystals are so small that the cleavage may not be observable, except with the use of a microscope.

Cleavage is usually described according to perfection (perfect, good, fair, poor) and the number and

direction of cleavage planes. The number and direction of planes may be given as the name of the parallel form (cubic cleavage, octahedral cleavage, rhombohedral cleavage, etc.). (See Figure 5-7.) For example, in cubic cleavage, planes are parallel to opposite faces of a cube, and have three directions at right angles. The directions of cleavage may be expressed by the indices of the planes; for example, a cleavage plane parallel to the base of a monoclinic crystal would be (001), or (0001) for a hexagonal mineral.

A mineral frequently has cleavage parallel to different forms. Barite, for example, exhibits perfect prismatic and basal cleavage. Feldspar (Plate 37,

Plate 11. **NATIVE ELEMENTS** ▪ **SEMIMETALS** (*A,B,C*), **AND NONMETALS** (*D,E,F*)

A   x 0.68

BISMUTH
Bi

Massive bismuth showing characteristic pale reddish-silver color and lamellar structure: Ontario, Canada.

B   x 1.0

ANTIMONY
Sb

Silver white radiating massive antimony; granular texture: New Brunswick, Canada.

C   x 0.67

ARSENIC
As

Concentric shells of arsenic with characteristic gray-black tarnish: Andreasberg, Harz Mts., Germany.

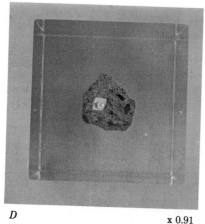

D   x 0.91

DIAMOND
C

Diamond crystal in kimberlite rock: Premier Mine, near Pretoria, Transvaal, South Africa. (Specimen imbedded in a clear plastic block; courtesy of Mahlon Miller–Diamond Distributors, N.Y.C.)

E   x 0.65

SULFUR
S

Sulfur occurring with white celestite: Sicily, Italy.

F   x 0.46

GRAPHITE
C

Section of a narrow vein of graphite showing foliated structure: Ticonderoga, New York.

Plate 12. SULFIDES ▪ PYRITE, PYRRHOTITE, MARCASITE, CINNABAR

A   x 0.53

PYRITE
FeS$_2$

Common pyrite crystals; pyritohedron (Leadville, Colorado) and striated cube (Coahuila, Mexico).

B   x 0.50

PYRITE

Aggregate of striated cubic crystals. Pale yellow-brass color is distinctive.

C   x 1.0

PYRRHOTITE
Fe$_{1-x}$S

Tabular six-sided crystals. Bronze color is distinctive. Potosi Mine, Chihuahua, Mexico.

D   x 0.61

MARCASITE
FeS$_2$

Twinned crystals showing characteristic "cockscomb" structure.

E   x 0.54

CINNABAR
HgS

Bright red earthy cinnabar: Terlingua, Texas.

F   x 0.54

CINNABAR

Massive bright red cinnabar in white calcite showing brilliant luster: Excelsior Mine, W. Terlingua, Mexico.

facing p. 219) shows a blocky cleavage fragment formed by the intersection of two cleavage planes—basal (001) and side pinacoid (010)—at almost right angles.

Parting

Parting resembles cleavage but is a limited effect. A mineral that exhibits parting breaks across weak atomic bond planes that are usually caused by deformation or twinning. Such planes are LIMITED to certain specimens of a mineral species and certain parallel planes. The mineral corundum frequently shows rhombohedral parting caused by twinning. Garnet has no cleavage, but some specimens show flat parting planes.

Fracture

Fracture represents a significant mineral break other than along cleavage or parting planes. The break may occur in any direction, and is usually localized around some defect or inclusion in the crystal. Usually, most minerals show uneven or irregular fracture. Thus, fracture ordinarily is not a useful property for mineral identification. A possible exception is a type of fracture, shown in Figure 5-8, called CONCHOIDAL (shell-like); they consist of smoothly curved surfaces. Glass, obsidian, chalcedony, and opal (Plate 36, facing p. 218) are examples of materials that show conchoidal fracture. Native metals like gold, silver, and copper exhibit a HACKLY fracture, which is a ragged surface with sharp points and edges.

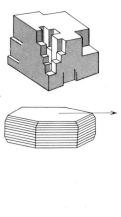

CUBIC {001}—3 planes at right angles. Examples: galena, halite, sylvite.

RECTANGULAR {100}, {010}, and {001}—3 planes. Example: anhydrite.

BASAL {001} or {0001}—one plane parallel to base or basal pinacoid. Examples: mica (Plate 41(*B*), facing p. 243), graphite, topaz (Plate 46(*B*), facing p. 290), molybdenite (Plate 16(*D*), facing p. 99).

PINACOIDAL—one plane parallel to either front (100), side (010), or basal (001) pinacoid. Example: stibnite (perfect side pinacoidal cleavage (010)).

RHOMBOHEDRAL {10$\bar{1}$1}—3 planes parallel to opposite faces of a rhombohedron. Examples: calcite, siderite (Plate 24(*C, F*), facing p. 147), dolomite, rhodochrosite.

OCTAHEDRAL {111}—4 planes parallel to each pair of faces of an octahedron. Example: fluorite (Plate 23(*E*), facing p. 146).

PRISMATIC {110}—2 planes (3 for hexagonal prism). Examples: spodumene, rhodonite, hornblende, augite.

DODECAHEDRAL {001}—6 planes parallel to each pair of faces of a dodecahedron. Example: sphalerite.

Fig. 5-7. Types of cleavage.

Fig. 5-8. Obsidian (volcanic glass) showing conchoidal fracture.

Specific Gravity *(Relative Density)*

The specific gravity of a mineral represents its weight with respect to an equal volume of pure water. Thus, a mineral with a specific gravity of 3 is three times as heavy as water. This is a characteristic property and depends upon the atomic weight of the atoms and their structural arrangement.

Specific gravity is determined by the weight of a mineral in air, divided by the weight of an equal volume of water at 4°C. Since the weight of an equal volume of water is equal to the mineral's loss in weight when weighed in water, the

Specific gravity (Sp.Gr.)

$$= \frac{\text{Weight of a mineral in air}}{\text{Weight in air} - \text{Weight in water}}$$

Temperature affects values in the third place of decimals, but may be neglected for routine work.

Various types of balances as well as heavy liquids are used to measure specific gravity. However, experience in handling minerals will often enable an observer to estimate approximately the specific gravity of a mineral. For example, a metallic mineral like pyrite has a specific gravity of about 5. If one lifts a specimen of copper (Sp.Gr. 8.9), gold (Sp.Gr. 19.3), or silver (Sp.Gr. 10.5), each will feel heavy in comparison. Typical nonmetallic minerals like quartz, feldspar, calcite, and aragonite have specific gravities from 2.6 to 2.95. In contrast, nonmetallic minerals like barite, scheelite, or cinnabar feel heavy, and such minerals as sulfur and halite feel light.

■ EXAMPLES OF NONMETALLIC MINERALS

LOW SP.GR. ($<$2.2)		HIGH SP.GR. ($>$3.5)	
Borax	1.7	Topaz	3.49–3.57
Sulfur	2.07	Corundum	4.0–4.1
Halite	2.16	Barite	4.5
Opal	2.0–2.2	Scheelite	6.1
		Anglesite	6.38
AVERAGE SP.GR. (2.6–3.0)		Cerussite	6.55
Quartz	2.65	Cassiterite	7.0
Feldspar	2.56–2.75	Cinnabar	8.09
Calcite	2.71		
Aragonite	2.94		

■ EXAMPLES OF METALLIC MINERALS

LOW SP.GR.		HIGH SP.GR. ($>$7)	
Graphite	2.09–2.23	Galena	7.57
		Wolframite	7.12–7.51
AVERAGE SP.GR. ($\sim$5)		Niccolite	7.78
Marcasite	4.88	Tantalite	7.9
Pyrite	5.01	Copper	8.95
Magnetite	5.2	Bismuth	9.7–9.83
		Silver	10.5
		Uraninite	10.0±
		Platinum	14–19
		Gold	19.3

In general metallic minerals have higher specific gravities than nonmetallic minerals. Thus, metallic minerals may be best compared with other metallic minerals, and nonmetallics compared with each other. If a mineral specimen is porous or impure the specific gravity will often appear to be lower than the true value.

UNIQUE PHYSICAL PROPERTIES

Chatoyancy

Chatoyancy describes the wavy, silky sheen that is sometimes concentrated in a narrow band of light; it is commonly shown by minerals with a fibrous structure, or by those with myriads of parallel microscopic needle-like inclusions or needle-like cavities. The chatoyant effect is produced by reflection of light from parallel fibers, cavities, or inclusions, and occurs at right angles to the direction of elongation. Peculiarities of structure may also cause a chatoyant sheen. Numerous needle-like parallel etch cavities may be produced in crystals by etch solutions that dissolve holes in the crystal at the centers of those growth spirals (cores of screw dislocations) which are chemically more reactive than the more perfect areas of the crystal.

When chatoyant gem minerals called "cat's eyes" are cut with a rounded polished surface (cabochon cut), the reflected light appears to be concentrated in a narrow band of light that extends across the surface of the stone (Figure 5-9), and shifts from side to side as the stone is turned; hence, the name cat's eyes. A number of minerals like chrysoberyl, quartz, sillimanite, scapolite, orthoclase, albite, and varieties of beryl may at times occur as "cat's eye" gems. Chrysoberyl is one of the most valuable. Light passing through a chrysoberyl cat's eye is reflected by thousands of microscopic needle-like cavities in parallel position, or parallel needle-like crystals. Quartz cat's eyes may at times resemble chrysoberyl but are much less valuable. Inclusions of asbestos-like fibers in cat's eye quartz frequently cause the chatoyant effect. Fibrous gypsum (satin-spar) and tiger's eye (fibrous crocidolite replaced by quartz) also exhibit chatoyancy.

Asterism

STAR CRYSTALS are those minerals that show a peculiar star of light when viewed in transmitted or reflected light. This multiple chatoyant effect is called ASTERISM. Star crystals commonly contain numerous

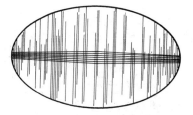

Fig. 5-9. Schematic diagram showing the chatoyant effect of a cat's eye. Band of reflected light occurs at right angles to microscopic parallel needle-like inclusions (courtesy of Ralph J. Holmes[7]).

microscopic oriented needle-like inclusions that are formed by a process called EXSOLUTION (unmixing of a solid solution). At high temperatures, foreign ions are dissolved in the solid host crystal, but at lower temperatures the host structure can no longer tolerate the foreign ions. The misfit ions thus migrate through the solid host structure and finally group themselves locally to form long oriented needle-like crystals within the host (see exsolution, unmixing of a solid solution, p. 55, Figure 3-27). The exsolved needle-like crystals are related to the atomic structure of the enclosing host crystal because the low-temperature host structure has room only in certain directions for the rejected impurity ions to segregate and form their own crystals.

When light passes through a cabochon-cut star gem, it is reflected by the inclusions. Each ray of light is at right angles to the direction of elongation of a set of inclusions (Figure 5-10). The direction and number of rays depend upon the structural geometry of the host crystal, since the inclusions are aligned only in certain crystallographic directions.

Asterism is sometimes caused by oriented needle-like cavities or by such peculiarities of structure as symmetrically arranged twinning lamellae.

In a star ruby or sapphire, the tiny needles of rutile that have been exsolved from solid solution are oriented in three directions at 120° to each other. These included needles reflect light so that a spot source on a rounded polished surface reflects six rays (Figure 5-11). The inclusions in corundum can also be minute oriented needle-like cavities.

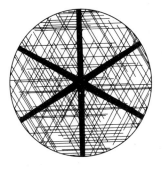

Fig. 5-10. Schematic diagram showing three sets of parallel inclusions that intersect at 120°C and produce a six-rayed star of light. The direction of each ray is at right angles to a parallel set of inclusions (courtesy of Ralph J. Holmes[7]).

Rose quartz frequently shows a six-rayed star of light, caused by exsolved rutile needles.

Some minerals exhibit asterism in transmitted light. For example, when a point source of light is viewed through a thin cleavage sheet of phlogopite mica, a star-like effect may be produced; this effect is caused by symmetrically arranged microscopic inclusions that deflect light.

Luminescence

Numerous minerals are luminescent; they glow in the dark when exposed to ultraviolet radiation (Plate 9, facing p. 66). Energy is absorbed from ultraviolet light, x-rays, visible light, infrared, or cathode rays, after which it is released in the form of visible or near-visible light. The absorption of energy is EXCITATION, and the release of energy is LUMINESCENCE. Luminescence includes both FLUORESCENCE and PHOSPHORESCENCE. As ordinarily observed, fluorescence stops as the excitation ceases, but a phosphorescent mineral may continue to emit light after the exciting source is removed.

Luminescence is caused by the motion of electrons. Energy absorbed during excitation raises electrons to higher energy states. Subsequently, an excited electron, on returning to a lower energy state, loses energy by emitting a light photon.

Technically, if light emission occurs within one one-hundred millionth of a second (10^{-8} sec.) of excitation, the process is called FLUORESCENCE; but if the crystal takes one ten-millionth of a second (10^{-7} sec.) or longer to emit light, the process is called PHOSPHORESCENCE and the crystal called a PHOSPHOR. It is difficult to distinguish between long fluorescence and short phosphorescence (D. Curie, 1963[3]). However, a distinction can often be made by studying the relationship of luminescence to temperature. Under ordinary conditions, fluorescence is not usually dependent upon temperature, but the duration of phosphorescence (after-glow) is strongly temperature-dependent, since thermal energy is necessary to liberate electrons from traps.

In a fluorescent mineral, emission takes place by one or more spontaneous transitions (Figure 5-12), but in long-persistence phosphorescent solids (Figure 5-15) some internal mechanism, such as trapping centers, stores the excitation energy and releases is slowly over a period of time that may range from microseconds to hours, depending upon the type of phosphor and the temperature.

Fig. 5-11. Star Ruby and Sapphires. *Left:* The Edith Haggin de Long Star Ruby: Burma; 100 carats. *Center:* The Star of India Star Sapphire: Ceylon; 563.35 carats. *Right:* The Midnight Star Sapphire: Ceylon; 116.75 carats (courtesy of the American Museum of Natural History).

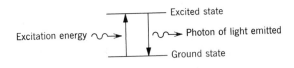

Fig. 5-12. An example of fluorescence.

Luminescence frequently depends upon small amounts of atomic impurities called ACTIVATORS, although in some cases other structural defects furnish activation. The properties of a luminescent mineral—energy required to become excited, the color and brightness of the light emitted, and the duration of after-glow—depend primarily upon the chemical nature and atomic structure of the host crystal and the type of impurities or structural defects it contains. The only general requirement for luminescense is an upper energy state which electrons of a certain energy state can occupy, and a lower state to which they can return with the spontaneous emission of light photons.

Two distinct forms of luminescence usually require activator atoms—the quasi-atomic system and the "ionized" activator system.

1. In THE QUASI-ATOMIC SYSTEM (nonphotoconducting), excited electrons are associated with the activator ions. Absorption of an appropriate quantum of energy raises an electron to a higher energy state CHARACTERISTIC OF THE ACTIVATOR ION. The excited states are unstable, hence the electrons will quickly return to a lower energy state with the spontaneous emission of a photon of energy that may be seen if it is within the visible spectrum.

2. THE "IONIZED" ACTIVATOR SYSTEM (photoconducting). In a photoconducting phosphor the excitation energy dissociates or "ionizes" electrons from the activator system and raises them to a conduction band. Here they become free electrons that can wander about the conduction band (Figure 5-15). When a free electron recombines with a positive hole, energy is released, frequently in the form of a visible light photon.

■ **LUMINESCENT MINERALS**

Where luminescence is caused by small amounts of impurity atoms, only mineral specimens containing the impurity will luminesce.

Willemite ($ZnSiO_4$) emits a bright yellowish-green light when it contains small amounts of manganese. Willemite from Franklin, New Jersey, that contains manganese activator atoms is highly luminescent, but from other localities it is manganese-free

and nonluminescent. The luminescent willemite from Franklin occurs with white calcite that emits a bright red glow under ultraviolet light (Plate 9, facing p. 66), and is also activated by manganese.

Scheelite ($CaWO_4$) is luminescent and glows bright bluish-white under short-wave ultraviolet light (Plate 9). The tungstate ion WO_4^{-2} is believed to be responsible for the light emission. If molybdenum ions replace some of the tungsten ions, scheelite gives off a yellowish-white glow.

Most luminescent minerals will fluoresce or phosphoresce under short-wave ultraviolet light (2537 Å), but a few, such as *scapolite, hackmanite, amblygonite,* and *pectolite,* luminesce better under long-wave ultraviolet (3660 Å).

Table 7-10, p. 148, lists minerals that are sometimes luminescent, with the colors most commonly observed. However, luminescent color may vary in different parts of a single specimen, or in different specimens of the same mineral species.

Thallium-Activated KCl. Thallium-activated potassium chloride crystals, KCl(Tl), furnish an example of a quasi-atomic emitting system. Small amounts of thallium ions, about 0.01% or less, randomly replace potassium ions (K^+) in the crystal structure. When ultraviolet light approximating 2500 Ångströms shines on the phosphor, each thallium ion absorbs a photon, becomes excited, and is temporarily raised to a higher energy state. The excitation raises the thallium ion from the unexcited or GROUND STATE (electronic configuration of lowest possible energy) by changing its electronic configuration (Figure 5-13), but does not "ionize" (remove) electrons from the activator system.

The energy rise is represented schematically by the length of the arrow *AB* in Figure 5-14. Since the electronic configuration differs in the excited state, the forces between the thallium activator ion and the 6 surrounding chlorine ions that comprise the ACTIVATOR SYSTEM also differ from the ground state. In the excited state, the surrounding chlorine ions are closer to the thallium ion. The activator system assumes various configurations, caused by the thermal vibrations of the chlorine ions around their equilibrium positions in the crystal lattice.[11] The CONFIGURATION

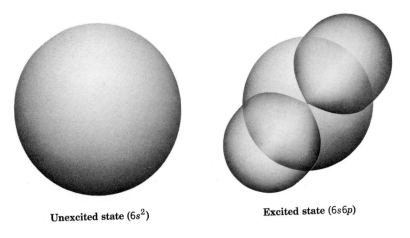

Unexcited state ($6s^2$)   Excited state ($6s6p$)

Fig. 5-13. Electron cloud of a thallium ion (Tl^+) differs in the unexcited state from that in the excited state. The model shows the difference in probable electron distributions around the nucleus (after General Electric Research Laboratory).

COORDINATE represents the distance between the thallium ion and its neighboring chlorine ions.

Since the higher energy state is unstable, the excited thallium ion will almost immediately revert to its lower energy state with the spontaneous emission of a photon of light (3050 Å). The arrow *CD* in Figure 5-14 represents the energy of the emitted photon of light. The unequal lengths of the arrows *AB* and *CD* indicate that the peaks of absorption and emission spectra come at different wavelengths. However, both are in the invisible ultraviolet range. The difference in energy forms heat in the crystal. When the thallium ion returns to the ground state, the activator system adjusts itself to the ground state coordination of ions, and the thallium ion is now ready to start a new cycle by absorbing another photon of ultraviolet light. A cycle is completed in about a millionth of a second.

Excitation with ultraviolet light in the 1960 Å band excites the thallium ion to a higher energy state represented by arrow *AE*. On returning to the ground state, the thallium ion emits ultraviolet radiation centered at 2470 Å, and the activator system adjusts itself to the ground state coordination of ions. However, the 2470 Å emission falls within the 2500 Å absorption band; thus, the emitted light is strongly reabsorbed.

Emission of visible blue light (4750 Å) is also observed, but its origin is still unexplained.

Figure 5-14 shows three energy states ($^1S_0$, $^3P_1{}^0$, $^1P_1{}^0$)* of a thallium activator ion (Tl^+) in potassium chloride. These serve as a function of the configuration coordinate. The thallium ion in the ground state is close to point *A*. When irradiated with ultraviolet light near 2500 Å the thallium ion absorbs a photon of light and its energy is raised from *A* to *B*. After the transition, the neighboring chlorine ions in the activator system adjust themselves to the equilibrium position of the excited state; the system thus ends at equilibrium position *C*, and the energy difference *B–C* is dissipated in the crystal as heat (thermal vibrations of atoms). The distance between the chlorine and thallium ions in equilibrium is about 0.26 Ångström units smaller in the excited state than in the ground state. On returning to the ground state from *C*, the thallium ion emits a photon of light ($C \rightarrow D$) in the 3050 Å band; the activator system also readjusts itself to its equilibrium position *A*, and again the rearrangement energy $D \rightarrow A$ is given to the crystal as heat.

When irradiated with ultraviolet light near 1960 Å, the energy of the thallium ion is raised from *A* to *E*. Shorter wavelengths have more energy than longer wavelengths; thus, absorption of energy around 1960 Å raises the ion to a higher energy state than absorption of the 2500 Å light. Arrow *FG* represents emission of light centered at 2470 Å.

* $^1S_0$ is the ground state with two $6s$ electrons ($6s^2$). $^3P_1{}^0$ and $^1P_1{}^0$ are energy levels in the excited or *P* state ($6s6p$).

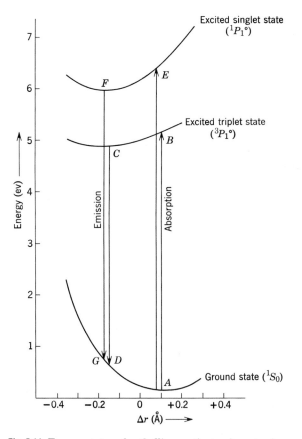

Fig. 5-14. Energy states of a thallium activator in potassium chloride (after Johnson and Williams[8]). Here Δr is the displacement in Ångström units of the neighboring Cl^- ions from their mean positions in the perfect KCl structure

2500 Å absorption—$^1S_0 \rightarrow {}^3P_1^0$

3050 Å emission—$^3P_1^0 \rightarrow {}^1S_0$

1960 Å absorption—$^1S_0 \rightarrow {}^1P_1^0$

2470 Å emission—$^1P_1^0 \rightarrow {}^1S_0$

Photoconducting Phosphorescent Crystals. It is unlikely that all phosphorescent phosphors can be described in terms of any one mechanism; hence, a hypothetical energy-level diagram for a zinc sulfide type phosphor (Figure 5-15) as explained by J. N. Shive[13] is given as one example.

Electrons in crystals are said to move in two broad energy bands, the **VALENCE BAND** with available

energy states occupied by valence electrons,* and the **CONDUCTION BAND** occupied by free electrons. The two bands are separated by a **FORBIDDEN REGION** which electrons normally do not occupy. If an electron is to jump across the forbidden region to the conduction band it must absorb a unit of energy, such as a photon of light. This must provide enough energy to carry it across the forbidden gap. Otherwise, the electron must simultaneously absorb the energy of a photon plus the energy of a particular crystal lattice thermal vibration to make up the required total energy. However, certain impurity atoms present in small amounts may introduce localized energy levels into the forbidden region (Figure 5-15).

Atoms that donate electrons to the conduction band are called **DONORS**; if they cause luminescence in the crystal they are also **ACTIVATORS**. A donor atom frequently has more valence electrons than the atom it replaces in the crystal, and since the extra electron is not needed to complete the bonding between the impurity atom and its neighbors, the "orphaned" electron will revolve around the impurity atom in a wide, loosely bound orbit of its own. As long as the electron remains on this level, it "belongs" to the impurity atom; in some cases, however, only a small amount of thermal energy is needed to displace it into the upper conduction band.

In the zinc sulfide type of phosphor (Figure 5-15) the activator energy level is too far below the upper conduction band for small amounts of thermal energy to elevate electrons to the upper band; thus, the phosphor must absorb larger units of excitation energy.[13] When the external excitation energy shines on the phosphor, some of the activator electrons are transferred to the conduction band, along with some electrons from the lower band that leave positive holes behind.

To explain the mechanism for long-persistence phosphors, provision must be made to store the excitation energy necessary to permit the phosphor to emit light after the excitation energy is shut off. The

---

* Valence electrons are outer electrons that form bonds between atoms.

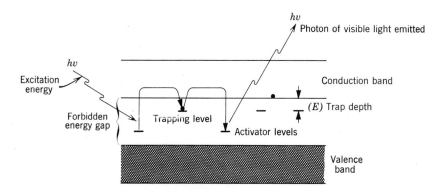

Fig. 5-15. Schematic energy level diagram of a zinc sulfide type of phosphor. An electron absorbs excitation energy and is liberated from an activator level into the conduction band; from there it falls into a trapping level of an acceptor-like impurity. Later, the trapped electron absorbs thermal energy (E), which again raises it to the conduction band. On returning to the activator level without being recaptured by a trap, the electron emits a photon of light. A transition from a trap to an activator level or valence band does not ordinarily occur. The trapping center is the agency chiefly responsible for delay in the emission of light that enables the phosphor to glow when the excitation energy is removed (after John N. Shive[13]).

electrons in such phosphors are believed to be trapped for a time in trapping centers.

ELECTRON TRAPS in photoconducting crystals are crystal imperfections (impurities, vacancies, dislocations, etc.) that introduce localized metastable energy levels in the crystal that are capable of "capturing" electrons and holding them for significant periods of time. The TRAP DEPTH is the thermal energy needed to liberate a trapped electron. Hole traps also occur. When a crystal imperfection captures a hole, and then an electron, it is called a RECOMBINATION CENTER. As a free electron falls into a trapped hole a photon is emitted (Figure 5-16).

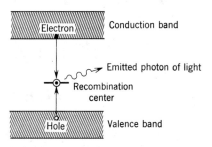

Fig. 5-16. A recombination center.

In the zinc sulfide type of phosphor, the trapping center has a normally empty energy level some distance below the bottom of the upper band. A free electron, liberated from the activator atom or from the lower valence band and wandering about in the upper band, may by chance encounter a trapping center and become trapped by falling down onto its energy level, otherwise called a TRAPPING LEVEL. Since the trapping level lies some distance below the bottom of the upper band, it may take some time before the electron is thermally liberated into the upper band. A phosphor whose trapping center energies lie far below the upper band will trap electrons longer than one whose trapping center energies lie closer, and thus need less energy to reach the upper band. This trapping interval is chiefly responsible for the delay of long-persistence phosphors.

When the excitation energy is removed, the traps slowly liberate the electrons to the upper band. An electron may then recombine with a positive hole and give off energy, may become trapped again, or, if it escapes entrapment, may recombine with an activator (luminescent center) and emit radiation rather than recombine with a positive hole in the lower band.

A x 0.74

NICCOLITE
NiAs

Massive niccolite with white quartz. Pale copper color is distinctive: Wolfgang, Saxony.

B x 0.64

PENTLANDITE IN PYRRHOTITE
$(Fe, Ni)_9S_8$

Massive bronze-colored pyrrhotite intermixed with pentlandite: Sudbury, Ontario.

C x 0.72

MILLERITE
NiS

Needle-like radiating crystals of millerite on hematite: Antwerp, New York.

D x 0.74

COBALTITE
CoAsS

Cobaltite crystals: Tunaberg, Sweden.

E x 0.9

PROUSTITE
Ag_3AsS_3

Dark red striated crystals with brilliant luster: Chile.

F x 0.68

ARGENTITE
Ag_2S

Aggregates of distorted argentite crystals with dull blackish-gray tarnish: Sonora, Mexico.

A   x 0.87

GALENA
PbS

Groups of lead gray galena crystals showing characteristic cubic habit and bright
metallic luster: Robinson Mine, Kansas.

B   x 0.72

BOURNONITE
$PbCuSbS_3$

Aggregate of twinned crystals showing
characteristic cog-wheel shape: Cornwall, England.

C   x 0.54

SPHALERITE
ZnS

Massive amber colored sphalerite showing resinous luster: Picos de Europa,
Santander, Spain.

D   x 0.50

SPHALERITE, GALENA

Concentrically banded intergrowth of
sphalerite and dark gray galena.

E   x 0.66

GREENOCKITE   SPHALERITE
CdS   ZnS

Sphalerite occurring with lemon yellow
greenockite: Marion County, Arkansas.

F   x 0.54

BOULANGERITE
$Pb_5Sb_4S_{11}$

Boulangerite ("feather ore") showing fibrous feathery texture: Stevens County,
Washington.

Silver or copper are common activator atoms in zinc sulfide, but the introduction of iron, nickel, or cobalt atoms will "poison" the luminescence. They form KILLER CENTERS (centers in which the probability of light emission is much smaller than for nonradiative emission). Luminescent centers are the opposite of killer centers.

The color of the light the phosphor emits usually depends upon the nature of the impurities and the locations of their energy levels in the forbidden region, as well as upon the composition and structure of the host crystal.

Thermoluminescence. Certain phosphors excited at low temperatures will not release the absorbed energy until the temperature of the crystal is raised. As the temperature is raised, electrons are liberated from traps, the shallow traps emptying first. Such minerals are THERMOLUMINESCENT and will usually phosphoresce when heated at low temperatures (below red heat). A pink to rose color variety of fluorite called *chlorophane* phosphoresces green after being gently heated. The luminescence is best observed in total darkness. The minerals celestite, barite, apatite, lepidolite, spodumene, and diamond are sometimes thermoluminescent.

Triboluminescence. Certain specimens of some minerals become luminescent when ground, scratched, rubbed, or struck with a hammer. This reaction is called TRIBOLUMINESCENCE, and is commonly observed in fluorite (chlorophane variety), sphalerite, and lepidolite; sometimes it also occurs in corundum and willemite.

Magnetic Properties

When a small hand magnet is used to separate so called "magnetic" from "nonmagnetic" minerals it actually separates FERROMAGNETIC minerals from those that are PARAMAGNETIC and DIAMAGNETIC. Paramagnetic substances are weakly magnetic and diamagnetic ones are repelled by a strong magnetic field. Two common minerals are ferromagnetic: magnetite, Fe_3O_4, and pyrrhotite, $Fe_{1-x}S$. Some specimens of magnetite called *lodestone* are themselves natural magnets and have the power of attracting iron filings.

Magnetic properties are caused by the motions of electrons. Atoms consist of small nuclei surrounded by clouds of electrons. The moving electrons are electric currents and create magnetic fields; this occurs as a result of their orbital motion and their spins, which can be clockwise or counterclockwise. When the spins of two electrons are in opposite directions, the net magnetic moment is zero, but WHEN THE SPINS ARE IN THE SAME DIRECTION THE MAGNETIC EFFECT IS POSITIVE.

In such transition elements as iron, cobalt, and nickel the atoms have an incomplete d shell of electrons; this shell is more stable when it contains the maximum number of electrons spinning in one direction and a smaller number spinning in the other direction. Hence, these atoms will exhibit magnetism under certain conditions.

In some crystal structures, at certain distances between atoms, the electron spins of unpaired electrons in each atom assume parallel orientation; however, if the optimum interatomic distance is decreased, an antiparallel spin may be favored. Manganese crystals, for example, are not ferromagnetic, but some compounds containing manganese atoms spaced farther apart are ferromagnetic. Hence, magnetic properties are structure-sensitive, and alloys that contain neither iron, cobalt, nor nickel—the common magnetic metals—can be ferromagnetic.

The mineral pyrrhotite, $Fe_{1-x}S$, is a good example of the way a change in interatomic spacing influences the magnetic properties. The iron content in pyrrhotite ranges from specimen to specimen, and some that contain a smaller number of iron atoms are more magnetic than those with a greater number. It has been shown that the higher the iron deficit in the pyrrhotite crystal the smaller the atomic cell spacing will be. At a certain interatomic distance, the forces favoring parallel orientation of electron spins will be strongest, and the pyrrhotite structure that contains the proper number of iron atoms for producing this favorable interatomic distance will be strongly ferromagnetic. Thus, some specimens of pyrrhotite are ferromagnetic and others are paramagnetic.

It is believed that ferromagnetic substances are composed of a number of small regions called DOMAINS. The domains are about 0.001 in. in size and are com-

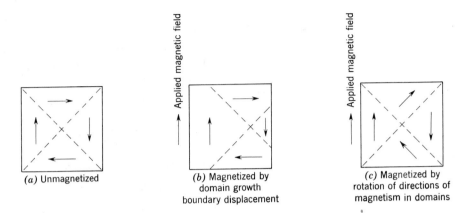

Fig. 5-17. Schematic diagram of the domains and their directions of magnetism in an unmagnetized crystal (*a*) and in magnetized crystals (*b*) and (*c*)—after C. Kittel[9].

posed of groups of atoms. Within each domain, the magnetization is constant in magnitude and direction, and thus each domain is roughly equivalent to a permanent magnet; however, since the magnetic moments (spins) of different domains point in different directions, their effects cancel out each other when there is no external magnetic force acting upon them (Figure 5-17a). However, under the action of an applied magnetic field, domains that are favorably oriented with respect to the field "grow" either by migration of their boundaries at the expense of unfavorably oriented domains (Figure 5-17b) or by rotation of the directions of magnetism in different domains toward the direction of the field (Figure 5-17c). It should be noted that the domains themselves, which are composed of groups of atoms fixed in the crystal structure, are not rotated, but only the direction of their magnetism. Thus, the spins of the electrons in the domains change direction, but the individual atoms in the domain do not change their positions.

Electrical Properties

Piezoelectric Effect. Pressure applied in appropriate directions to certain crystals produces an electric current, and the crystal develops positive and negative charges. As the pressure is applied, the crystal changes shape by the motion of its ions; and as some of the negative and positive ions move in OPPOSITE

DIRECTIONS they produce an electric current. When the pressure is released, the ions return to their original positions, producing an electric current in the reverse direction. Hence, a series of directed blows would produce an alternating current. The current flows only when the crystal is changing shape (ions are moving). This effect, that of a mechanical deformation producing an electric current, is called the PIEZOELECTRIC EFFECT. However, to produce a current, the crystal must have a type of atomic arrangement that lacks a center of symmetry.* When ions are arranged so that opposite charges move in the same direction their motions cancel out electrically, and no current is produced. The piezoelectric effect is direction-sensitive, since the arrangement of atoms varies in different directions. Two minerals that readily show the piezoelectric effect are quartz and tourmaline.

The converse piezoelectric effect is produced when a crystal is mechanically deformed by placing it in an electric field. As voltage is applied, the negative and positive ions move to opposite electrodes, thus changing the shape of the crystal.

Pyroelectric Effect. Temperature changes cause some crystals to develop electric charges at opposite

---

* A crystal is said to have a center of symmetry if an imaginary line can be drawn from any point on the surface through the center to a corresponding point on the opposite side of the crystal (see p. 59, Chapter 4).

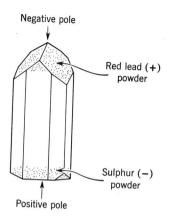

Negative pole

Red lead (+) powder

Sulphur (−) powder

Positive pole

Fig. 5-18. Pyroelectric effect shown by tourmaline.

ends. This property is called PYROELECTRICITY, and, like piezoelectricity, can only occur in minerals that lack a center of symmetry. The pyroelectric effect can be demonstrated by first heating a tourmaline crystal. Then, while cooling, the crystal is dusted with a mixture of sulfur and red lead (Pb_3O_4) that has been previously electrified by being blown through a silk screen. The mixture of sulfur and red lead powder can be placed in a rubber bulb with a silk screen over the mouth. As the particles are blown through the silk, the sulfur receives a negative charge and settles on the positive end of the charged tourmaline crystal and the positively charged red lead settles on the negative end of the crystal (Figure 5-18). The faces at the opposite poles of the vertical axis of a tourmaline crystal differ in atomic arrangement. Such crystals with different forms at opposite ends of an axis are called HEMIMORPHIC from the Greek *hemi* meaning half and *morphe* for form.

IDENTIFICATION OF MINERALS BY X-RAY DIFFRACTION

X-rays

X-rays are produced when a target metal such as tungsten, copper, molybdenum, or iron is bombarded in a vacuum with high-speed electrons induced by high voltage. These bombarding electrons knock out inner electrons from the atoms of the target metal. As electrons from the outer shells of the target atoms fall into the vacant inner-electron positions, energy is released in the form of characteristic x-rays. These x-rays are characteristic radiation, and have wavelengths characteristic of the atoms in the target metal. X-rays are a form of electromagnetic radiation that is similar to visible light but of extremely short wavelengths (about 10^{-8} or 10^{-9} cm).

■ X-RAY DIFFRACTION

THE DISTANCES BETWEEN PLANES OF ATOMS IN A MINERAL AND THE KINDS AND DISTRIBUTION OF ATOMS IN THESE PLANES ARE CHARACTERISTIC FOR EACH PARTICULAR MINERAL SPECIES. By using x-ray diffraction techniques, it is possible to calculate the distance between atomic planes and to identify unknown minerals with precision.

Since the wavelength of x-rays is approximately equal to distance between atoms in a crystalline mineral, the mineral will diffract the x-rays similarly to the way in which a ruled grating diffracts visible light. Although the diffraction effect involves the concept of each atom acting as a center of wave propagation, it is simpler to consider planes of atoms acting as mirrors "reflecting" the incident x-ray beam.

It was shown by Sir William Bragg that a thin ribbon of x-rays would be reflected at a UNIQUE ANGLE that would in turn depend upon the interatomic spacing of the planes of atoms parallel to the reflecting crystal face. However, when a beam of characteristic x-rays strikes a set of parallel crystal planes at some ARBITRARY ANGLE, there will usually be no reflected beam; this is because the x-rays reflected from the parallel planes travel paths of different lengths, and the reflected rays from each plane are out of phase with each other and cancel each other out. Only when each ray is out of phase by exactly one wavelength, or two, or three, or more, will the reflected beam consist of rays that are in phase again. The angle at which reflection occurs is known as the Bragg angle, θ (theta).

In Figure 5-19 entering and emerging rays are parallel, although the path along *CFD* is longer than *AOB*. The two sets of waves will be in phase if the

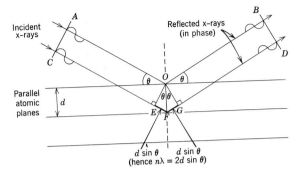

Fig. 5-19. Reflection of an x-ray beam by parallel atomic planes in a crystal (d represents the distance between atomic planes). The incident x-rays strike the crystal planes at the proper angle θ and the reflected rays are in phase.

path difference $EF + FG$ is equal to a whole number of wavelengths ($n\lambda$). In this case, λ is the wavelength and n can have the value 1,2,3, etc. EF and FG form equal legs of right triangles where both equal $d \sin \theta$. Thus, $EF + FG = 2d \sin \theta$, where d is the distance between consecutive atomic planes and θ is the angle of reflection. Thus, the condition for the production of a reflected x-ray beam is $n\lambda = 2d \sin \theta$, which is known as Bragg's law.

When a powdered mineral specimen consisting of thousands of minute crystals is placed in a beam of characteristic x-rays, numerous reflections from different sets of parallel atomic planes are produced; this occurs because the x-rays will strike the randomly oriented crystals at various angles, some of which will be correct to produce x-rays. Different sets of atomic planes will cause different reflections.

These reflected beams are recorded on a strip of photographic film, or on a paper strip chart with a pen that is actuated by the electric impulses from a scanning geiger tube that picks up the reflected x-ray beams. The films, because of their value in mineral recognition, have been called "mineral finger prints."

The distances between lines on the film or the peaks on the chart are measured, and the values con-

verted into d values, the distances between planes of atoms in the mineral. The intensities of the lines on the film or the peak heights on the chart are determined by the kinds and distribution of atoms in an atomic plane. Since the distances between atomic planes and the intensities are characteristic for each particular mineral species, for an unknown mineral to be identified the d spacings and the intensities of the reflected x-rays are compared with x-ray diffraction cards on which there are recorded d values and intensities for hundreds of minerals and other compounds.[*]

REFERENCES

1. L. V. Azároff, *Introduction to Solids,* McGraw-Hill, New York, 1960, pp. 404–406.
2. B. D. Cullity, *Elements of X-ray Diffraction,* Addison-Wesley Publishing Company, Reading, Mass., 1956.
3. D. Curie, *Luminescence in Crystals,* John Wiley and Sons, New York, 1963.
4. P. J. Darragh and J. V. Sanders, "The Origin of Color in Opal Based on Electron Microscopy," *Lapidary Journal,* Dec., 1965, p. 1052.
5. P. E. Desautels, "Interaction between Light and Minerals," *Natural History,* Oct., 1965, p. 53.
6. A. Holden and P. Singer, *Crystals and Crystal Growing,* Doubleday and Company, New York, 1960 (Piezoelectric effect, Chapter 11).
7. R. J. Holmes, "Synthetic and Other Man-made Gems," *Foote Prints,* 32, No. 1, 1960, pp. 11–13.
8. P. D. Johnson and F. E. Williams, "Simplified Configurational Coordinate Model for KCl(Tl)," *Phys. Rev.,* 117, 1960, p. 964.
9. C. Kittel, *Introduction to Solid State Physics,* John Wiley and Sons, New York, 1956, pp. 415, 491–502.
10. E. W. Lee, *Magnetism,* Penguin Books, 1963.
11. J. S. Prener and D. B. Sullenger, "Phosphors," *Scientific American,* October, 1954.
12. K. Przibram, *Irradiation Colors and Luminescence,* Pergamon Press, London, 1956.
13. J. N. Shive, *The Properties, Physics, and Design of Semiconductor Devices,* D. Van Nostrand Company, Princeton, N.J., 1959, pp. 415–417.

[*] The American Society for Testing Materials has issued cards giving x-ray data for many minerals.

6.

Chemical Tests ✓

In any procedure directed toward mineral recognition, it must be kept in mind that minerals are chemicals found in nature. Thus, a chapter devoted to chemical tests is essential, notwithstanding the great emphasis on physical features found elsewhere in the text.

The tests selected are, for the most part, simple. As far as possible, they are designed to provide unique data for indicating the presence or absence of specific elements or combinations of elements. These tests may indirectly indicate specific minerals, may discriminate between a few alternate minerals, or may confirm a mineral identification suggested by other criteria. They supplement the identification tables of Chapter 7.

Since the chemical tests usually destroy a portion of the mineral tested, they should be used with caution, and reagents should be applied to small amounts. The tests yield the most information after the physical properties of a specimen have been established. If a chemical test is to be applied in a significant way, the chemical composition of the mineral sought must be kept in mind.

TESTING EQUIPMENT

Agate mortar and pestle
Blowpipe
Bunsen burner, alcohol lamp, or candle. (A small commercial blowtorch is useful.)
Charcoal blocks ($4'' \times 1'' \times \frac{1}{2}''$)
Filter paper
Glass tubing (about $\frac{1}{4}$-in. outside diameter)
Hammer and anvil (small)
Magnet (Alnico)
Platinum-tipped forceps
Platinum wire (No. 27). (A small metal holder or a holder made by fusing the wire in a small bore glass tube is desirable.)
Pyrex test tubes and holder
Triangular file

REAGENTS

Wet

Alcohol—95 percent ethyl alcohol used to dissolve dimethylglyoxime
Ammonium hydroxide, NH_4OH (ammonia). Mix one volume of concentrated ammonium hydroxide with two volumes of water.

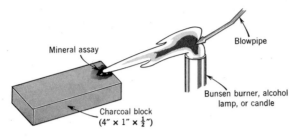

Fig. 6-1. Fusion of a mineral on charcoal.

Hydrochloric acid, HCl. Mix two parts of concentrated acid with three parts of water.

Hydrogen peroxide, H_2O_2

Nitric acid, HNO_3. One part concentrated acid with two parts water.

Sulfuric acid, H_2SO_4. One part of concentrated acid to four parts of water. Pour acid slowly into water. DO NOT ADD WATER TO ACID.

Dry

Ammonium molybdate, $(NH_4)_2MoO_4$. Mix 10 grams of MoO_3 with 40 ml of water and 8 ml of concentrated NH_4OH. Then with constant stirring slowly pour solution into a mixture of 40 ml of concentrated nitric acid and 60 ml of water. Let stand for several days and filter before using.

Borax, $Na_2B_4O_7 \cdot 10H_2O$. Used dry.

Cobalt nitrate, $Co(No_3)_2 \cdot 6H_2O$. Dissolve 7 grams of salt in 100 ml of water.

Dimethylglyoxime. Dissolve 1 gram in 100 ml of 95 percent ethyl alcohol.

Potassium bisulfate, $KHSO_4$. Dry reagent.

Potassium ferricyanide, $K_3Fe(CN)_6$. Dissolve $5\frac{1}{2}$ grams in 100 ml of water.

Potassium ferrocyanide, $K_4Fe(CN)_6 \cdot 3H_2O$. Dissolve 20 grams in 100 ml of water.

Potassium iodide, KI. Used dry. Also dissolve 8 grams in 100 ml of water.

Salt of phosphorus (microcosmic salt), $HNaNH_4PO_4 \cdot 4H_2O$. Used dry.

Silver nitrate, $AgNO_3$. Dissolve 4 grams in 100 ml of water. Keep in dark opaque bottle.

Sodium carbonate, Na_2CO_3. Used dry.

Sulfur. Dry reagent. Mixed with potassium iodide, forms the bismuth or iodide flux.

Tin, metallic

Turmeric paper

Water. Distilled water is implied in order to obtain pure reagents and reactions.

Zinc, metallic

Reagents may be secured from a chemical supply house.

THE BLOWPIPE

The blowpipe is used to produce an intensely hot flame (about 1200°–1500°C) that can be directed against a mineral fragment or powder (assay). (See Figure 6-1.) The simple blowpipe is a brass tube about 9 in. long, bent at one end, and used in conjunction with a Bunsen burner, alcohol lamp, or candle. The tip of the blowpipe is placed close to, or just within the flame, and air is blown through the opposite end. To produce a steady blast, the blower should keep his cheeks puffed out while breathing naturally through the nose. This is a simple procedure, but often requires a little practice. Widely available are more sophisticated blowpipes with small gas pressure tanks capable of producing a hot flame.

Reducing Flame

Intense burning takes place in the BLUE portion of the flame, termed the REDUCING FLAME. A mineral frag-

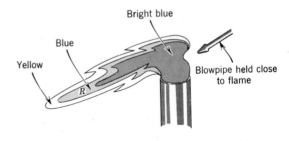

Fig. 6-2. *R*—Reducing flame (removes oxygen).

ment placed in the outer tip of the blue flame *loses oxygen* to the burning gases and is thus reduced. Reduction is a process by which electrons are *gained.* For example, the copper ions (Cu^+) in cuprite (Cu_2O) gain electrons when cuprite is reduced to metallic copper (Cu^0).

The best reducing flame is produced by a gentle blast of air from the blowpipe held a little outside the flame. The substance to be reduced is placed within the blue cone at *R* and is surrounded by a yellow flame (Figure 6-2). Do not place the specimen too far into the blue cone, or it may become coated with soot, since the interior is relatively cool.

Oxidizing Flame

The outer ALMOST INVISIBLE HEAT CONE is the OXIDIZING FLAME. A specimen placed here reacts easily with the oxygen of the surrounding air at elevated temperature. In oxidation, electrons are *lost,* with a change occurring, such as ferrous iron (Fe^{2+}) changing to ferric iron (Fe^{3+}).

The best oxidizing flame is produced by a strong blast of air from a blowpipe held just within the flame (Figure 6-3). This produces a narrow flame that contains no yellow. The specimen to be oxidized is held in the invisible heat cone at *O*, well beyond the blue reducing cone.

FUSIBILITY

Fusibility represents the relative ease with which a mineral melts. In testing fusibility, a thin fragment

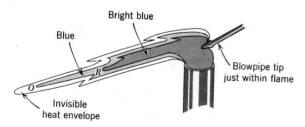

Fig. 6-3. *O*—Oxidizing flame (adds oxygen).

may be held, with forceps, in the hottest part of the flame, just outside the tip of the blue flame.

Metallic minerals are usually fused on charcoal, for they frequently contain arsenic or antimony which can alloy with the tip of the platinum forceps. The nonmetallic arsenic minerals, orpiment (yellow) and realgar (orange), should also be heated on charcoal.

Scale of fusibility*

SCALE	MINERAL	BEHAVIOR
1.	Stibnite	Fuses easily in any flame; about 550°C.
2.	Chalcopyrite	Fuses with some difficulty in a gas flame; about 800°C.
3.	Almandine garnet	Fuses easily in a blowpipe flame; about 1050°C.
4.	Actinolite	Thin edges fuse easily in a blowpipe flame; about 1200°C.
5.	Orthoclase	Fuses on edges with difficulty in a blowpipe flame; about 1300°C.
6.	Enstatite (bronzite)	Blowpipe flame round only the thinnest edges; about 1400°C.
7.	Quartz	Infusible in ordinary blowpipe flame.

Such relative terms as "fuses easily," "fusible," "fuses with difficulty," and "infusible" are often used in the description of minerals. When testing fusion, it is not only the ease of fusion that should be noted, but also the flame coloration (see flame tests) and any unusual behavior of the mineral when heated. For example, such minerals as the zeolites, scapolite, and epidote swell and bubble up when heated, as the result of released gases (INTUMESCE). Others, like pyrophyllite, vermiculite, and wavellite, spread out like leaves of a

---

* Applies to a gas burner and simple blowpipe. A blow torch or compressed air-gas blowpipe yields higher temperatures.

book (**EXFOLIATE**). Still others, like barite and serpentine, fly apart (**DECREPITATE**).

REDUCTION TO A METALLIC GLOBULE ON CHARCOAL

Glowing charcoal is a strong reducing agent. Some minerals can be reduced to a metallic globule by just heating them on charcoal. Others, however, must be mixed with a reducing mixture before they can be reduced. **REDUCING MIXTURE:** sodium carbonate and powdered charcoal mixed in equal proportions.

ELEMENT	METALLIC GLOBULE
Gold, Au	Golden yellow; soft, malleable globule; does not tarnish (heat with reducing mixture).
Silver, Ag	Silver-white; soft, malleable globule; remains bright when cooled. Reducing mixture is usually necessary.
Copper, Cu	Copper-red; malleable globule; surface of globule turns black on cooling. Roast copper-bearing minerals to remove sulfur, antimony, and arsenic; then heat with a reducing mixture.
Lead, Pb	Gray; soft, malleable globule.
Tin, Sn	Silver-white; soft, malleable globule. On cooling, becomes coated with a white oxide film.
Bismuth, Bi	Dark gray; brittle globule.

BEAD TESTS

When certain elements are fused in a borax, salt of phosphorus, or sodium carbonate flux, they impart a characteristic color to the flux. The flux may be held in a small ($\frac{1}{8}$-in. diameter) loop of a platinum wire (Figure 6-4).

The platinum loop is first heated in the flame, then dipped into the powdered flux and reheated until a clear bead forms. The process is repeated until a rounded clear bead fills the loop of the platinum wire.

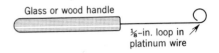

Glass or wood handle

$\frac{1}{8}$-in. loop in platinum wire

Fig. 6-4. Platinum loop for bead tests.

While still hot, the bead should be touched to the powdered mineral and heated first in the oxidizing flame, and the color observed, and then in the reducing part of the flame (Figures 6-2 and 6-3). The color should be observed not only when the bead is hot, but also when it is cold. Care must be taken not to let too much powdered mineral adhere to the bead, or it will turn black. A few minute grains are usually enough to color the bead. Table 6-1 lists bead tests for borax, $Na_2B_4O_7 \cdot 10H_2O$, salt of phosphorus (microcosmic salt), $HNaNH_4PO_4 \cdot 4H_2O$, and sodium carbonate, Na_2CO_3.

FLAME TESTS

Ignition of minerals in a Bunsen flame or blowpipe flame imparts a color that is characteristic of certain elements. However, the characteristic flame color of an element may be masked by the presence of another element. To remove the strong yellow color of sodium, the flame may be viewed through a cobalt blue glass.

A simple flame test is made with a platinum wire. First, the wire is cleaned by dipping it in hydrochloric acid and heating it in the Bunsen flame; the clean wire is then dipped into the powdered mineral previously moistened with hydrochloric acid, and held in the flame. The color of the flame is at times indicative of particular elements.

OPEN AND CLOSED TUBES

Open and closed tubes, made from glass tubing of about $\frac{1}{4}$ in. in outside diameter, are used primarily to test minerals containing **ARSENIC, ANTIMONY, MERCURY,** or **WATER;** these minerals form volatile sublimates, and condense on the upper cooler walls of the

A   x 0.47

CHALCOPYRITE

$CuFeS_2$

Massive chalcopyrite with quartz. Deep golden brass yellow color is distinctive: Ore Knob, North Carolina.

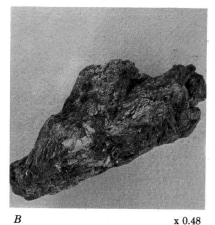

B   x 0.48

COVELLITE

CuS

Massive covellite showing characteristic iridescent blue color and platy structure: Colorado.

C   x 0.80

CHALCOCITE

Cu_2S

Striated chalcocite crystals coated with black tarnish. The two lower crystals are geniculate twins (knee-like).

D   x 0.60

BORNITE

Cu_5FeS_4

Massive bornite showing reddish bronze color on a fresh surface and characteristic iridescent blue tarnish: Chile.

E   x 0.74

ENARGITE

Cu_3AsS_4

Prismatic iron black enargite crystals: Montana.

F   x 0.78

TETRAHEDRITE

$(Cu, Fe, Zn, Ag)_{12}(Sb, As)_4S_{13}$

Massive gray tetrahedrite occurring with chalcopyrite and quartz: Cornwall, England.

A   x 0.70

ORPIMENT
As$_2$S$_3$

Massive lemon yellow orpiment showing characteristic foliated structure: Kurdistan, Turkey.

B   x 0.50

REALGAR
AsS

Massive orange-red realgar showing typical resinous luster, granular texture, and yellow alteration: Manhattan, Nevada.

C   x 0.80

ARSENOPYRITE
FeAsS

Striated arsenopyrite crystals.

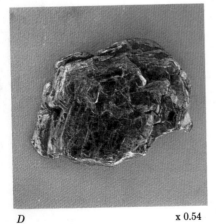

D   x 0.54

MOLYBDENITE
MoS$_2$

Molybdenite showing characteristic *bluish* lead gray color and foliated structure: Okanogan County, Washington.

E   x 0.25

STIBNITE
Sb$_2$S$_3$

Aggregate of steel gray striated bladed stibnite crystals: Ichinokwa, Iyo, Japan.

F   x 0.68

BISMUTHINITE
Bi$_2$S$_3$

Dark lead gray bismuthinite showing characteristic fibrous texture: Kingsgate, New South Wales.

Table 6-1

Bead Tests

Element	Oxidizing Flame	Reducing Flame
Borax Bead		
Chromium, Cr	green*	green
Cobalt, Co	blue	blue
Copper, Cu	blue-green	red, opaque
Iron, Fe	yellow	pale bottle green
Manganese, Mn	violet	colorless
Molybdenum, Mo	colorless	dark brown
Nickel, Ni	reddish-brown	gray, opaque
Titanium, Ti	colorless	brownish-violet
Tungsten, W	colorless	yellow
Uranium, U	yellow to yellow-brown	pale green (luminesces bright green under ultra-violet light)
Vanadium, V	yellow-green	green
Salt of Phosphorus Bead		
Chromium, Cr	green	green
Cobalt, Co	blue	blue
Columbium, Cb (niobium)	colorless	red-brown
Copper, Cu	greenish-blue	red, opaque
Iron, Fe	yellow to colorless	colorless
Manganese, Mn	violet	colorless
Molybdenum, Mo	colorless	green
Nickel, Ni	yellow to yellow-brown	yellow to yellow-brown
Silica, Si	insoluble skeleton	insoluble skeleton
Titanium, Ti	colorless	violet
Tungsten, W	colorless	blue
Uranium, U	almost colorless	green
Vanadium, V	greenish-yellow	green
Sodium Carbonate Bead		
Chromium, Cr	opaque yellow	
Manganese, Mn	opaque bluish-green	
Silica, Si	insoluble SiO_2 skeleton	

*Colors refer to cold beads.

Table 6-2

Flame Coloration

Color	Element
Red	
crimson	Lithium, Li
crimson	Strontium, Sr
Violet	Potassium, K
Orange	Calcium, Ca
Yellow	
bright yellow	Sodium, Na
Blue	
azure blue	Copper chloride (copper mineral moistened with HCl)
dark blue	Selenium, Se
light blue	Arsenic, As
whitish-blue	Lead, Pb
Blue-green	Antimony, Sb
	Phosphorus, P
	Zinc, Zn
Yellow-green	Barium, Ba
	Boron, B
pale yellow-green	Molybdenum, Mo
Green	
emerald green	Copper, Cu
grass green	Tellurium, Te
light green	Antimony, Sb
whitish-green	Bismuth, Bi

tube. The open tube is a 4-in. length of tubing, open at both ends. The mineral assay is placed about 1 in. from the end of the tube and gently heated (Figure 6-5).

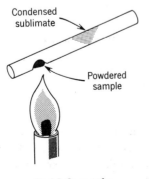

Fig. 6-5. Open tube.

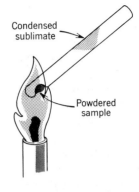

Fig. 6-6. Closed tube.

To make two closed tubes, one should heat an 8-in. length of tubing in the center, pull it apart quickly, and fuse the drawn ends into a closed terminal. The assay is placed in the bottom of the tube and gently heated (Figure 6-6).

TESTS FOR INDIVIDUAL ELEMENTS AND COMMON ANIONS (CO_3, PO_4, SO_4, H_2O)

Aluminum, Al

INFUSIBLE light-colored aluminum-bearing minerals can be tested by intensely heating a fragment in the oxidizing flame. After heating, moisten with a few drops of cobalt nitrate, $Co(NO_3)_2$, and reheat. If aluminum is present, the fragment will turn deep blue.

Ex. Gibbsite, cliachite (bauxite), alunite, topaz. (Al-bearing silicates are apt to fuse slightly and produce a blue cobalt glass, but this is not a test.)

Antimony, Sb

When heated moderately in the open tube, antimony minerals produce dense white fumes of antimony oxide, Sb_2O_4, on the underside of the tube.

Dissolve the powdered mineral in concentrated hydrochloric acid; then pour a few drops of the solution into a test tube containing a solution of potassium iodide (KI). A bright orange precipitate of antimony iodide results. (Lead minerals produce a lemon yellow precipitate of lead iodide, and arsenic minerals produce a yellow precipitate of arsenic iodide.)

The heating of antimony minerals on charcoal in the oxidizing flame produces a volatile white sublimate of antimony oxide near the assay, and dense, odorless white fumes.

Ex. Stibnite, native antimony.

Arsenic, As

When heated moderately in the open tube with a good draft, arsenic-bearing minerals produce a volatile white sublimate of arsenic oxide, As_2O_3, and fumes of arsine, AsH_3, with garlic odor.

The garlic odor, characteristic of arsenic-bearing minerals, is often produced by powdering the mineral specimen.

Heated in the closed tube, arsenic-bearing minerals produce a brilliant gray "arsenic mirror" of crystalline arsenic and a bright black coating of arsenic sulfide nearby. Garlic odor is present.

When heated on charcoal, most arsenic minerals give a volatile white coating (As_2O_3) some distance from the assay; they also give fumes with a garlic odor.

Ex. Arsenopyrite, native arsenic, realgar, orpiment.

Barium, Ba

Yellowish-green flame test.

Ex. Barite, witherite.

Beryllium, Be

Beryllium is not easily tested qualitatively. A radiation device making use of neutron absorption and known as a "berylometer" has been widely used. The emission spectrograph is also satisfactory. Beryl and other beryllium minerals are frequently identified optically as well as by physical tests.

Ex. Beryl, phenakite.

Bismuth, Bi

Bismuth minerals heated on charcoal with potassium iodide and sulfur produce a bright red coating of bismuth iodide with a yellow ring near the fused mass.

A powdered bismuth mineral fused on charcoal with sodium carbonate can be reduced to a brittle metallic globule of bismuth. Also, coating of bismuth oxide, Bi_2O_3, which is orange-yellow when hot and yellow when cold, forms near the fused mass.

Ex. Bismuthinite, bismuth.

Boron, B

Borates produce a yellow-green flame, similar to the flame of barium-bearing minerals.

Ex. Borax, kernite, colemanite, datolite.

Bromine, Br

$AgNO_3$ solution added to a solution of a bromine mineral in dilute nitric acid will precipitate white silver bromide (AgBr).

Carbon dioxide (carbonate, CO₃)

Carbonates effervesce in acids. A range of effervescence exists, depending on the acid, the temperature, and the concentration. See Table 7-9, p. 147.

Carbon dioxide in contact with lime water causes a turbid solution.

Ex. Calcite.

Chlorine, Cl

Dissolve the chlorine-bearing mineral in dilute nitric acid. The addition of silver nitrate ($AgNO_3$) forms a curdy white precipitate of silver chloride (AgCl), which darkens on exposure to light. Bromine and iodine give similar reactions.

A powdered Cl mineral heated with sulfuric acid and manganese dioxide yields fumes of chlorine.

Ex. Carnallite, $KMgCl_3 \cdot 6H_2O$, halite, sylvite.

Chromium, Cr

Chromium-bearing minerals will give an emerald green color to a borax bead in either the reducing or the oxidizing flame.

Ex. Chromite.

Cobalt, Co

A small amount of cobalt will color a borax bead blue. An excess will alloy with the platinum wire that contains the bead, and make it brittle.

Ex. Cobaltite.

Columbium, Cb, (niobium)

Test: Fuse the columbium-bearing mineral with sodium carbonate, and dissolve the fusion in concentrated hydrochloric acid. Add a piece of metallic tin, and boil. If columbium is present, then the solution will turn dark blue and change slowly to brown. The color disappears entirely with the addition of water. (Tungsten-bearing minerals also turn blue when dissolved in HCl and boiled with tin.)

Ex. Columbite.

Copper, Cu

Test: Dissolve copper-bearing mineral in acid. The solution turns dark blue when neutralized with ammonium hydroxide.

Copper gives an emerald green flame. If mineral is moistened with hydrochloric acid an azure blue flame is produced.

Ex. Chalcocite, chalcopyrite, enargite.

Fluorine, Fl

Fluorides dissolve in concentrated sulfuric acid and liberate hydrofluoric acid (HF) that etches glass. Test: Coat a glass plate with paraffin and remove a small area to expose the glass. Add concentrated sulfuric acid (H_2SO_4) to the exposed area. Then add the powdered mineral to the acid, and allow the test to stand. The evolution of hydrofluoric acid will etch the glass. When the glass is cleaned, a rough etched surface will be observed.

Ex. Fluorite, cryolite.

Gold, Au

Fused with sodium carbonate on charcoal, gold compounds are reduced to a malleable golden yellow globule.

Ex. Calaverite.

Iodine, I

When heated in the closed tube with $KHSO_4$, iodine-bearing minerals yield violet vapors.

The addition of $AgNO_3$ to solutions of iodides in dilute nitric acid yields a white precipitate of AgI that is almost insoluble in ammonia.

Iron, Fe

Minerals bearing ferric iron, when dissolved in HCl, form a red-brown precipitate of iron hydroxide, if the acid solution is neutralized with ammonium hydroxide.

Iron-bearing minerals become magnetic when heated in the reducing flame. (Nickel and cobalt minerals also become magnetic.)

Test: Dissolve the iron mineral (not an iron sulfide) in hydrochloric acid. (If solution occurs with difficulty, previous fusion with sodium carbonate may be necessary.) Divide the cool dilute solution into two portions. Test one with a few drops of potassium ferricyanide. If the dissolved iron is ferrous iron, then a dark blue precipitate results. Test the second portion of the solution with a few drops of potassium ferrocyanide. It will turn dark blue if ferric ions are present.

Ex. Hematite, magnetite, limonite.

Lead, Pb

Test: Dissolve the powdered mineral in concentrated hydrochloric acid. Pour a few drops into a test tube containing a solution of potassium iodide (KI). A lemon yellow precipitate of lead iodide (PbI_2) forms. Arsenic produces a similar reaction, and antimony compounds produce an orange precipitate.

Fused on charcoal with sodium carbonate, lead minerals may be reduced to a malleable lead globule. A yellow coating forms near the assay, and a white coating forms farther away.

Ex. Cerussite, anglesite, galena.

Lithium, Li

Lithium-bearing minerals impart to a flame a crimson color that is similar to the flame color of strontium.

Ex. Spodumene.

Magnesium, Mg

A solution* containing magnesium may be tested qualitatively and sequentially for iron, calcium,

---

* Magnesium-bearing silicates must be fused with sodium carbonate prior to testing.

and magnesium. Test: Dissolve in nitric acid to which a small amount of hydrochloric acid has been added. Precipitate iron with ammonia, then filter. Add ammonium oxalate to the filtrate, and boil. Filter away calcium oxalate if calcium is present; cool. Liquid salt of phosphorus will slowly precipitate magnesium in a cool solution from which all calcium has been removed.

Manganese, Mn

This mineral will color a borax bead violet, and a sodium carbonate bead pale blue.
Ex. Pyrolusite, manganite.

Mercury, Hg

Heated in the closed tube with soda, mercury minerals give a black sublimate of mercury sulfide, and drops of mercury are liberated.

When heated gently in the open tube, sulfur is oxidized, and minute gray metallic globules of mercury adhere to the walls of the tube.
Ex. Cinnabar.

Molybdenum, Mo

Test: On a charcoal block, fuse the powdered mineral in a large sodium metaphosphate bead. Place bead in a filter paper and put it on a watch glass. Dissolve in very dilute hydrochloric acid and add some metallic tin. A blue solution is then formed, which turns brown on heating.
Ex. Molybdenite.

Nickel, Ni

Test: Dissolve powdered mineral in nitric acid, neutralize with ammonium hydroxide, and use filter to remove brown ferric hydroxide. Add a few drops of dimethylglyoxime. A bright rose precipitate indicates the presence of nickel.
Ex. Pentlandite.

Phosphorus, P (phosphates, PO₄)

Test: Dissolve mineral in nitric acid. Add a few drops of the solution to a solution of ammonium molybdate. A yellow precipitate results.
Ex. Pyromorphite, apatite.

Potassium, K

This element gives violet flame, often masked by yellow sodium flame, but is visible through a blue glass or Merwin's flame-color screen.

Potassium-bearing minerals show a weak radioactivity that is detected with highly sensitive scintillation counters.
Ex. Sylvite.

Selenium, Se

When fused on charcoal, selenium gives a disagreeable horse radish odor and a volatile grayish deposit with a red border.
Ex. Native selenium.

Silicon, Si

Test: Boil mineral in hydrochloric acid and evaporate almost to dryness. If insoluble, fuse first with sodium carbonate. A jelly-like mass results, which indicates silica.
Ex. Sodalite, olivine.

Silver, Ag

Test: Dissolve mineral in nitric acid. Add a few drops of hydrochloric acid. A white, curdy precipitate of silver chloride, which darkens on exposure to light, indicates silver. Precipitate is soluble in ammonium hydroxide.
Ex. Silver.

Fuse silver-bearing mineral intensely on charcoal with sodium carbonate to obtain a malleable silver globule.
Ex. Proustite, pyrargyrite.

Strontium, Sr

Strontium-bearing minerals give a crimson flame test, which is similar to the flame of lithium.
Ex. Strontianite.

Sulfates (SO₄)

The addition of barium chloride to a solution of a soluble sulfate produces a heavy and dense white precipitate of barium sulfate.
Ex. Gypsum.

Sulfur, S

Test: Fuse powdered mineral on charcoal with sodium carbonate and charcoal dust. Place fused mass on a bright silver coin, moisten with a drop of water, and crush. A blackish stain of silver sulfide will form on the coin.

When heated in the open tube, sulfides give off suffocating fumes of sulfur dioxide, SO_2.

Ex. Pyrite, pyrrhotite.

Tellurium, Te

A powdered tellurium mineral added to hot concentrated sulfuric acid produces a deep crimson color.

Ex. Native tellurium.

Tin, Sn

Test: Heat a lump of cassiterite (SnO_2) that is wrapped with zinc shavings in dilute hydrochloric acid. The hydrogen evolved reduces the oxide, and cassiterite will become coated with metallic tin. The latter may be polished by rubbing it on cloth.

Powdered tin-bearing minerals that are fused on charcoal with a reducing mixture of sodium carbonate and powdered charcoal produce (with difficulty) a soft, white, malleable globule of metallic tin.

Titanium, Ti

Test: Fuse mineral with sodium carbonate on charcoal. Dissolve fusion in concentrated sulfuric acid. Add the acid solution *very slowly* to an equal volume of water. A few drops of hydrogen peroxide, H_2O_2, added to the cold dilute solution turns it yellow-amber, indicating titanium.

Fuse with sodium carbonate, and dissolve fusion in hydrochloric acid. Heat the solution with metallic tin. The solution turns pale violet, indicating titanium.

Ex. Rutile.

Tungsten, W

Test: Fuse powdered mineral with sodium car-

bonate on charcoal, and dissolve fusion in hydrochloric acid. Boil the solution with metallic zinc or tin. If tungsten is present, then the solution will turn deep blue and then brown. (Columbium (niobium) gives a similar test.)

Ex. Wolframite.

Uranium, U

A borax or sodium fluoride bead fused with a uranium mineral luminesces brilliant yellow under ultraviolet light.

Uranium minerals are radioactive as indicated by a Geiger counter, electroscope, or the exposure of a sealed photographic negative.

Ex. Uraninite, carnotite.

Vanadium, V

A brownish-red solution is obtained when a vanadium mineral is dissolved in concentrated hydrochloric acid. The addition of a small amount of water decolorizes the solution, and the red-brown color is restored upon the addition of a few drops of hydrogen peroxide.

In the reducing flame, vanadium minerals turn the borax or salt of phosphorus bead to a color between brownish and dirty green; when the bead cools it turns green.

Ex. Vanadinite.

Water, H_2O

When minerals containing water are heated in the closed tube, they will deposit drops of water on the cool walls of the upper portion of the tube.

Ex. Gypsum, stilbite, heulandite.

Zinc, Zn

A fragment of a zinc mineral that is moistened with a few drops of cobalt nitrate solution and intensely heated will turn bright green, or yield a green sublimate on charcoal. Zinc silicates should be sintered with sodium carbonate. Willemite, a common zinc silicate, often luminesces in ultraviolet light.

Periodic Table of the Elements

KEY TO CHART

- Atomic Number → 50
- Symbol → Sn
- Atomic Weight → 118.69
- Electron Configuration → -18-18-4
- Oxidation States → +2 +4

Transition Elements — Group 8

At. No.	Sym.	Oxidation States	Atomic Weight	Electron Config.	Orbit
1	H	+1, -1	1.00797	1	K
2	He	0	4.0026	2	K
3	Li	+1	6.939	2-1	K-L
4	Be	+2	9.0122	2-2	K-L
5	B	+3	10.811	2-3	K-L
6	C	+2 +4 -4	12.01115	2-4	K-L
7	N	+1 +2 +3 +4 +5 -2 -3	14.0067	2-5	K-L
8	O	-2	15.9994	2-6	K-L
9	F	-1	18.9984	2-7	K-L
10	Ne	0	20.183	2-8	K-L
11	Na	+1	22.9898	2-8-1	K-L-M
12	Mg	+2	24.312	2-8-2	K-L-M
13	Al	+3	26.9815	2-8-3	K-L-M
14	Si	+2 +4 -4	28.086	2-8-4	K-L-M
15	P	+3 +5 -3	30.9738	2-8-5	K-L-M
16	S	+4 +6 -2	32.064	2-8-6	K-L-M
17	Cl	+1 +5 +7 -1	35.453	2-8-7	K-L-M
18	Ar	0	39.948	2-8-8	K-L-M
19	K	+1	39.102	-8-8-1	-L-M-N
20	Ca	+2	40.08	-8-8-2	-L-M-N
21	Sc	+3	44.956	-8-9-2	-L-M-N
22	Ti	+2 +3 +4	47.90	-8-10-2	-L-M-N
23	V	+2 +3 +4 +5	50.942	-8-11-2	-L-M-N
24	Cr	+2 +3 +6	51.996	-8-13-1	-L-M-N
25	Mn	+2 +3 +4 +7	54.9380	-8-13-2	-L-M-N
26	Fe	+2 +3	55.847	-8-14-2	-L-M-N
27	Co	+2 +3	58.9332	-8-15-2	-L-M-N
28	Ni	+2 +3	58.71	-8-16-2	-L-M-N
29	Cu	+1 +2	63.54	-8-18-1	-L-M-N
30	Zn	+2	65.37	-8-18-2	-L-M-N
31	Ga	+3	69.72	-8-18-3	-L-M-N
32	Ge	+2 +4	72.59	-8-18-4	-L-M-N
33	As	+3 +5 -3	74.9216	-8-18-5	-L-M-N
34	Se	+4 +6 -2	78.96	-8-18-6	-L-M-N
35	Br	+1 +5 -1	79.909	-8-18-7	-L-M-N
36	Kr	0	83.80	-8-18-8	-L-M-N
37	Rb	+1	85.47	-18-8-1	-M-N-O
38	Sr	+2	87.62	-18-8-2	-M-N-O
39	Y	+3	88.905	-18-9-2	-M-N-O
40	Zr	+4	91.22	-18-10-2	-M-N-O
41	Nb	+3 +5	92.906	-18-12-1	-M-N-O
42	Mo	+6	95.94	-18-13-1	-M-N-O
43	Tc		(99)	-18-13-2	-M-N-O
44	Ru	+4 +6 +8	101.07	-18-15-1	-M-N-O
45	Rh	+3	102.905	-18-16-1	-M-N-O
46	Pd	+2 +4	106.4	-18-18-0	-M-N-O
47	Ag	+1	107.870	-18-18-1	-M-N-O
48	Cd	+2	112.40	-18-18-2	-M-N-O
49	In	+3	114.82	-18-18-3	-M-N-O
50	Sn	+2 +4	118.69	-18-18-4	-M-N-O
51	Sb	+3 +5 -3	121.75	-18-18-5	-M-N-O
52	Te	+4 +6 -2	127.60	-18-18-6	-M-N-O
53	I	+1 +5 +7 -1	126.9044	-18-18-7	-M-N-O
54	Xe	0	131.30	-18-18-8	-M-N-O
55	Cs	+1	132.905	-18-8-1	-N-O-P
56	Ba	+2	137.34	-18-8-2	-N-O-P
57*	La	+3	138.91	-18-9-2	-N-O-P
72	Hf	+4	178.49	-32-10-2	-N-O-P
73	Ta	+5	180.948	-32-11-2	-N-O-P
74	W	+6	183.85	-32-12-2	-N-O-P
75	Re	+4 +6 +7	186.2	-32-13-2	-N-O-P
76	Os	+3 +4 +6 +8	190.2	-32-14-2	-N-O-P
77	Ir	+3 +4	192.2	-32-15-2	-N-O-P
78	Pt	+2 +4	195.09	-32-16-2	-N-O-P
79	Au	+1 +3	196.967	-32-18-1	-N-O-P
80	Hg	+1 +2	200.59	-32-18-2	-N-O-P
81	Tl	+1 +3	204.37	-32-18-3	-N-O-P
82	Pb	+2 +4	207.19	-32-18-4	-N-O-P
83	Bi	+3 +5	208.980	-32-18-5	-N-O-P
84	Po	+2 +4	(210)	-32-18-6	-N-O-P
85	At	-1	(210)	-32-18-7	-N-O-P
86	Rn	0	(222)	-32-18-8	-N-O-P
87	Fr		(223)	-18-8-1	-O-P-Q
88	Ra	+2	(227)	-18-8-2	-O-P-Q
89**	Ac	+3	(227)	-18-9-2	-O-P-Q

*Lanthanides

At. No.	Sym.	Oxidation States	Atomic Weight	Electron Config.
58	Ce	+3 +4	140.12	-19-9-2
59	Pr	+3 +4	140.907	-20-9-2
60	Nd	+3	144.24	-22-8-2
61	Pm	+3	(145)	-23-8-2
62	Sm	+2 +3	150.35	-24-8-2
63	Eu	+2 +3	151.96	-25-8-2
64	Gd	+3	157.25	-25-9-2
65	Tb	+3	158.924	-26-9-2
66	Dy	+3	162.50	-28-8-2
67	Ho	+3	164.930	-29-8-2
68	Er	+3	167.26	-30-8-2
69	Tm	+3	168.934	-31-8-2
70	Yb	+2 +3	173.04	-32-8-2
71	Lu	+3	174.97	-32-9-2

Orbit: -N-O-P

**Actinides

At. No.	Sym.	Oxidation States	Atomic Weight	Electron Config.
90	Th	+4	232.038	-19-9-2
91	Pa	+5 +4	(231)	-20-9-2
92	U	+3 +4 +5 +6	238.03	-21-9-2
93	Np	+3 +4 +5 +6	(237)	-22-9-2
94	Pu	+3 +4 +5 +6	(242)	-23-9-2
95	Am	+3 +4 +5 +6	(243)	-24-9-2
96	Cm	+3	(245)	-25-9-2
97	Bk	+3 +4	(249)	-26-9-2
98	Cf	+3	(249)	-28-8-2
99	Es		(254)	-29-8-2
100	Fm		(252)	-30-8-2
101	Md		(256)	-31-8-2
102	No		(254)	-32-8-2
103	Lw			-32-9-2

Orbit: -O-P-Q

Numbers in parentheses are mass numbers of most stable isotope of that element.

(Courtesy of The Chemical Rubber Company, *Handbook of Chemistry and Physics*, 45th edition, 1964–1965.)

CHEMICAL ELEMENTS AND THEIR SYMBOLS

ELEMENT	SYMBOL	ELEMENT	SYMBOL
Actinium	Ac	Germanium	Ge
Aluminum	Al	Gold	Au
Americium	Am	Hafnium	Hf
Antimony	Sb	Helium	He
Argon	Ar	Holmium	Ho
Arsenic	As	Hydrogen	H
Astatine	At	Indium	In
Barium	Ba	Iodine	I
Berkelium	Bk	Iridium	Ir
Beryllium	Be	Iron	Fe
Bismuth	Bi	Krypton	Kr
Boron	B	Lanthanum	La
Bromine	Br	Lawrencium	Lw
Cadmium	Cd	Lead	Pb
Calcium	Ca	Lithium	Li
Californium	Cf	Lutetium	Lu
Carbon	C	Magnesium	Mg
Cerium	Ce	Manganese	Mn
Cesium	Cs	Mendelevium	Mv
Chlorine	Cl	Mercury	Hg
Chromium	Cr	Molybdenum	Mo
Cobalt	Co	Neodymium	Nd
Columbium	Cb	Neon	Ne
(Niobium, Nb)		Neptunium	Np
Copper	Cu	Nickel	Ni
Curium	Cm	Nitrogen	N
Dysprosium	Dy	Nobellium	No
Einsteinium	Es	Osmium	Os
Erbium	Er	Oxygen	O
Europium	Eu	Palladium	Pd
Fermium	Fm	Phosphorus	P
Fluorine	F	Platinum	Pt
Francium	Fr	Plutonium	Pu
Gadolinium	Gd	Polonium	Po
Gallium	Ga	Potassium	K

ELEMENT	SYMBOL	ELEMENT	SYMBOL
Praseodymium	Pr	Tantalum	Ta
Promethium	Pm	Technetium	Tc
Protactinium	Pa	Tellurium	Te
Radium	Ra	Terbium	Tb
Radon	Rn	Thallium	Tl
Rhenium	Re	Thorium	Th
Rhodium	Rh	Thulium	Tm
Rubidium	Rb	Tin	Sn
Ruthenium	Ru	Titanium	Ti
Samarium	Sm	Tungsten	W
Scandium	Sc	Uranium	U
Selenium	Se	Vanadium	V
Silicon	Si	Xenon	Xe
Silver	Ag	Ytterbium	Yb
Sodium	Na	Yttrium	Y
Strontium	Sr	Zinc	Zn
Sulfur	S	Zirconium	Zr

Zirconium, Zr

Test: Fuse the mineral with sodium carbonate and dissolve in hydrochloric acid. The acid solution will give an orange color to a piece of turmeric paper.

Zircon may luminesce orange under ultraviolet light.

REFERENCES

1. Dennen, W. H., *Principles of Mineralogy,* Ronald Press Company, New York, 1960, Chapter 5.
2. Hurlbut, C. S., and E. S. Dana, *Manual of Mineralogy,* 17th edition, John Wiley and Sons, New York, 1959, Chapter 4.
3. Smith, O. C., *Identification and Qualitative Chemical Analysis of Minerals,* 2nd edition, D. Van Nostrand Company, Princeton, N.J., 1953.
4. *Handbook of Chemistry and Physics,* 45 edition, 1964–1965, The Chemical Rubber Company, Cleveland, Ohio.

7.

Mineral Identification Tables

The tables that follow are intended to supply a key to mineral recognition. Experience shows that condensed and comprehensively tabulated data considerably accelerate mineral recognition. The tables also provide a convenient guide for anyone interested in becoming acquainted with common minerals.

RECOGNITION OF MINERALS

Each mineral species is unique in two fundamentals: CHEMICAL COMPOSITION and CRYSTALLIZATION. A precise knowledge of both is essential if one is to be able to establish the identity of a mineral.

Notwithstanding the importance of these fundamentals, there are many indirect observations that provide convenient short cuts to mineral recognition. Consider, for example, the case of common salt (the mineral halite): Its chemical composition of sodium chloride can be verified by chemical analysis, and its isometric crystallization can be established either by crystal measurement or by x-ray reflection. However, if one wants merely to identify halite, a reasonably good sense of taste is convenient for rapid recognition.

A number of minerals have unusual properties that assist in recognition, some examples being the magnetism of magnetite and pyrrhotite; the radio-activity of uraninite or carnotite; and the luminescence of scheelite and willemite.

There are a goodly number of common minerals, and many not so common, that frequently occur in specimens exhibiting distinctive properties; these can be recognized, with reasonable accuracy, by observation alone. A KNOWLEDGE OF CHARACTERISTIC PHYSICAL PROPERTIES, A FAMILIARITY WITH MINERAL ASSOCIATIONS, AND A KNOWLEDGE OF WHICH MINERALS ARE LIKELY TO BE CONFUSED WITH EACH OTHER SHOULD BE ACQUIRED AS AIDS IN RAPID RECOGNITION. Where sight identification is inadequate, special x-ray or optical equipment and supplementary chemical tests may be required.

Observational Sequence in Recognition

A systematic observational sequence greatly assists in mineral recognition. While observing an unknown specimen, it is desirable for one to conduct a detailed mental audit of these several properties: color, luster, streak, hardness, mode of aggregation, crystal habit, cleavage, tenacity, specific gravity, magnetism, radioactivity, water solubility, taste, behavior in acid, luminescence.

In many fortunate instances, just this outline, or even a small portion of it coupled with some reference to the Mineral Tables, will provide a key to mineral identification. There will be cases, however, where two, three, or more alternative criteria will be suggested for the specimen under examination. In these cases, the individual mineral descriptions should be consulted for more complete data. If uncertainties still exist, then the mineral under examination may well be beyond the range of the introductory procedure covered in this text. In such cases, recourse must be made to more advanced texts on mineralogy, supplemented by established laboratory methods.

A study of physical properties not only provides a vast array of useful knowledge of minerals, but also furnishes the basis for many observations that may be widely utilized in simplifying the process of identification. Reference to the illustrative and descriptive material of this text, a small hand lens ($6\times$ to $10\times$), a pen knife to test relative hardness, a streak plate, a small strong magnet, and a comparison of relative weights by heft, combine to provide much progress in terms of minerals recognized.

RECOGNITION TABLES

The following tables provide an observational sequence for the rapid recognition of mineral specimens:

Table 1. Color; luster; streak; hardness
  2. Mineral aggregates
  3. Crystal habit
  4. Cleavage
  5. Specific gravity
  6. Magnetism
  7. Radioactivity
  8. Water solubility—taste
  9. Effervescence in acid
  10. Luminescence

Table 7-1
Color, Luster, Streak, and Hardness

Color, luster, streak, and hardness are among the most distinctive and easily recognizable mineral properties.

Mineral colors range across the spectrum. Although the same mineral may often occur in a number of different colors, a few minerals are essentially invariable in color (idiochromatic). Some examples are cinnabar (vermillion), malachite (green), orpiment (yellow), and covellite (blue). Minerals with essentially invariable color are italicized in Table 7-1.

Luster may be metallic (lustrous, opaque, and metal-like), submetallic (essentially opaque, partially metallic), and nonmetallic (transparent in fine particles immersed in liquid). All surfaces examined for luster should be clean and free of surface films or other contamination. A freshly broken surface is often desirable. Nonmetallic minerals may be adamantine (brilliantly reflecting as a diamond), subadamantine (less brilliant), and vitreous (bright and shiny as broken glass). Other terms like earthy, waxy, resinous, pearly, silky, greasy, and dull are largely selfexplanatory. Luster is a less variable property than color.

Streak represents the smear of a mineral powder on a surface of unglazed porcelain. Hematite (red-brown), limonite (yellow-brown), and pyrolusite (black) are easily distinguished in this way. Dark colored metallic or submetallic minerals are more likely to yield informative streaks. Streaks for such soft minerals as graphite and molybdenite may be obtained on a glazed paper surface.

A simple hardness test consists in using a penknife to make a small scratch on a smooth, fresh surface of a mineral; this process will distinguish minerals with a hardness of above $5\frac{1}{2}$ from those whose hardness is less. The process may be reversed by trying to scratch the knife blade with the edge of the mineral.

DEEP RED

COLOR	LUSTER	STREAK	HARDNESS	REMARKS	MINERAL
Bright scarlet to red-brown	Adamantine to submetallic	Scarlet	$2\frac{1}{2}$	Heavy (Sp. Gr. 8.09)	*Cinnabar*
Deep ruby red	Adamantine	Red	$2-2\frac{1}{2}$	Ruby silver; striated crystals	*Proustite*
Dark purplish-red to almost black	Submetallic	Purplish-red	$2\frac{1}{2}$	Dark ruby silver	*Pyrargyrite*
Ruby red, orange-red, reddish-brown	Adamantine	Yellowish	3	Often in small six-sided crystals	VANADINITE
Bluish-red to almost black	Subadamantine	Red	$3\frac{1}{2}-4$	Associated with other copper minerals	*Cuprite*
Brick red	Earthy	Brick red	Soft	Earthy masses; often öolitic	HEMATITE
Ruby red; also brown, yellow, black	Adamantine	Yellowish	$3\frac{1}{2}-4$	Note complex dodeca-hedral cleavage	SPHALERITE (ruby, zinc)
Deep red to orange-yellow	Subadamantine	Orange-yellow	$4-4\frac{1}{2}$	Associated with wille-mite and franklinite	ZINCITE
Deep red, reddish-brown, black	Adamantine to submetallic	Pale brown	$6-6\frac{1}{2}$	Knee-like twins; verti-cally striated crystals	RUTILE
Deep red to purplish-red	Vitreous	White	$6\frac{1}{2}-7\frac{1}{2}$	Dodecahedrons; trapezohedrons	GARNET (pyrope, almandine, spessartite)
Dark red (opaque)	Waxy to dull	White	7	Microcrystalline	CHALCEDONY (jasper)
Dark red to pink	Vitreous	White	7	Six-sided crystals	QUARTZ (ferruginous)
Ruby red	Vitreous	White	$7\frac{1}{2}-8$	Octahedron	SPINEL (ruby spinel)
Ruby red	Vitreous	White	9	Six-sided crystals	CORUNDUM (ruby)

COPPER-RED

COLOR	LUSTER	STREAK	HARDNESS	REMARKS	MINERAL
Copper	Metallic	Copper (shining)	$2\frac{1}{2}-3$	Sectile	*Copper*
opper	Metallic	Gray-black	3	Often shows bluish-purple tarnish	*Bornite*
Copper	Metallic	Brownish-black	$5-5\frac{1}{2}$	Brittle, massive	*Niccolite*
Copper	Metallic	Reddish-brown	$5\frac{1}{2}$	Resembles niccolite	*Breithauptite*

ORANGE-RED

COLOR	LUSTER	STREAK	HARDNESS	REMARKS	MINERAL
Orange-red	Resinous	Orange-red	$1\frac{1}{2}$–2	Often associated with yellow orpiment	*Realgar* ✓
Deep orange-red	Adamantine	Orange-yellow	$2\frac{1}{2}$–3	Striated, elongated crystals; often hollow	*Crocoite*
Orange-red, yellow, gray	Adamantine	Yellowish	3	Tabular crystals	WULFENITE
Orange-red	Vitreous	White	5–6	Cryptocrystalline	OPAL (fire opal)
Orange-red	Waxy	White	7	Microcrystalline (translucent)	CHALCEDONY (carnelian)
Orange-red	Vitreous	White	$6\frac{1}{2}$–7	Thin wedge-shaped crystals	AXINITE

PINK: ROSE

COLOR	LUSTER	STREAK	HARDNESS	REMARKS	MINERAL
Pink to crimson	Vitreous, pearly, earthy	Pink	$1\frac{1}{2}$–2	Cobalt bloom	ERYTHRITE
Pink (more frequently white)	Vitreous to pearly	White	2–$2\frac{1}{2}$	A mica	MUSCOVITE
Pink, lilac, less frequently yellow or white	Pearly to vitreous	White	$2\frac{1}{2}$–4	A lithium-bearing mica	LEPIDOLITE
Flesh red, brick red, gray, white	Vitreous	White	3–$3\frac{1}{2}$	Bitter taste	POLYHALITE
Pink (many other colors and white)	Vitreous	White	3	Effervescent in cold HCl	CALCITE
Pale pink, white, light tints	Vitreous to pearly	White	$3\frac{1}{2}$–4	Curved saddle-shaped crystals	DOLOMITE
Pink to deep rose; sometimes gray or brown	Vitreous	White	$3\frac{1}{2}$–$4\frac{1}{2}$	Rhombohedral cleavage; effervescent in hot HCl	RHODOCHROSITE
Pink, violet, white, gray	Vitreous to pearly	White	$3\frac{1}{2}$–5	A brittle mica associated with corundum	MARGARITE
Reddish; commonly white	Vitreous to pearly	White	$3\frac{1}{2}$–4	Sheaf-like aggregates	STILBITE
Reddish, yellow; commonly white	Vitreous to pearly	White	$3\frac{1}{2}$–4	Coffin-shaped crystals	HEULANDITE

PINK: ROSE *(continued)*

Deep rose to amethyst	Porcelain-like	White	$3\frac{1}{2}$–4	Botryoidal crusts and small crystals, associated with iron phosphates	STRENGITE
Pink; many other colors	Vitreous	White	4	Cubic crystals; octahedral cleavage	FLUORITE
Flesh color, white, gray	Vitreous to pearly	White	4	Usually massive	ALUNITE
Pink, white, green, blue, yellow, brown	Vitreous	White	4–$4\frac{1}{2}$	Botryoidal masses; effervescent in cold HCl	SMITHSONITE
Pink, white, red, brown, yellow	Vitreous	White	4–5	Pseudocubic crystals	CHABAZITE
Pink, pale green or blue; white	Vitreous to pearly	White	$4\frac{1}{2}$–5	Pseudocubic crystals; pearly base	APOPHYLLITE
Pink, white; many colors	Vitreous	White	5–6	Massive; conchoidal fracture	OPAL
Pink, pale yellow, gray, violet, white	Vitreous	White	5–6	Often luminesces yellow-orange under long-wave ultraviolet light	SCAPOLITE
Pink, white	Vitreous	White	$5\frac{1}{2}$–6	May fade in daylight and turn pink again on exposure to ultraviolet radiation; massive	SODALITE (hackmanite)
Pink to deep rose; grayish	Vitreous	White	$5\frac{1}{2}$–6	Tabular, rough crystals; may contain black veinlets of pyrolusite	RHODONITE
Pink	Vitreous	White	6	Columnar crystals radiating needle-like aggregates	ZOISITE (thulite)
Pink, white, yellowish, brownish	Vitreous	White	6	A feldspar	ORTHOCLASE
Pale pink, other pale tints; commonly colorless or white	Vitreous to pearly	White	6	Resembles feldspar; fuses easily	AMBLYGONITE
Pink, usually emerald to pale green; also white, lilac, yellowish, brownish	Vitreous	White	$6\frac{1}{2}$–7	Tough, compact masses	JADEITE
Pink, lilac	Vitreous	White	$6\frac{1}{2}$–7	Flat, striated prismatic crystals; blocky cleavage	SPODUMENE (kunzite)

PINK: ROSE *(continued)*

Color	Luster	Streak	Hardness	Remarks	Mineral
Rarely pink; pale green and other tints	Vitreous	White	$6\frac{1}{2}-7\frac{1}{2}$	Dodecahedral and trapezohedral crystals	GARNET (grossularite)
Pink, violet, blue	Vitreous	White	7	Fibrous masses	DUMORTIERITE
Pink, rose; many other colors; white	Vitreous	White	7	Pink quartz in pegmatites is usually massive	QUARTZ
Pink, gray, white, brown, reddish-brown greenish	Vitreous	White	$7\frac{1}{2}$	Blocky crystals; nearly square cross section	ANDALUSITE
Pink, rose	Vitreous	White	$7-7\frac{1}{2}$	Rounded; triangular cross section	TOURMALINE (rubellite)
Pink to rose	Vitreous	White	$7\frac{1}{2}-8$	Usually massive; six-sided crystals (uncommon)	BERYL (morganite)
Pink; many other colors	Vitreous	White	$7\frac{1}{2}-8$	Octahedral crystals	SPINEL
Pink, pinkish-brown, lilac, yellowish, brownish-yellow, brown, pale blue, white	Vitreous	White	8	Perfect basal cleavage	TOPAZ
Pink; many other colors	Vitreous	White	9	Six-sided crystals	CORUNDUM (pink sapphire)
Pink, usually colorless or pale tints; occasionally deeply colored	Greasy adamantine	White	10	Rounded octahedrons	DIAMOND (pink, rare)

AMETHYST, VIOLET

COLOR	LUSTER	STREAK	HARDNESS	REMARKS	MINERAL
Pale violet, pink; less frequently white or yellow	Vitreous to pearly	White	$2\frac{1}{2}-4$	A lithium-bearing mica	LEPIDOLITE
Amethyst to deep rose	Porcelain-like	White	$3\frac{1}{2}-4$	Botryoidal crusts; small crystals associated with iron phosphates	STRENGITE
Amethyst to violet; many other colors are also common	Vitreous	White	4	Cubic crystals; octahedral cleavage	FLUORITE
Violet (rare); commonly green or brown; also blue	Vitreous	White	5	Six-sided prisms; loses violet color on heating	APATITE

AMETHYST, VIOLET *(continued)*

Violet, reddish	Vitreous	White	5–6	Bladed aggregates	**TREMOLITE** (hexagonite)
Violet, white, gray, pale yellow, pink	Vitreous	White	5–6	May luminesce yellow-orange under longwave ultraviolet radiation	**SCAPOLITE**
Amethyst	Vitreous	White	$6\frac{1}{2}$–$7\frac{1}{2}$	Dodecahedrons	**GARNET**
Pale violet (rare); usually green; sometimes white, yellowish, brownish	Vitreous	White	$6\frac{1}{2}$–7	Tough, compact masses	**JADEITE**
Violet-brown	Vitreous	White	$6\frac{1}{2}$–7	Thin, wedge-shaped crystals	**AXINITE**
Deep amethyst to pale violet	Vitreous	White	7	Six-sided prisms	**QUARTZ** (amethyst)
Violet, pink, blue	Vitreous	White	7	Fibrous masses	**DUMORTIERITE**
Violet; numerous other colors	Vitreous	White	$7\frac{1}{2}$–8	Octahedrons	**SPINEL**
Amethyst, violet; many other colors	Vitreous	White	9	Six-sided crystals with sets of intersecting striations	**CORUNDUM**

YELLOW

COLOR	LUSTER	STREAK	HARDNESS	REMARKS	MINERAL
Lemon yellow	Resinous to pearly	Pale yellow	$1\frac{1}{2}$–2	Associated with orange-red realgar	*Orpiment*
Bright yellow; may be amber, reddish, or blackish when contaminated	Resinous to greasy	White	$1\frac{1}{2}$–$2\frac{1}{2}$	Ignited with a match; blue flame	*Sulfur*
Bright yellow	Earthy	White	Soft	Powdery masses; radioactive	*Carnotite*
Lemon yellow	Vitreous to pearly	Yellowish	2–$2\frac{1}{2}$	Scaly aggregates; radioactive	*Autunite*
Yellow-ocher, yellow	Pearly	Yellowish	$2\frac{1}{2}$–3	Scaly masses; metallic taste	**COPIAPITE**
Lemon yellow, orange-yellow, orange-red, gray, white	Adamantine	White	3	Thin, tabular crystals	**WULFENITE**

YELLOW *(continued)*

Yellow, white, pale tints	Vitreous	White	3	Scalenohedrons, rhombohedrons; effervescent in cold HCl	CALCITE
Yellow, orange, brick red	Resinous	Yellow-orange	$3-3\frac{1}{2}$	Associated with sphalerite	GREENOCKITE
Yellow-ocher, brown, black	Dull	Yellow-ocher	$1-5\frac{1}{2}$	Massive	LIMONITE
Yellow, red-brown, brown, black, white, green	Adamantine	Yellowish-white	$3\frac{1}{2}-4$	Dodecahedral cleavage	SPHALERITE
Yellow, white, pale tints	Vitreous	White	$3\frac{1}{2}-4$	Effervescent in cold HCl	ARAGONITE
Yellow; many other colors	Vitreous	White	4	Cubic crystals; octahedral cleavage	FLUORITE
Yellow; colored by greenockite; also green, blue, white, gray, pink, brownish	Vitreous	White	$4-4\frac{1}{2}$	Botryoidal aggregates; effervescent in cold HCl	SMITHSONITE
Yellow, greenish-yellow, green, brownish-red, white	Vitreous	White	$5\frac{1}{2}$	Often luminesces bright green under ultraviolet radiation	WILLEMITE
Yellow, green, gray, brown, black	Adamantine	White	$5-5\frac{1}{2}$	Wedge-shaped crystals	SPHENE
Yellow; numerous other colors	Vitreous	White	5–6	Conchoidal fracture; cryptocrystalline	OPAL
Honey-yellow, white, reddish, light blue, gray	Vitreous	White	5–6	A feldspathoid associated with blue sodalite	CANCRINITE
Yellow, white, violet, pink, gray, greenish, bluish	Vitreous	White	5–6	May luminesce yellow-orange in long wave ultraviolet	SCAPOLITE
Yellow to red-brown	Vitreous	Nearly white	$6-6\frac{1}{2}$	Grains in crystalline limestone	CHONDRODITE
Yellow, white, pink, lilac, green, colorless	Vitreous	White	$6\frac{1}{2}-7$	Flat elongated striated crystals	SPODUMENE
Yellow to yellow-orange	Vitreous	White	$6\frac{1}{2}-7$	Thin wedge-shaped crystals	AXINITE
Honey-yellow	Silky	White	7	Yellow quartz pseudomorph after fibrous crocidolite	TIGER'S EYE
Yellow to brownish-yellow	Vitreous	White	7	Conchoidal fracture	QUARTZ (citrine)

114

Plate 17. **OXIDES** ▪ CUPRITE, ZINCITE, CORUNDUM

A   x 1.0

CUPRITE
Cu₂O

Aggregate of loosely matted hair-like crystals "chalcotrichite": Globe, Arizona.

B   x 0.4

CUPRITE

Massive dark red cuprite with alteration of green malachite: Chuquicamata, Chile.

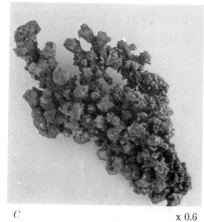

C   x 0.6

CUPRITE

Branching aggregate of small blackish-red octahedral crystals: Globe, Arizona.

D   x 0.7

ZINCITE
ZnO

Massive dark red zincite with black franklinite and gray tephroite: Franklin, New Jersey.

E   x 0.48

CORUNDUM
Al₂O₃

Rough rounded crystals, with characteristic barrel shape: Transvaal, South Africa.

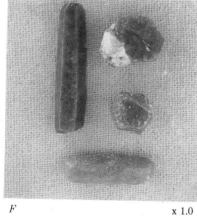

F   x 1.0

CORUNDUM

Six-sided corundum crystals: Transvaal, South Africa.

A   x 0.50

Black botryoidal hematite (kidney ore): Cumberland, England.

B   x 0.75

Lead gray foliated mass of brilliantly reflecting micaceous hematite: Hawley, Massachusetts.

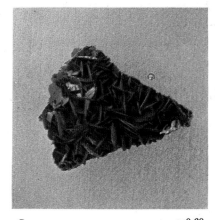

C   x 0.60

Aggregate of thin divergent hematite plates: Rio, island of Elba.

D   x 0.66

Massive oölitic hematite: Clinton, New York.

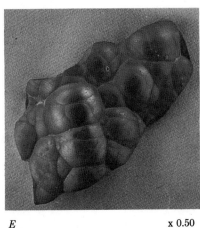

E   x 0.50

Red botryoidal hematite (kidney ore): Lancashire, England.

F   x 0.8

Thin hematite plates grouped in rosette forms (iron roses): St. Gotthard, Switzerland.

YELLOW (continued)

Color	Luster	Streak	Hardness	Remarks	Mineral
Yellow	Vitreous	White	$7\frac{1}{2}$–8	Six-sided crystals	BERYL (golden beryl)
Pale yellow, yellowish-brown, pinkish-brown, pink, lilac, pale blue, colorless, white	Vitreous	White	8	Perfect basal cleavage	TOPAZ
Yellow, greenish-yellow, gray, blue-green, brown	Vitreous	White	$8\frac{1}{2}$	Flat pseudohexagonal crystals	CHRYSOBERYL
Yellow; numerous other colors	Vitreous	White	9	Six-sided crystals	CORUNDUM (yellow sapphire)
Deep yellow (rare); usually colorless or yellowish	Greasy adamantine	White	10	Rounded octahedrons	DIAMOND

GOLD, BRASS

COLOR	LUSTER	STREAK	HARDNESS	REMARKS	MINERAL
Golden yellow	Metallic	Gold (shining)	$2\frac{1}{2}$–3	Sectile	*Gold*
Brass-yellow to silver-white	Metallic	Yellowish to greenish-gray	$2\frac{1}{2}$	Brittle; elongated striated crystals	CALAVERITE
Brass-yellow with a greenish tinge	Metallic	Greenish-black	3–$3\frac{1}{2}$	Hair-like crystals	*Millerite*
Deep golden brass-yellow	Metallic	Greenish-black	$3\frac{1}{2}$–4	Color distinctive	*Chalcopyrite*
Pale brass-yellow	Metallic	Greenish-black	6–$6\frac{1}{2}$	Striated cubes; pyritohedrons	*Pyrite*
Pale brass-yellow with greenish cast	Metallic	Greenish-black	6–$6\frac{1}{2}$	Cockscomb structure	*Marcasite*

BROWN: RED-BROWN

COLOR	LUSTER	STREAK	HARDNESS	REMARKS	MINERAL
Violet-brown, gray, greenish-gray	Waxy	White to gray	2–$2\frac{1}{2}$	Very sectile, wax-like masses	CERARGYRITE
Light to dark brown	Vitreous to pearly	White	$2\frac{1}{2}$–3	A brown mica	*Phlogopite*
Dark brown to black	Vitreous	White	$2\frac{1}{2}$–3	A mica	BIOTITE
Red-brown, brown, yellow-brown; commonly ruby red	Subadamantine	Nearly white	3	Small six-sided crystals	VANADINITE

BROWN: RED-BROWN *(continued)*

Light to dark brown; gray	Vitreous	White	3½–4	Rhombohedral cleavage; effervescent in hot HCl	**SIDERITE**
Brown, yellow-brown, deep red, green, black	Vitreous	Orange to brownish-red	3½	Crusts of minute crystals often associated with vanadinite and wulfenite	**DESCLOIZITE**
Brown; usually green; also gray, yellow, white	Subadamantine	Nearly white	3½–4	Small six-sided crystals	**PYROMORPHITE**
Yellow-brown, brown, black	Vitreous to dull	Yellowish-brown	Variable to 5½	Streak is characteristic	**GOETHITE** **LIMONITE**
Red-brown, red, black	Earthy to metallic	Red to red-brown	Variable 1–6½	Streak is characteristic	**HEMATITE**
Bronze	Metallic	Black	4	Bronze color; usually massive; may be magnetic	*Pyrrhotite*
Brown, white, green, blue, yellow, pink	Dull to vitreous	White	4–4½	"Dry bone ore"; dull, brownish, honey-combed masses	**SMITHSONITE**
Brown, green, blue, yellow, violet, white	Vitreous	White	5	Six-sided prisms; often mottled	**APATITE**
Reddish-brown to yellow-brown	Resinous	White	5–5½	Small flat monoclinic crystals	**MONAZITE**
Brown, green, yellow, gray	Resinous to adamantine	Pale yellow-brown	5–5½	Flat wedge-shaped crystals	**SPHENE**
Dark brown	Metallic	Red-brown	5–5½	Bladed crystals	*Huebnerite*
Reddish-brown, gray	Vitreous	White	5½	Six-sided crystals; may luminesce green under ultraviolet	**WILLEMITE** (troostite)
Bronze-brown	Submetallic to silky	Gray	5½–6	Bronze, fibrous aggregates	**BRONZITE**
Brown, gray, pink	Vitreous	White	6	Elongated crystals	**ZOISITE**
Red-brown, deep red, black	Submetallic	Pale brown	6–6½	Knee-shaped twins; striated elongated crystals	**RUTILE**
Red-brown to yellow	Vitreous	Nearly white	6–6½	Grains in metamorphic limestones	**CHONDRODITE**
Red-brown, dark brown, yellow-brown, black	Adamantine to greasy	Nearly white	6–7	High specific gravity (6.99)	**CASSITERITE**
Brown, green, yellow, blue	Vitreous	White	6½	Square prisms	**IDOCRASE**

116

BROWN: RED-BROWN *(continued)*

COLOR	LUSTER	STREAK	HARDNESS	REMARKS	MINERAL
Brown, violet-brown, gray, greenish, yellow-orange (rare)	Vitreous	White	$6\frac{1}{2}$–7	Thin wedge-shaped crystals	AXINITE
Smoky brown to almost black	Vitreous	White	7	Six-sided crystals	SMOKY QUARTZ
Brown, black, rose, green, pink, blue, white	Vitreous	White	7–$7\frac{1}{2}$	Crystals common, triangular cross section	TOURMALINE
Reddish-brown to brownish black	Vitreous to dull	White	7–$7\frac{1}{2}$	Cross-twins	STAUROLITE
Commonly brown; also colorless, gray, green, reddish, bluish	Adamantine	White	$7\frac{1}{2}$	Square prisms	ZIRCON
Reddish-brown, pink, olive-green, gray, white	Vitreous	White	$7\frac{1}{2}$	Rough, nearly square prisms	ANDALUSITE
Yellowish-brown, pinkish-brown, pale yellow, pink, light blue, white, colorless	Vitreous	White	8	Perfect basal cleavage	TOPAZ
Brown, yellowish to greenish-brown; dark green, red, violet, blue, white	Vitreous	White	8	Octahedral crystals	SPINEL
Brown; numerous other colors	Vitreous	White	9	Six-sided crystals, often barrel-shaped	CORUNDUM

BLUE

COLOR	LUSTER	STREAK	HARDNESS	REMARKS	MINERAL
Dark blue; purplish to blackish tarnish	Metallic	Lead-gray to black	$1\frac{1}{2}$–2	Color, basal cleavage, brittle	*Covellite*
Pale blue-green, bright azure blue; blackish-blue, greenish; colorless when unaltered	Vitreous, pearly, earthy	Bluish, but turns dark blue on exposure to light	$1\frac{1}{2}$–2	Flexible cleavage lamellae, sectile	VIVIANITE
Light blue to blue-green; pale green	Pearly	White	2	Scaly micaceous aggregates	AURICHALCITE
Blue (uncommon); usually colorless or white	Vitreous	White	$2\frac{1}{2}$	Cubic crystals, salty taste	HALITE

BLUE *(continued)*

Light blue (not common); white and numerous other tints	Vitreous	White	3	Rhombohedral cleavage; effervescent in cold HCl	**CALCITE**
Sky blue, bluish-green, green	Vitreous to waxy, or dull	White	2–4 (harder when impregnated with quartz)	Massive; resembles turquoise	**CHRYSOCOLLA**
Bluish-purple tarnish	Metallic	Gray-black	3	"Peacock ore," purple-blue tarnish on copper-red bornite	**BORNITE** (tarnish)
Pale blue, colorless, reddish, yellowish, brownish	Vitreous	White	3–3½	Tabular crystals; good cleavage in 3 directions; heavy	**BARITE**
Pale blue, colorless, white, reddish, brownish	Vitreous	White	3–3½	Resembles barite	**CELESTITE**
Azure to dark blue	Vitreous to earthy	Blue	3½–4	Associated with green malachite; effervescent in HCl	*Azurite*
Pale blue; commonly white or pale tints	Vitreous	White	3½–4	Effervescent in cold HCl; pseudohexagonal prisms	**ARAGONITE**
Blue; numerous other colors	Vitreous	White	4	Cubic crystals; octahedral cleavage	**FLUORITE**
Blue, white, green, yellow, pink	Vitreous	White	4½–5	Botryoidal aggregates, effervescent in cold HCl	**SMITHSONITE**
Blue, brown, green, yellow, violet, white	Vitreous	White	5	Six-sided prisms; often mottled with fused edges	**APATITE**
Bright azure blue	Vitreous	White	5–5½	Associated with pyrite	**LAZURITE** (*lapis Lazuli*)
Blue; numerous other colors	Vitreous	White	5–6	Conchoidal fracture; massive	**OPAL**
Sky blue, greenish-blue, green	Waxy	White	5–6	Commonly in compact masses	**TURQUOISE**
Blue, green, colorless; sometimes pink, yellow, black	Vitreous	White	5–7	Bladed crystals; uneven color distribution	**KYANITE**
Blue (usual color); also white, gray, pink, green, violet	Vitreous	White	5½–6	Usually massive	**SODALITE**

118

BLUE *(continued)*

COLOR	LUSTER	STREAK	HARDNESS	REMARKS	MINERAL
Blue to white	Vitreous	White	$6\frac{1}{2}$	Associated with neptunite; uneven blue and white color distribution	BENITOITE
Blue (cyprine), usually green or brown	Vitreous	White	$6\frac{1}{2}$	Square prisms	IDOCRASE
Blue-green iridescence	Vitreous	White	6	Blue-green iridescent play of colors; polysynthetic twinning	LABRADORITE (bluish reflections)
Milky blue (uncommon); usually white, numerous other colors	Vitreous	White	7	Conchoidal fracture	QUARTZ
Blue-gray, white, brown, gray, yellowish, red, green	Waxy	White	7	Microcrystalline masses	CHALCEDONY
Blue, violet, pink	Vitreous	White	7	Fibrous masses	DUMORTIERITE
Blue; numerous other colors	Vitreous	White	$7-7\frac{1}{2}$	Triangular cross section	TOURMALINE
Violet-blue, gray	Vitreous	White	$7-7\frac{1}{2}$	Commonly in grains which change from violet-blue to yellowish-gray when viewed through different directions	CORDIERITE
Blue, blue-green; numerous other colors	Vitreous	White	$7\frac{1}{2}-8$	Six-sided prisms common	BERYL
Blue; numerous other colors	Vitreous	White	8	Octahedral crystals	SPINEL
Blue; numerous other colors	Vitreous	White	8	Basal cleavage	TOPAZ
Blue; numerous other colors	Vitreous	White	9	Six-sided crystals; sets of intersecting parting striations	CORUNDUM (sapphire)
Bluish-white, deep blue (rare); usually colorless	Greasy adamantine (polished)	White	10	Rounded Octahedrons	DIAMOND

GREEN

COLOR	LUSTER	STREAK	HARDNESS	REMARKS	MINERAL
Pale green, dark green, white, gray, blackish gray	Greasy to pearly	White	$1-1\frac{1}{2}$	Greasy feel; micaceous cleavage; sectile	TALC (soapstone)

Greenish, usually blue	Vitreous, pearly, earthy	Greenish, bluish	$1\frac{1}{2}$–2	Flat crystals; radiating aggregates; sectile	**VIVIANITE**
Light green; usually pale blue-green; also light blue	Pearly	White	2	Scaly aggregates	**AURICHALCITE**
Emerald to yellowish-green	Pearly to vitreous	Light green	2–$2\frac{1}{2}$	Thin green square plates; scaly aggregates; radioactive	*Torbernite*
Greenish-gray, gray-brown, violet-brown	Waxy	White to gray	2–$2\frac{1}{2}$	Wax-like masses; very sectile	**CERARGYRITE**
Green; usually colorless; also yellowish, rose, amber	Vitreous to pearly	White	2–$2\frac{1}{2}$	Micaceous cleavage	**MUSCOVITE**
Usually in various shades of green; also rose, yellow, white	Vitreous to pearly	White, pale green	2–$2\frac{1}{2}$	Micaceous cleavage	**CHLORITE**
Green to whitish	Earthy	Greenish	2–3	Earthy masses, incrustations	**GARNIERITE**
Pale apple-green, gray	Vitreous to earthy	White	$2\frac{1}{2}$–3	Earthy crusts on nickel-cobalt minerals	**ANNABERGITE**
Light and dark shades of green; also white, red, yellow, black, brown	Waxy or silky	White	2–5	Massive; fibrous asbestos	**SERPENTINE**
Emerald green, blackish-green, pale green	Vitreous	Pale green	$3\frac{1}{2}$	Cross fiber veinlets	*Antlerite*
Emerald green, blackish-green, pale green	Vitreous	Pale green	$3\frac{1}{2}$–4	Resembles antlerite and malachite	*Brochantite*
Bright emerald green to blackish-green	Vitreous	Pale green	$3\frac{1}{2}$–4	Effervescent in cold HCl	*Malachite*
Green, white, yellow, brown	Vitreous	White	$3\frac{1}{2}$–4	Radiating aggregates	**WAVELLITE**
Green, brown, gray, yellow, white	Subadamantine	Nearly white	$3\frac{1}{2}$–4	Small hollow six-sided crystals	**PYROMORPHITE**
Pale green, blue-green, colorless	Waxy	White	$3\frac{1}{2}$–$4\frac{1}{2}$	Dense masses; resembles turquoise	**VARISCITE**
Green; numerous other colors	Vitreous	White	4	Cubic crystals; octahedral cleavage	**FLUORITE**
Light green, white, numerous other colors	Vitreous	White	4–$4\frac{1}{2}$	Botryoidal aggregates; effervescent in cold HCl	**SMITHSONITE**

GREEN *(continued)*

Color	Luster	Streak	Hardness	Characteristics	Mineral
Pale green, pale blue, colorless, white, pink	Vitreous to pearly	White	$4\frac{1}{2}$–5	Pseudocubic crystals; pearly on base	**APOPHYLLITE**
Green, brown, blue, yellow, reddish-brown, violet, white	Vitreous	White	5	Six-sided prisms; often mottled	**APATITE**
Emerald green	Vitreous	Pale green	5	Short six-sided prisms terminated by rhombohedrons	*Dioptase*
Green, yellow-green, yellow, brown, gray	Adamantine	White	5–$5\frac{1}{2}$	Flat wedge-shaped crystals	**SPHENE**
Greenish tint, colorless, white	Vitreous	White	5–$5\frac{1}{2}$	Complex crystals; in cavities in basalt	**DATOLITE**
Yellowish-green, brownish-red, yellow, white	Vitreous	White	$5\frac{1}{2}$	Luminesces green; occurs with franklinite and zincite	**WILLEMITE**
Yellowish-green to yellow	Vitreous	White	$5\frac{1}{2}$	Complex monoclinic crystals	**BRAZILIANITE**
Dark green	Waxy	White	5–6	Tough compact masses	**NEPHRITE**
Light green to dark green	Vitreous	White	5–6	Fibrous aggregates	**ACTINOLITE**
Light green to white	Vitreous	White	5–6	Prismatic crystals in crystalline limestones	**DIOPSIDE**
Dark green to black	Vitreous	White	5–6	Crystals with rectangular cross section	**AUGITE**
Green; numerous other colors	Vitreous	White	5–6	Conchoidal fracture; massive	**OPAL**
Light green, pale blue-green, sky blue	Waxy	White	5–6	Pale bluish-green; dense masses	**TURQUOISE**
Green, blue, colorless, pink, yellow, black	Vitreous	White	5–7	Flat bladed crystals; uneven color distribution	**KYANITE**
Bright green, bluish-green	Vitreous	White	6	Perthite structure; blocky cleavage	**MICROCLINE** *(amazonite)*
Yellowish-green, apple-green, white	Vitreous	White	6–$6\frac{1}{2}$	Botryoidal aggregates in cavities in basalt	**PREHNITE**
Green, yellow-green, brown, yellow, blue	Vitreous	White	$6\frac{1}{2}$	Square prisms; granular masses	**IDOCRASE** *(vesuvianite)*
Pistachio-green to blackish-green; gray-brownish	Vitreous	White	6–$6\frac{1}{2}$	Elongated striated prisms	**EPIDOTE**

GREEN *(continued)*

COLOR	LUSTER	STREAK	HARDNESS	REMARKS	MINERAL
Light to bright green; also white, violet, yellowish, brownish	Vitreous	White	$6\frac{1}{2}$–7	Tough compact masses	**JADEITE**
Bottle green, olive green, greenish-gray, brown	Vitreous	White	$6\frac{1}{2}$–7	Granular masses (dunite) grains in basic igneous rocks	**OLIVINE**
Emerald green to yellow-green	Vitreous	White	$6\frac{1}{2}$–7	Flat striated crystals; blocky cleavage	**SPODUMENE** (*hiddenite*)
Light green to emerald green	Vitreous	White	$6\frac{1}{2}$–$7\frac{1}{2}$	Dodecahedrons	**GARNET** (*uvarovite; demantoid; grossularite*)
Dark green	Waxy	White	7	Microcrystalline quartz	*Prase*
Green with spots of red	Waxy	White	7	Green chalcedony with spots of red jasper	**BLOODSTONE**
Apple-green	Waxy	White	7	Apple-green, translucent calcedony	*Chrysoprase*
Green	Vitreous	White	7	Quartz with included scales of green mica	**ADVENTURINE**
Green (rare), white, colorless; numerous other colors	Vitreous	White	7	Conchoidal fracture	**QUARTZ**
Green; numerous other colors	Vitreous	White	7–$7\frac{1}{2}$	Triangular cross section	**TOURMALINE**
Green, blue-green; numerous other colors	Vitreous	White	$7\frac{1}{2}$–8	Six-sided prisms	**BERYL**
Dark green; numerous other colors	Vitreous	White	$7\frac{1}{2}$–8	Octahedral crystals	**SPINEL**
Yellowish-green, bluish-green, yellow, brownish	Vitreous	White	$8\frac{1}{2}$	Flat pseudohexagonal crystals	**CHRYSOBERYL**
Green; numerous other colors	Vitreous	White	9	Six-sided crystals; sets of intersecting parting striations	**CORUNDUM**

SILVER

COLOR	LUSTER	STREAK	HARDNESS	REMARKS	MINERAL
Silver	Metallic	Gray	$1\frac{1}{2}$–2	Brittle; branching forms "graphic tellurium"	*Sylvanite*

Plate 19. **OXIDES** ▪ CASSITERITE, RUTILE, URANINITE, CURITE, SCHOEPITE

A   x 0.52

CASSITERITE
SnO_2

Compact reddish-brown masses of cassiterite, some showing botryoidal structure (wood tin): Durango, Mexico.

B   x 1.0

CASSITERITE

Brown cassiterite crystals showing characteristic adamantine luster: Zinnwald, Bohemia.

C   x 0.5

CURITE, SCHOEPITE, AND SODDYITE (Silicate)

Polished specimen of secondary uranium minerals. Curite (orange), schoepite and soddyite (yellow): Shinkolobwe Mine, Katanga, Congo.

D   x 0.9

URANINITE
UO_2

Massive uraninite showing narrow alteration veins of orange curite.

E   x 0.5

URANINITE (WITH ALTERATION MINERALS)

Massive black uraninite with green metatorbernite, and yellow-orange mixture of curite, soddyite, and schoepite: Shinkolobwe Mine, Katanga, Congo.

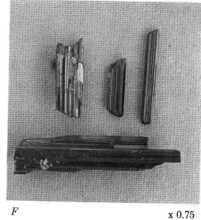

F   x 0.75

RUTILE
TiO_2

Vertically striated prismatic rutile crystals: White Plains, North Carolina.

A   x 0.7

MAGNETITE
Fe_3O_4

Black octahedral magnetite crystals: Mineville, New York.

B   x 0.5

FRANKLINITE
$(Zn, Fe, Mn)(Fe, Mn)_2O_4$

Black octahedral franklinite crystal in white calcite: Franklin, New Jersey.

C   x 0.7

GAHNITE
(Zinc spinel), $ZnAl_2O_4$

Octahedral gahnite crystal and pyrrhotite in white calcite: Ogdensburg, New Jersey.

D   x 0.6

CHROMITE
$FeCr_2O_4$

Black granular chromite and serpentine: Banat, Hungary.

E   x 1.0

SPINEL
$MgAl_2O_4$

Octahedral spinel crystal and yellow chondrodite occurring in crystalline limestone: Sparta, New York.

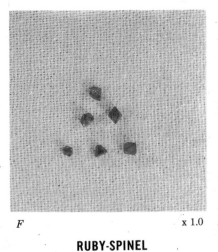

F   x 1.0

RUBY-SPINEL

Tiny rose red octahedral spinel crystals: Ceylon.

SILVER *(continued)*

	LUSTER	STREAK	HARDNESS	REMARKS	MINERAL
Pale reddish-silver	Metallic	Silver (shining)	2–2½	Reddish tinge; iridescent tarnish; sectile	*Bismuth*
Silver to gold	Metallic	Gray	2½	Brittle; deeply striated crystals	CALAVERITE
Silver (tarnishes dark)	Metallic	Silver (shining)	2½–3	Sectile; **gray-black** tarnish	*Silver*
Bright silver-white	Metallic	Gray	3–3½	Very brittle; radiating masses with a granular texture	*Antimony*
Silver-gray	Metallic	Shiny gray	4–4½	Malleable	*Platinum*
Silver with slightly pinkish cast	Metallic	Black	5½	Striated cubes, pyritohedrons	*Cobaltite*
Silver	Metallic	Black	5½–6	Commonly massive	*Skutterudite*
Silver-gray	Metallic	Black	5½–6	Garlic odor when struck with a hammer	*Arsenopyrite*
Bright silver (readily tarnishes gray-black)	Metallic	Black	5½–6	Botryoidal masses with gray-black tarnish	*Arsenic*

LEAD-GRAY to BLACK

	LUSTER	STREAK	HARDNESS	REMARKS	MINERAL
	Metallic	Greenish-gray	1–1½	Greasy feel; bluish tinge; micaceous cleavage	MOLYBDENITE
	Metallic	Black	1–2	Greasy feel; micaceous cleavage	GRAPHITE
	Metallic	Gray-black	2	Striated bladed crystals with cross striations	STIBNITE
	Metallic	Lead-gray	2	Thin striated bladed crystals; resembles stibnite	BISMUTHINITE
	Metallic	Blackish-gray (shining)	2–2½	Very sectile; found with other silver minerals	ARGENTITE
	Submetallic	Purplish-red	2–2½	Dark ruby silver	PYRARGYRITE
	Metallic	Black	2–2½	Stains fingers sooty black	PYROLUSITE
	Metallic	Black	2–3	Six-sided plates with triangular markings	POLYBASITE

Luster	Color	Hardness	Characteristics	Mineral
Metallic	Black	2–2½	Short pseudohexagonal prisms; associated with other silver minerals	STEPHANITE
Metallic	Black	2–3	Fibrous masses with a feather-like appearance	JAMESONITE
Metallic	Lead-gray	2½	Cubic cleavage	GALENA
Vitreous	White	2½–3	Micaceous cleavage	BIOTITE
Metallic	Gray-black	2½–3	"Cogwheel crystals"	BOURNONITE
Metallic	Gray-black (shining)	2½–3	Subsectile	CHALCOCITE
Metallic	Gray-black	3	Bladed masses; brittle	ENARGITE
Adamantine	Brownish	3½–4	Rounded, distorted crystals	SPHALERITE
Metallic	Dark gray to black	3½–4	Tetrahedrons; brittle	TETRAHEDRITE
Metallic	Dark brown	4	Groups of vertically striated prismatic crystals	MANGANITE
Metallic	Steel-gray	4–5	Malleable	IRON
Adamantine-metallic	Yellow-brown	5–5½	Yellowish streak	GOETHITE
Metallic	Black to dark brown	5–5½	Bladed crystals	WOLFRAMITE
Metallic	Black to brownish-black	5–6	Botryoidal to stalactitic masses	PSILOMELANE
Metallic	Black to brownish-red	5–6	Tabular crystals; massive	ILMENITE
Vitreous	White, gray	5–6	Prismatic crystals with a diamond-shaped cross section	HORNBLENDE
Metallic	Brownish black	5½	Commonly massive; orange, yellow, green alteration; strongly radioactive	URANINITE
Metallic	Brown	5½	Occurs with serpentine; usually massive	CHROMITE
Metallic	Red-brown	5½–6½	Botryoidal masses; tabular crystals	HEMATITE
Metallic	Black to dark brown	6	Tabular crystals	COLUMBITE-TANTALITE

Metallic	Black	6	Octahedrons; strongly magnetic	MAGNETITE
Metallic-adamantine	Light brown	6–6½	Slender, vertically striated crystals	RUTILE
Adamantine	Pale brown to white	6–7	Twinned crystals	CASSITERITE
Vitreous	White	6½–7½	Dodecahedrons; trapezohedrons	GARNET
Vitreous	White	7–7½	Striated prisms with triangular cross section	TOURMALINE
Vitreous	White	7	Six-sided crystals	QUARTZ
Vitreous	White	8	Octahedrons	SPINEL
Vitreous	White	9	Sets of intersecting striations; six-sided crystals	CORUNDUM
Adamantine	White	10	Great hardness	DIAMOND (carbonado)

COLORLESS; WHITE; LIGHT GRAY; PALE TINTS

LUSTER	STREAK	HARDNESS	REMARKS	MINERAL
Silky	White	1	Rounded fibrous masses; "cotton balls"	ULEXITE
Greasy to pearly	White	1–1½	Foliated masses; greasy feel	TALC
Greasy to pearly	White	1–2	Fine-grained masses; radiating crystals	PYROPHYLLITE
Dull	White	2–2½	Earthy white masses	KAOLINITE
Vitreous	White	2	Often transparent fibrous with silky luster; fine-grained mass crystals	GYPSUM
Dull	White to reddish	1–3	Earthy rock composed of pisolitic grains	BAUXITE
Waxy	White to gray	2–2½	Wax-like pearl gray masses; very sectile (readily darkens to violet-brown)	CERARGYRITE
Vitreous to silky	White	2	Silky tufts of needle-like crystals; cooling taste	NITER

Vitreous	White	$1\frac{1}{2}$–2	Crusts; cooling taste	SODA NITER
Vitreous	White	2	Bitter, salty taste; cubic cleavage	SYLVITE
Vitreous	White	2–$2\frac{1}{2}$	Sweetish astringent taste; water soluble	BORAX
Vitreous	White	2–$2\frac{1}{2}$	Cotton-like fibrous crusts; botryoidal masses; water soluble; bitter taste; translucent, almost disappears in water	EPSOMITE
Vitreous	White	$2\frac{1}{2}$	Pseudocubic parting; associated brown siderite	CRYOLITE
Vitreous	White	$2\frac{1}{2}$	Cubic crystals and cleavage; salty taste	HALITE
Vitreous to pearly	White	$2\frac{1}{2}$	Foliated masses; sectile	BRUCITE
Vitreous to pearly	White	2–$2\frac{1}{2}$	Micaceous cleavage	MUSCOVITE
Vitreous to pearly	White	$2\frac{1}{2}$–4	Micaceous cleavage; associated with spodumene, amblygonite, pink tourmaline	LEPIDOLITE
Vitreous	White	3	Long splintery cleavage fragments; soluble in hot water	KERNITE
Vitreous	White	3	Rhombohedral cleavage; effervescent in cold dilute HCl	CALCITE
Adamantine to dull	White	3	Heavy prismatic crystals; "woody anglesite" associated with galena	ANGLESITE
Vitreous	White	3–$3\frac{1}{2}$	Pseudohexagonal pyramids; effervescent in cold HCl	WITHERITE
Adamantine	White	3–$3\frac{1}{2}$	Grid-like aggregates; associated with galena; effervescent in hot HNO_3	CERUSSITE
Vitreous	White	3–$3\frac{1}{2}$	Heavy tabular crystals	BARITE

Vitreous	White	3–3½	3 cleavages at right angles	ANHYDRITE
Vitreous	White	3½–4	Heavy tabular crystals; often associated with yellow sulfur	CELESTITE
Vitreous	White	3½–4	Effervescent in cold dilute HCl; poor cleavage	ARAGONITE
Vitreous	White	3½–4	Rhombohedral crystals often curved	DOLOMITE
Subadamantine	White	3½–4	Small six-sided crystals, often rounded or barrel-shaped	PYROMORPHITE, MIMETITE
Vitreous	White	3½–4	Radiating aggregates; effervescent in cold HCl	STRONTIANITE
Vitreous to dull	White	3½–5	Usually compact white masses; effervescent in hot HCl	MAGNESITE
Vitreous	White	4	Cubic crystals; octahedral cleavage	FLUORITE
Vitreous to dull	White	4–4½	Botryoidal masses; effervescent in cold dilute HCl	SMITHSONITE
Vitreous	White	4–4½	Complex monoclinic crystals associated with borax, ulexite, and kernite	COLEMANITE
Vitreous	White	4–5	Pseudocubic crystals associated with other zeolites and calcite	CHABAZITE
Vitreous to pearly	White	4½–5	Pseudocubic crystals, pearly on base; associated with stilbite	APOPHYLLITE
Vitreous	White	4½–5	Radiating groups; botryoidal masses; "segmented worms"	HEMIMORPHITE
Vitreous	White	4½–5	Pyramidal crystals; granular masses; luminesces pale blue in short-wave ultraviolet radiation	SCHEELITE

127

Vitreous	White	5	Six-sided prisms	APATITE
Vitreous to silky	White	5	Radiating needle-like crystals; associated with prehnite and zeolites	PECTOLITE
Vitreous	White	5–5½	Radiating needle-like crystals; associated with other zeolites and prehnite	NATROLITE
Vitreous to silky	White	5–5½	Prismatic crystals; fibrous masses; resembles tremolite	WOLLASTONITE
Vitreous	White	5–5½	Trapezohedrons; associated with prehnite, calcite, and zeolites	ANALCITE
Vitreous	White	5–5½	Complex monoclinic crystals with greenish tinge; associated with native copper and zeolites; cryptocrystalline masses	DATOLITE
Vitreous	White	5–6	Botryoidal colorless masses (hyalite); earthy chalk-like masses (diatomite); massive; conchoidal fracture	OPAL
Vitreous	White	5–6	Fibrous; prismatic crystals	TREMOLITE
Vitreous	White	5–6	Square prisms; commonly found in metamorphic limestones	DIOPSIDE
Vitreous	White	5–7	Long-bladed crystals; commonly associated with staurolite and garnet in metamorphic rocks	KYANITE
Vitreous	White	5½	Associated with franklinite, zincite, and calcite; luminesces green	WILLEMITE
Vitreous to greasy	White	5½–6	Greasy luster; associated with sodalite and leucite	NEPHELINE

Vitreous to dull	White	5–6	Large prismatic crystals with a dirty, white, woody surface	SCAPOLITE
Vitreous	White	5½–6	Usually massive; often luminesces	SODALITE
Vitreous	White	5½–6	Trapezohedral crystals	LEUCITE
Vitreous to pearly	White	6	Cleavable masses; resembles feldspar; fuses easily	AMBLYGONITE
Vitreous	White	6	Blocky cleavage; parallel twinning striations on plagioclase feldspars	FELDSPARS
Vitreous	White	6–6½	Botryoidal masses with a crystalline surface	PREHNITE
Vitreous	White	6–7	Long, slender, striated crystals; found in gneiss and schist	SILLIMANITE
Vitreous	White	6½–7	Flat vertically striated crystals; often rough wood-like surfaces	SPODUMENE
Vitreous	White	7	Six-sided prisms; massive; conchoidal fracture	QUARTZ
Waxy	White	7	Microcrystalline quartz	CHALCEDONY
Vitreous to dull	White	7½	Rough nearly square prisms	ANDALUSITE
Vitreous	White	7½–8	Flat colorless crystals; resembles quartz	PHENAKITE
Vitreous	White	7½–8	Six-sided prisms	BERYL
Vitreous	White	8	Prismatic crystals; basal cleavage	TOPAZ
Vitreous	White	9	Six-sided crystals; often barrel shaped	CORUNDUM
Greasy adamantine	White	10	Rounded octahedrons	DIAMOND

129

Table 7-2

Mineral Aggregates*

Minerals frequently found in significant aggregates, together with illustrative photographs, are included in this table. References to illustrations found elsewhere in the text are also given.

Rounded to Radial Growth Forms

CONCENTRIC (Fig. 7-1)—circular layers arranged about a center (onion-like structure).

■ **EXAMPLES**

Chalcedony (agate), Pl. 35(*E*) facing p. 211.
Goethite, Pl. 21(*B*), facing p. 130.
Malachite
Rhodochrosite, Pl. 1(*B*), facing p. 2, Pl. 25(*D*), facing
   p. 154.
Smithsonite

COLLOFORM (Fig. 7-2)—general term denoting rounded masses that often possess a radial inner structure (BOTRYOIDAL, RENIFORM, GLOBULAR).

*Figures 7-2, 7-5, 7-7, 7-8, 7-11, 7-12, 7-13, 7-15, and 7-20 by courtesy of the American Museum of Natural History, New York. Figures 7-1, 7-3, 7-4, 7-9, 7-10, 7-18, 7-19, and 7-21 by courtesy of the United States National Museum, Washington, D.C.

Fig. 7-2. *Botryoidal* malachite, a *colloform* structure; Aurbriz, Africa.

■ **EXAMPLES**

Azurite, Pl. 27(*B,D*), facing p. 163.
Cassiterite, Pl. 19(*A*), facing p. 122.
Chalcedony, Pl. 35(*B*), facing p. 211.
Collophane
Goethite
Hematite, Pl. 18(*A,E*), facing p. 115.
Hemimorphite, Pl. 1(*D*), facing p. 2.
Limonite, Pl. 21(*E*), facing p. 130.
Malachite, Pl. 27(*A*), facing p. 163.
Mimetite
Opal (hyalite), Pl. 36(*B*), facing p. 218.
Pitchblende
Prehnite, Pl. 40(*F*), facing p. 242.
Psilomelane
Pyromorphite
Rhodochrosite
Smithsonite, Pl. 25(*A,B*), facing p. 154.

OOLITIC (Fig. 7-3)—composed of small, rounded particles the size of fish roe (0.25 to 2.00 mm in diameter).

■ **EXAMPLES**

Calcite
Hematite, Pl. 18(*D*), facing p. 115.
Limonite
Opal (hyalite)

Fig. 7-1. *Concentric* malachite (polished); Clifton, Arizona.

Plate 21. **HYDROXIDES** ▪ **GOETHITE, LIMONITE, DIASPORE**

A x 0.5

GOETHITE
FeO(OH)

Black stalactitic goethite.

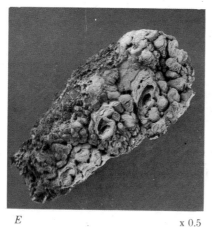

B x 0.5

GOETHITE

Radiating fibrous and concentric inner structure of stalactitic goethite with brown and yellow color bands perpendicular to the fibers: Lake Superior.

C x 0.35

GOETHITE

Stalactitic goethite with black varnish-like luster: Roxbury, Connecticut.

D x 0.7

LIMONITE

Massive yellow ocher limonite; loose porous structure.

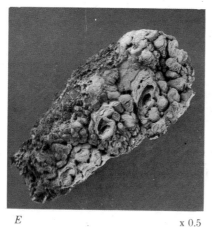

E x 0.5

LIMONITE

Yellow ocher limonite with colloform structure: Ariege.

F x 0.75

DIASPORE
AlO(OH)

Aggregate of thin bladed grayish-lavender diaspore crystals: Massachusetts.

Plate 22. **HYDROXIDES ▪ BAUXITE, GIBBSITE, BRUCITE, MANGANITE**

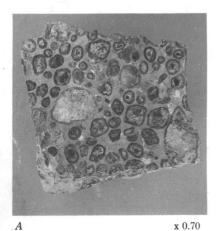

A   x 0.70

BAUXITE

Bauxite formed by the alteration of a porphyritic rock. The structure of the original rock is largely preserved: Saline County, Arkansas.

B   x 0.50

GIBBSITE

Intertwined stalactites of gibbsite with a concentric radial inner structure: Minas Gerais, Brazil.

C   x 0.50

GIBBSITE

Al(OH)$_3$

Stalactitic aggregate with smooth surface: Richmond, Massachusetts.

D   x 0.70

BAUXITE

Bauxite showing typical pisolitic structure.

E   x 0.63

BRUCITE

Mg(OH)$_2$

Aggregate of plate-like crystals of brucite showing pearly luster: Lancaster County, Pennsylvania.

F   x 0.55

MANGANITE

MnO(OH)

Groups of vertically striated prismatic crystals: Ilfeld, Harz Mountains, Germany.

Fig. 7-3. *Oölitic* opal; Tateyma, Japan.

Fig. 7-5. *Radiating* wavellite aggregate; Garland County, Arkansas.

PISOLITIC (Fig. 7-4)—composed of pea-size rounded particles exceeding 2.00 mm in diameter.

■ EXAMPLES

Bauxite, Pl. 22(*D*), facing p. 131.
Calcite

RADIAL (Fig. 7-5)—divergent groups of crystals (stellated—star-like shape).

■ EXAMPLES

Clinozoisite
Erythrite, Pl. 32(*E*), facing p. 194.
Goethite, Pl. 21(*B*), facing p. 130.
Malachite.
Marcasite
Millerite, Pl. 13(*C*), facing p. 90.
Natrolite
Pectolite, Pl. 41(*E*), facing p. 243.
Pyrophyllite, Pl. 40(*D*), facing p. 242.
Tourmaline
Wavellite, Pl. 32(*B*), facing p. 194.

Banded Growth

BANDED (Fig. 7-6)—composed of alternating layers differing in color and texture.

■ EXAMPLES

Aragonite
Barite
Calcite, Pl. 24(*A*), facing p. 147.
Chalcedony, Pl. 35(*E,F*), facing p. 211.
Fluorite
Malachite, Pl. 27(*C*), facing p. 163.
Rhodochrosite, Pl. 1(*B*), facing p. 2.

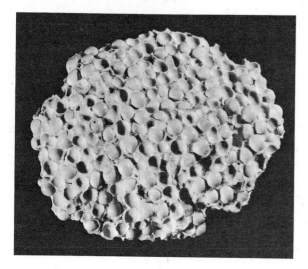

Fig. 7-4. *Pisolitic* calcite; Carlsbad, Bohemia, Germany.

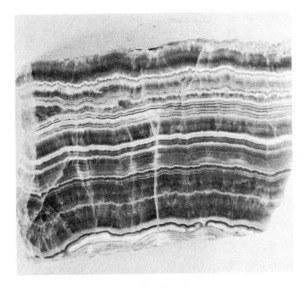

Fig. 7-6. *Banded* calcite.

Fig. 7-7. *Stalactitic* wavellite; Chester County, Pennsylvania.

Icicle-shaped Growth Forms

STALACTITIC (Fig. 7-7)—icicle-shaped masses.

■ **EXAMPLES**

Aragonite
Calcite
Gibbsite, Pl. 22(*C*), facing p. 131.
Goethite, Pl. 21(*A,C*), facing p. 130.
Limonite
Marcasite
Psilomelane
Rhodochrosite
Wavellite

Crystal Groups or Single Crystals

BLADED (Fig. 7-8)—flat elongated crystals (aggregates or single crystals).

■ **EXAMPLES**

Crocoite, Pl. 30(*C*), facing p. 186.
Epidote
Gypsum

Fig. 7-8. *Bladed* stibnite; Japan.

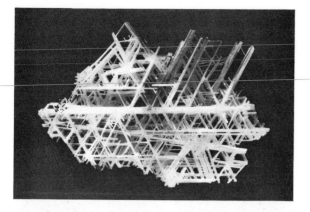

Fig. 7-9. *Reticulated* cerussite; Tintic District, Utah.

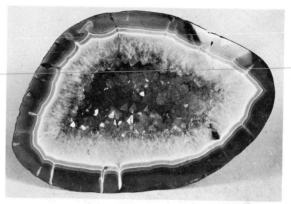

Fig. 7-11. *Geode,* agate lined with quartz crystals; Uruguay.

Kyanite, Pl. 49(*A*), facing p. 307.
Rutile, Pl. 19(*F*), facing p. 122.
Staurolite
Stibnite, Pl. 2(*B*), facing p. 3.
Zoisite

RETICULATED (Fig. 7-9)—aggregate of slender crystals forming a net-like pattern.

■ EXAMPLES

Cerussite, Pl. 26(*D*), facing p. 162.
Rutile
Stibnite

Fig. 7-10. *Sheaf structure,* stilbite; Cape Blomidon, Nova Scotia.

SHEAF-LIKE (Fig. 7-10)—groups of crystals resembling a highly reduced sheaf of wheat.

■ EXAMPLE

Stilbite, Pl. 39(*D,E*), facing p. 227.

GEODE (Fig. 7-11)—a rock cavity, often lined with crystals but not completely filled.

■ EXAMPLE

Amethyst geode, Pl. 34(*E*), facing p. 210.

COLUMNAR (Fig. 7-12)—thick column-like prismatic crystals (groups or single crystals).

■ EXAMPLES

Amphiboles
Andalusite, Pl. 49(*C*), facing p. 307.
Apatite, Pl. 6(*B,C*), facing p. 35.
Beryl, Pl. 6(*A*), facing p. 35.
Corundum, Pl. 17(*E,F*), facing p. 114.
Microcline
Pyroxenes, Pl. 42(*B*), facing p. 258.
Quartz
Scapolite, Pl. 38(*C*), facing p. 226.
Tourmaline, Pl. 45(*E*), facing p. 275.

TABULAR (lamellar) (Fig. 7-13)—book-like or plate-like shapes (groups or single crystals).

Fig. 7-12. *Columnar* beryl; Portland, Connecticut.

■ **EXAMPLES**

Albite (cleavelandite), Pl. 37(*F*), facing p. 219.
Barite, Pl. 28(*A,B,C*), facing p. 178.
Calcite
Celestite
Columbite-tantalite
Hematite
Ilmenite
Rhodonite, Pl. 41(*D*), facing p. 243.
Siderite, Pl. 24(*E*), facing p. 147.
Spodumene
Wulfenite, Pl. 30(*A,B*), facing p. 186.

Flaky

FOLIATED (Fig. 7-14)—composed of thin, easily separated plates which are usually slightly warped.

■ **EXAMPLES**

Hematite, Pl. 18(*B*), facing p. 115.
Orpiment, Pl. 16(*A*), facing p. 99.

Fig. 7-13. *Tabular* calcite crystals; Andreasberg, Harz, Germany.

Talc, Pl. 40(*E*), facing p. 242.
Vermiculite

MICACEOUS (Fig. 7-15)—exceedingly thin sheets are easily separated. Mica flakes are frequently aligned in rocks to yield a schist structure. (See micaceous cleavage, p. 82. Groups or single crystals.)

■ **EXAMPLES**

Autunite, Pl. 33(*E*), facing p. 195.
Biotite, Pl. 41(*A*), facing p. 243.

Fig. 7-14. *Foliated* talc.

Fig. 7-15. Muscovite mica.

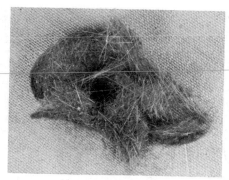

Fig. 7-17. *Capillary* millerite.

Chlorite
Lepidolite
Muscovite, Pl. 40(*C*), facing p. 242.
Phlogopite, Pl. 41(*B*), facing p. 243.
Torbernite, Pl. 33(*D*), facing p. 195.

Whisker Growths

ACICULAR (Fig. 7-16)—slender needle-like crystals. (Groups or single crystals.)

■ **EXAMPLES**

Erythrite, Pl. 32(*E*), facing p. 194.
Millerite, Pl. 13(*C*), facing p. 90.
Natrolite, Pl. 39(*A*), facing p. 227.
Pectolite, Pl. 41(*E*), facing p. 243.
Tourmaline

CAPILLARY (Fig. 7-17)—hair-like crystals.

■ **EXAMPLES**

Cuprite (chalcotrichite), Pl. 17(*A*), facing p. 114.
Jamesonite
Millerite

WIRE-LIKE (Fig. 7-18)—(Groups or single crystals.)

■ **EXAMPLES**

Copper
Gold
Silver, Pl. 10(*C*), facing p. 67.

Fig. 7-16. *Acicular* natrolite; Bergen Hill, New Jersey.

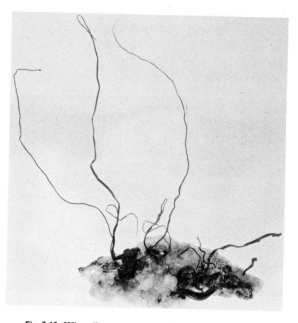

Fig. 7-18. *Wire* silver on quartz; Guanajuato, Mexico.

Fig. 7-19. *Fibrous* structure, asbestos; Thetford, Canada.

FIBROUS (Fig. 7-19)—composed of fibers that are often separable.

■ **EXAMPLES**

Actinolite, Pl. 43(*B*), facing p. 259.
Crocidolite, Pl. 43(*E*), facing p. 259.
Dumortierite, Pl. 49(*E*), facing p. 307.
Enstatite

Fig. 7-21. *Dendritic* copper aggregate.

Fig. 7-20. *Flos ferri* aragonite; Chihuahua, Mexico.

Rounded triangular cross section →

Fig. 7-22. Striated *tourmaline* crystal showing characteristic rounded triangular cross section.

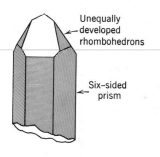

Fig. 7-23. Common quartz crystals.

Gypsum
Kernite
Serpentine (chrysotile asbestos), Pl. 2(*C*), facing p. 3.
Sillimanite
Strontianite
Tremolite (hexagonite), Pl. 43(*A*), facing p. 259.
Wollastonite, Pl. 41(*F*), facing p. 243.

Branching Growth

CORALLOIDAL (Fig. 7-20)—branching coral-like structure.

■ **EXAMPLES**

Aragonite, Pl. 26(*A*), facing p. 162.
Calcite

DENDRITIC (Fig. 7-21)—branching or tree-like shape. A dendritic crystal or (dendrite) is a *single* branching skeleton crystal. Dendritic aggregates may be formed by groups of dendrites, or by the deformation of a dendritic crystal during growth (see dendritic aggregates, p. 33).

■ **EXAMPLES**

Copper, Pl. 1(*A*), facing p. 2.
Gold
Pyrolusite
Silver, Pl. 10(*E*), facing p. 67.

CRYSTAL HABIT

In some respects, the crystals of minerals may be compared to individuals. Some are short, some are broad, others are long and narrow, and many others exhibit a distinctive geometrical development. It is customary to refer to a crystal's characteristic tendency to grow according to a distinctive pattern as **CRYSTAL HABIT.**

TOURMALINE (Figure 7-22) often grows in crystals with three curved, convex, striated sides, and recognition of this habit is enough to identify the mineral.

ELONGATED HEXAGONAL QUARTZ crystals (Figure 7-23) with six-sided prisms terminated by pyramidal faces (two rhombohedrons) follow a common crystal habit.

Table 7-3
Crystal Habits of Common Minerals

The idealized diagrams in this table illustrate habits frequently observed among crystals of ordinary minerals. The habit and, if possible, the accompanying form or forms should constitute the first observa-tion for a crystallized specimen. A mineral suggested by Table 7-3 should be confirmed by reference to the appropriate text description.

CUBE

Fluorite, facing p. 146
Galena, facing p. 91
Halite, facing p. 146
Pyrite (striated),
    facing p. 83
Sylvite, facing p. 146

OCTAHEDRON

Cuprite, facing p. 114
Diamond (rounded),
    facing p. 82
Franklinite, facing p. 123
Magnetite, facing p. 123
Pyrite (rare)
Spinel, facing p. 123

TETRAHEDRON

Tetrahedrite

DODECAHEDRON

Garnet, facing p. 18

TRAPEZOHEDRON

Analcite, facing p. 227
Garnet
Leucite, facing p. 226

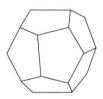

PYRITOHEDRON

Pyrite, facing p. 18

RHOMBOHEDRON

Calcite, facing p. 35
Dolomite
Rhodochrosite, facing p. 154
Siderite

SCALENOHEDRON

Calcite, facing p. 35

Table 7-3 *(continued)*

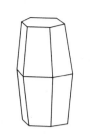

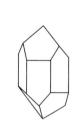

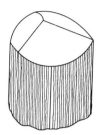

SIX-SIDED PRISMS

Apatite, facing p. 35
Beryl, facing p. 35
Corundum, facing p. 114
Pyromorphite, facing p. 187
Vanadinite, facing p. 187

BARREL-SHAPED

Corundum, facing p. 114

PRISMATIC HEXAGONAL

Quartz, facing p. 210

**PRISMATIC CONVEX
STRIATED SIDES**

Tourmaline, facing p. 275

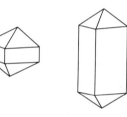

PRISMATIC TETRAGONAL

Idocrase, facing p. 34

PRISMATIC TETRAGONAL

Zircon, facing p. 290

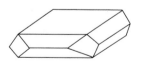

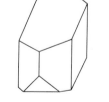

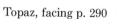

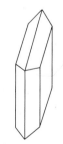

Barite, facing p. 178

Topaz, facing p. 290

Orthoclase, facing p. 50

Gypsum, facing p. 50

Table 7-3 *(continued)*

Common Twinned Crystals

Twinned
Cassiterite

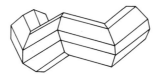

Knee-like twinned
Rutile, facing p. 51

Cross-twinned
Staurolite, facing p. 51

Carlsbad twin
Orthoclase, facing p. 51

Polysynthetic twinning
Plagioclase Feldspar, facing p. 51

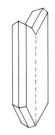

Fish-tail twin
Gypsum, facing p. 51

Table 7-4

Cleavage

Cleavage, especially in massive specimens, should be examined carefully with a hand lens. Reflections from small broken cleavage planes are often recognizable at low magnifications.

Crystals with Perfect Cleavage in One Direction

(Some may also have imperfect cleavage or parting in other directions)

BASAL CLEAVAGE

Topaz
Epidote
Bismuth
Antimony
Covellite
Apophyllite

SIDE PINACOIDAL CLEAVAGE

Gypsum	Stilbite
Diaspore	Wolframite
Zoisite	Stibnite
Heulandite	Bismuthinite
Colemanite	Vivianite
Sillimanite	Erythrite

FRONT PINACOIDAL CLEAVAGE

Kyanite

MICACEOUS CLEAVAGE (the mineral separates into thin sheets)

Muscovite	Brucite
Biotite	Aurichalcite
Phlogopite	Copiapite
Lepidolite	Autunite
Margarite	Torbernite
Chlorite	Graphite
Talc	Molybdenite
Pyrophyllite	Orpiment

Minerals with Prominent Cleavage in Two Directions

CLOSE TO 90°

Orthoclase
Microcline
Plagioclase feldspar
Spodumene

Pyroxene Group

OBTUSE AND ACUTE ANGLES

Amphibole Group 56° and 124°

Hornblende
Tremolite
Actinolite

Table 7-4   *(continued)*

Pyroxene Group
  Diopside
  Augite
Rhodonite
Wollastonite
Amblygonite (angle between
  (001) and (100) = 90°15′,
  also interrupted cleavage
  on other planes)

ELONGATE SPLINTERY CLEAVAGE FRAGMENTS
Kernite

Minerals with Prominent Cleavage in Three Directions

CUBIC CLEAVAGE (3)	RHOMBOHEDRAL (3)	2-PRISMATIC WITH 1-PINACOIDAL
Halite	Calcite	Barite
Sylvite	Dolomite	Celestite
Galena	Siderite	
	Magnesite	**3-PRISMATIC**
	Rhodochrosite	Cinnabar
	Proustite	
	Smithsonite	**3-PINACOIDAL**
	(cleavage	Anhydrite
	rarely seen)	(rectangular
	Corundum	cleavage)
	(no cleavage but	
	good rhombohedral	
	parting)	

Cleavage in Four Directions

OCTAHEDRAL CLEAVAGE
Fluorite
Diamond

Cleavage in Six Directions

DODECAHEDRAL CLEAVAGE
Sphalerite

Table 7-5
Minerals Listed According to Increasing Specific Gravity

Specific gravity is best determined with a delicate balance. In the absence of such a balance, several expedients may prove useful. A fair-sized specimen may be dropped repeatedly on one's hand until a fair sense of heft is obtained, particularly if compared with specimens of known specific gravity and similar size.

A small light 1-cc (or 10-cc) plastic container may be filled with the powdered mineral and weighed. The weight of the mineral in the container will give an approximation of the specific gravity. For 1-cc volume, the weight will approximate the specific gravity (for 10 cc it will be increased by a factor of 10).

MINERAL	SPECIFIC GRAVITY	MINERAL	SPECIFIC GRAVITY	MINERAL	SPECIFIC GRAVITY	MINERAL	SPECIFIC GRAVITY
Ice	0.917	**2.3–2.5**		**2.75–3.0**		**3.0–3.5**	
Epsomite	1.67	Gibbsite	2.4	Anorthite	2.76	Erythrite	3.1
Borax	1.7	Lazurite	2.4–2.45	Plagioclase	2.62–2.76	Autunite	3.1–3.2
Kernite	1.9	Colemanite	2.42	Pectolite	2.7–2.8	Chondrodite	3.1–3.2
Ulexite	1.95	Cancrinite	2.42–2.5	Phlogopite	2.76–2.90	Apatite	3.1–3.2
Sylvite	1.98	**2.5–2.75**		Talc	2.58–2.83	Spodumene	3.1–3.2
		Leucite	2.5	Turquoise	2.6–2.9	Andalusite	3.1–3.2
2.0–2.3		Serpentine	2.2–2.6	Chlorite	2.6–2.9	Lazulite	3.1–3.39
Sepiolite	2.0	Garnierite	2.5	Beryl	2.66–2.83	Fluorite	3.18
Opal	2.0–2.2	Scapolite	2.5–2.7	Polyhalite	2.78	Sillimanite	3.23
Sulfur	2.07	Orthoclase	2.56–2.59	Glauberite	2.81	Torbernite	3.2–3.7
Copiapite	2.08–2.17	Microcline	2.56–2.59	Pyrophyllite	2.65–2.90	Diopside	3.25–3.35
Niter	2.1	Chalcedony	2.57–2.64	Wollastonite	2.8–2.9	Jadeite	3.3–3.5
Chrysocolla	2.0–2.3	Albite	2.62	Dolomite	2.85	Augite	3.2–3.4
Bauxite	2.0–2.55	Nepheline	2.55–2.65	Lepidolite	2.8–2.9	Clinozoisite	3.21–3.38
Chabazite	2.05–2.1	Kaolinite	2.6	Muscovite	2.77–2.88	Axinite	3.26–3.36
Graphite	2.09–2.23	Cordierite	2.55–2.75	Prehnite	2.9–2.95	Dumortierite	3.3–3.4
Stilbite	2.1–2.2	Talc	2.58–2.83	Datolite	2.9–3.0	Enstatite	3.2–3.5
Halite	2.16	Oligoclase	2.65	Jarosite	2.91–3.26	Dioptase	3.3
Heulandite	2.1–2.2	Quartz	2.65	Biotite	2.7–3.3	Zoisite	3.15–3.36
Tridymite	2.2	Plagioclase	2.62–2.76	Hornblende	2.9–3.4	Olivine	3.2–4.3
Natrolite	2.2–2.26	(albite)	(anorthite)	Aragonite	2.94	Epidote	3.38–3.49
Variscite	2.2–2.5	Thenardite	2.67	Cryolite	2.97	Diaspore	3.3–3.5
Serpentine	2.2–2.6	Vivianite	2.68	Brazilianite	2.98	Hornblende	2.9–3.4
Soda Niter	2.25	Alunite	2.6–2.9	Anhydrite	2.98–3.0	Idocrase	3.4–3.5
Analcite	2.24–2.29	Turquoise	2.6–2.9	Eudialite	2.9–3.0	Hemimorphite	3.4–3.5
Sodalite	2.27–2.33	Chlorite	2.6–2.9	Danburite	2.97–3.02	Rhodonite	3.4–3.7
Chalcanthite	2.28	Collophane	2.6–2.9	Phenakite	3.0	Orpiment	3.49
		Pyrophyllite	2.65–2.90			Sphene	3.45–3.55
2.3–2.5		Beryl	2.66–2.83	**3.0–3.5**		Topaz	3.49–3.57
Gypsum	2.3	Pectolite	2.7–2.8	Danburite	2.97–3.02		
Christobalite	2.33	Calcite	2.71	Amblygonite	3.0–3.1	**3.5–3.75**	
Sodalite	2.27–2.33	Labradorite	2.71	Margarite	3.0–3.1	Diamond	3.50–3.53
Apophyllite	2.33–2.37	Bytownite	2.74	Magnesite	3.0–3.2	Realgar	3.56
Wavellite	2.36	Biotite	2.7–3.3	Tourmaline	3.0–3.25	Topaz	3.49–3.57
Brucite	2.4			Tremolite	3.0–3.3	Spinel	3.58–4.6

Table 7-5 (continued)

MINERAL	SPECIFIC GRAVITY	MINERAL	SPECIFIC GRAVITY	MINERAL	SPECIFIC GRAVITY	MINERAL	SPECIFIC GRAVITY
3.5–3.75		**4.0–4.5**		**5.0–5.5**		**6.0–7.0**	
Benitoite	3.6	Smithsonite	4.4	Millerite	5.5	Cassiterite	7.0
Rhodochrosite	3.7	Pyrolusite	4.4–5.0	**5.5–6.0**		**7.0–8.0**	
Kyanite	3.53–3.65	Enargite	4.4–4.5	Cerargyrite	5.56	Pyromorphite	7.04
Garnet	3.58–4.31	Barite	4.5	Proustite	5.57	Mimetite	7.24
Rhodonite	3.4–3.7	**4.5–5.0**		Chalcocite	5.5–5.8	Columbite-Tantalite	5.2–7.95
Staurolite	3.7–3.8	Spinel	3.58–4.6	Jamesonite	5.63	Wolframite	7.12–7.51
Strontianite	3.72	Stibnite	4.63	Zincite	5.68	Argentite	7.2–7.4
Chrysoberyl	3.75	Chromite	4.5–4.8	Arsenic	5.63–5.78	Low-Nickel Iron	7.3–7.87
3.75–4.0		Pyrrhotite	4.6–4.7	Columbite-Tantalite	5.2–7.95	Galena	7.57
Atacamite	3.76	Pyrolusite	4.4–5.0	Bournonite	5.83	Niccolite	7.78
Azurite	3.77	Zircon	4.6–4.7	Pyrargyrite	5.85	High-Nickel Iron	7.8–8.22
Staurolite	3.7–3.8	Covellite	4.6–4.76	**6.0–7.0**		**8.0–10.0**	
Goethite	3.3–4.3	Molybdenite	4.62–4.73	Crocoite	6.0–6.1	Sylvanite	8.1
Malachite	3.6–4.0	Pentlandite	4.6–5.0	Polybasite	6.0–6.2	Cinnabar	8.09
Antlerite	3.9±	Tetrahedrite-Tennantite	4.6–5.1	Arsenopyrite	6.07	Uraninite	8.0–10.9
Anatase	3.9	Monazite	4.6–5.4	Scheelite	6.1	Pitchblende	6.5–8.5
Siderite	3.9	Psilomelane	4.7	Cuprite	6.1	High-Nickel Iron	7.8–8.22
Brochantite	4.0±	Ilmenite	4.7	Tellurium	6.1–6.3	Breithauptite	8.23
Celestite	3.97	Selenium	4.8	Descloizite	6.2	Copper	8.95
Sphalerite	3.9–4.1	Marcasite	4.88	Stephanite	6.2	Calaverite	9.2
Willemite	3.9–4.2	Greenockite	4.9	Cobaltite	6.33	Bismuth	9.70–9.83
4.0–4.5		**5.0–5.5**		Anglesite	6.38	**>10.0**	
Sphalerite	3.9–4.1	Bornite	5.06–5.08	Skutterudite	6.5±	Silver	10.5
Corundum	4.0–4.1	Pyrite	5.01	Columbite-Tantalite	5.2–7.95	Mercury	13.59
Carnotite	4–5	Monazite	4.6–5.4	Cerussite	6.55	Platinum	14–19
Chalcopyrite	4.1–4.3	Magnetite	5.2	Antimony	6.61–6.72		(21.46 pure Pt)
Brookite	4.1	Columbite-Tantalite	5.2–7.95	Bismuthinite	6.8	Gold	19.3
Rutile	4.2	Hematite	5.26	Wulfenite	6.5–7.0		
Witherite	4.29	Linarite	5.35	Vanadinite	6.88		
Olivine	3.2–4.3						
Manganite	4.33						

Table 7-6
Magnetic Minerals

Powerful electro-magnets may be magnetically adjusted to separate minerals. Many common minerals are magnetic in such equipment. For ordinary use, a small alnico magnet is helpful.

* Chromite
  Franklinite
* Ilmenite (with included magnetite)
  Magnetite

  Nickel-iron
* Platinum (rich in iron)
* Pyrrhotite
* Ferberite

*Some specimens are magnetic

Table 7-7

Radioactive Minerals

Radioactive minerals (those of Uranium and Thorium) affect photographic film, illuminate a spinthariscope, cause the leaves of an electroscope to collapse, and activate either a geiger counter or scintillometer. Some are both radioactive and luminescent.

MINERAL	COLOR		MINERAL	COLOR
Autunite	yellow		Tuyamunite	yellow
Coffinite	black		Uraninite	black
Carnotite	yellow		Uranophane	yellow
* Monazite	brown		* Zircon	brown
Torbernite	green			

*Specimens are frequently radioactive

URANIUM	THORIUM
Oxides	
Uraninite (var. pitchblende), UO_2	Thorianite, ThO_2
Hydrous Oxides	
Schoepite, $4UO_3 \cdot 9H_2O$ (?)	
Gummite, $UO_3 \cdot nH_2O$	
Curite, $2PbO \cdot 5UO_3 \cdot 4H_2O$ (?)	
Becquerelite, $2UO_3 \cdot 3H_2O$	
Multiple Oxides	
Brannerite, $(U, Ca, Fe, Th, Y)_3Ti_5O_{16}$	Yttrocrasite, $(Y, Th, U, Ca)_2(Ti, Fe, W)_4O_{11}$
Davidite, $(Fe, Ce, U)(Ti, Fe, V, Cr)_3(O, OH)_7$	Zirkelite, $(Fe, Th, U, Ca)_2(Ti, Zr)O_5$
Silicates	
Coffinite, $U(SiO_4)_{1-x}(OH)_{4x}$	Thorite, $Th(SiO_4)$
Hydrous Silicates	
Uranophane, $CaO \cdot 2UO_3 \cdot 2SiO_2 \cdot 6H_2O$	Thorogummite, $Th(SiO_4)(OH)_{4x}$
Beta-uranotile, $CaO \cdot 2UO_3 \cdot 2SiO_2 \cdot 6H_2O$	
Skiodowskite, $MgO \cdot 2UO_3 \cdot 2SiO_2 \cdot 6H_2O$	
Soddyite, $5UO_3 \cdot 2SiO_2 \cdot 6H_2O$ (?)	

Table 7-7   (continued)

URANIUM	THORIUM

Hydrous Phosphates

Autunite, $CaO \cdot 2UO_3 \cdot P_2O_5 \cdot 8H_2O$
Torbernite, $CuO \cdot 2UO_3 \cdot P_2O_5 \cdot 8H_2O$

Anhydrous Phosphates

Monazite, $(Ce, Y, La, TH)PO_4$
Xenotime, $Y(PO_4) + Th$

Hydrous Arsenates

Zeunerite, $CuO \cdot 2UO_3 \cdot As_2O_5 \cdot 8H_2O$
Carnotite, $K_2O \cdot 2UO_3 \cdot V_2O_5 \cdot 3H_2O$
Tyuyamunite, $CaO \cdot 2UO_3 \cdot V_2O_5 \cdot 8H_2O$

Hydrous Sulphates

Zippeite, $2UO_3 \cdot SO_3 \cdot nH_2O$
Uranopilite, $6UO_3 \cdot SO_3 \cdot nH_2O$
Johannite, $CuO \cdot 2UO_3 \cdot 2SO_3 \cdot 7H_2O$

Carbonates

Schroeckingerite, $3CaO \cdot Na_2O \cdot UO_3 \cdot CO_2 \cdot SO_3F \cdot 10H_2O$

Organic Compounds

Urano-organic complexes—Mixture of hydrocarbon with U, Th, and rare earths.
Thucholite (Carburan)

Table 7-8
Water-Soluble Minerals—Taste

Only the smallest particle should be placed on the tip of the tongue, and it should be immediately washed away, for some minerals, like chalcanthite, are extremely poisonous if ingested in quantity.

MINERAL	TASTE	MINERAL	TASTE
Borax $Na_2B_4O_7 \cdot 10H_2O$	Sweetish-astringent	Kernite (soluble in hot water) $Na_2B_4O_7 \cdot 4H_2O$	—
Chalcanthite $CuSO_4 \cdot 5H_2O$	Sweetish-metallic	Niter (saltpeter) KNO_3	Cooling
Copiapite $(Fe,Mg)Fe_4(SO_4)_6(OH)_2 \cdot 20H_2O$	Metallic	Polyhalite $K_2Ca_2Mg(SO_4)_4 \cdot 2H_2O$	Bitter
Epsomite (epsom salt) $MgSO_4 \cdot 7H_2O$	Bitter	Soda niter (Chile saltpeter) $NaNO_3$	Cooling
Glauberite $Na_2Ca(SO_4)_2$	Bitter-salty	Sylvite KCl	Bitter-salty
Halite (common salt) $NaCl$	Salty	Thenardite Na_2SO_4	Salty

Plate 23. HALIDES ■ HALITE, SYLVITE, FLUORITE, CRYOLITE

A   x 0.46

HALITE
NaCl

Cubic halite crystals: Stassfurt, Prussia.

B   x 0.42

SYLVITE
KCl

Sylvite crystals; cubes modified by octahedrons: Stassfurt, Prussia.

C   x 4.0

HALITE

Halite with deep blue color localized in irregular patches.

D   x 0.52

FLUORITE
CaF_2

Groups of characteristic cubic fluorite crystals: Cumberland, England.

E   x 0.60

FLUORITE

Fluorite showing perfect octahedral cleavage and uneven color distribution: Rosiclare, Illinois.

F   x 0.57

CRYOLITE
Na_3AlF_6

Massive snow white cryolite occurring with brown siderite: Ivigtut, Greenland.

Plate 24. **CARBONATES** ▪ **CALCITE GROUP** CALCITE, SIDERITE

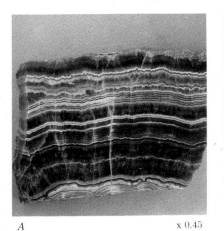

A   x 0.45

CALCITE
CaCO₃

"Mexican onyx"—Banded structure: Suisun, California.

B   x 0.30

CALCITE

Golden calcite crystal: Joplin, Missouri.

C   x 0.50

CALCITE

Cleavable mass of pink calcite with green actinolite: Canada.

D   x 0.50

CALCITE AND GALENA

Golden yellow calcite crystals with dark gray galena.

E   x 0.90

SIDERITE

Thin plate-like siderite crystals occurring with white calcite: Traversella, Peidmont, Italy.

F   x 0.46

SIDERITE
FeCO₃

Massive siderite showing perfect rhombohedral cleavage: Dauphiné, France.

Table 7-9
Minerals That Effervesce in Acid

Plastic acid bottles by means of which drops of cold dilute acid may be placed on a mineral or powder are convenient. Effervescence of carbonates is easily observed.

Cold Dilute Hydrochloric Acid (HCl)

MINERAL	COLOR	REMARKS
Aragonite, $CaCO_3$	Colorless, white; variously tinted	Pseudohexagonal prisms; poor cleavage; hardness $3\frac{1}{2}$–4
Azurite, $Cu_3(CO_3)_2(OH)_2$	Bright azure blue; crystals are dark blue	Hardness $3\frac{1}{2}$–4
Calcite, $CaCO_3$	Colorless, white; variously tinted	Perfect rhombohedral cleavage; hardness 3
Dolomite (powdered), $CaMg(CO_3)_2$	White; pale tints	Small pink curved crystals; hardness $3\frac{1}{2}$–4
Malachite, $Cu_2(CO_3)(OH)_2$	Green	Hardness $3\frac{1}{2}$–4
Smithsonite, $ZnCO_3$	White, green, blue, yellow, pink, gray, brown	Botryoidal aggregates; high specific gravity (4.4); hardness 4–$4\frac{1}{2}$
Strontianite, $SrCO_3$	White, gray, yellow, pink, greenish	Fibrous aggregates; high specific gravity (3.7); hardness $3\frac{1}{2}$–4
Witherite, $BaCO_3$	White	Twinned crystals resembling hexagonal pyramids or bipyramids; high specific gravity (4.3); hardness $3\frac{1}{2}$

Hot Dilute Hydrochloric Acid

MINERAL	COLOR	REMARKS
Aurichalcite $(Zn,Cu)_5(OH)_6(CO_3)_2$	Greenish-blue	Blue-green pearly scales; hardness 2
Dolomite, $CaMg(CO_3)_2$	White, pale tints	Small pink curved crystals; hardness $3\frac{1}{2}$–4
Hydrozincite, $Zn_5(OH)_6(CO_3)_2$	White	Dull white soft masses; usually luminesces pale blue to violet; hardness 2–$2\frac{1}{2}$
Magnesite, $MgCO_3$	White	Usually in dull compact masses; hardness $3\frac{1}{2}$–5
Rhodochrosite, $MnCO_3$	Pink to rose	Perfect rhombohedral cleavage; hardness $3\frac{1}{2}$–$4\frac{1}{2}$
Siderite, $FeCO_3$	Brown, gray	Perfect rhombohedral cleavage; hardness 3.7–4 2

Warm Dilute Nitric Acid (HNO_3)

MINERAL	COLOR	REMARKS
Cerussite, $PbCO_3$	White	Brilliant luster; high specific gravity (6.55); hardness 3–$3\frac{1}{2}$; grid-like aggregates

147

Table 7-10
Luminescent Minerals

Convenient portable long-wave and short-wave ultraviolet lamps may be used to examine the luminescence of many minerals in darkness. Minerals range widely in the resulting color, and the intensity of the visible light emitted ranges from strong to very weak. Effects should be observed for both long and short wavelengths. Luminescence may be observed in specimens of minerals from particular localities, but not observed in some from other localities. The luminescent color may also vary in different parts of a mineral specimen.

The mineral hackmannite exhibits a curious effect. The actual color of hackmannite (a pink variety of sodalite) is changed on exposure to ultraviolet light. Specimens of hackmannite from around Bancroft, Ontario, fade to white in daylight, but resume a pink hue when exposed to ultraviolet light. Hackmannite in darkness is also luminescent under long-wave ultraviolet radiation.

Minerals and Location	Red	Pink	Orange	Yellow-Orange	Yellow	Yellowish-Green	Green	Blue	Pale Blue	Brown	Violet	Bluish-Violet	White	Wavelength— S (short) L (long)
Adamite														
Laurium, Greece						×								S
Durango, Mexico					×									S
Anglesite														
Phoenixville, Penn.				×	×									L
Black Hills, S. Dak.					×									
Leadhills, Scotland					×									
Apatite														
Bohemia													×	S
Hull, Quebec, Canada					×									S
Ontario, Canada							×							S
Valyermo, Cal.				×										
Portland, Conn.					×									
Apophyllite														
Paterson, N.J.					×									S
Aragonite														
Death Valley, Cal.							×							S
Livingston, N.M.	×													
Crestmore, Cal.					×									S
La Junta, Colo.													×	S

Table 7-10   (continued)

Minerals and Location	Luminescent Color													Wavelength— S (short) L (long)
	Red	Pink	Orange	Yellow-Orange	Yellow	Yellowish-Green	Green	Blue	Pale Blue	Brown	Violet	Bluish-Violet	White	
Autunite														
Spruce Pine, N.C.						×								S,L
Spokane County, Wash.						×								S,L
Katanga District, Congo						×								S,L
Axinite														
Franklin, N.J.	×													
Barite														
Palos Verdes, Cal.					×									S
England							×							
Benitoite														
San Benito Co., Cal.								×						S,L
Beryl (emerald, especially synthetic emeralds)	×													
Calcite														
Franklin, N.J.	×													S
(numerous other locations)		×	×	×	×		×	×	×					
Calcium-larsenite														
Franklin, N.J.					×									S
Cerussite					×									L
Chalcedony (agate)														
San Diego, Cal.						×								S
Sweetwater Valley, Wyo.						×								S
Goldfield, Nev.						×								S
Colemanite														
Death Valley, Cal.													×	S
Corundum (blue sapphire)	×		×	×										
Siam (ruby)		×												
Burma (ruby)	×													
N. Carolina (ruby)	×													
Crocoite														
Dundas, Tasmania										×				
Ural Mts., U.S.S.R.										×				

149

Table 7-10 *(continued)*

Minerals and Location	Red	Pink	Orange	Yellow-Orange	Yellow	Yellowish-Green	Green	Blue	Pale Blue	Brown	Violet	Bluish-Violet	White	Wavelength— S (short) L (long)
Diamond														
South Africa									×					S-
Brazil			×						×					
Diaspore														
Chester, Mass.					×									
Diopside														
New York									×					S
Dolomite														
Baden, Germany								×						S
Dumortierite														
Lyons, France								×						S
San Diego Co., Cal.											×			S
Oreana, Nev.											×			
Epsomite														
Death Valley, Cal.									×					
Fluorite														
Durham, England								×				×		
Illinois-Kentucky region								×				×		
Clay Center, Ohio					×									
Duncan, Ariz.							×							
Glauberite														
Borax Lake, Cal.								×						
Gypsum					×									
Grand Rapids, Mich.							×							L
Halite														
Amboy, Cal.	×	×												
Kansas							×							
San Diego Co., Cal.	×	×												
Hemimorphite			×											L
Hexagonite														
Edwards, N.Y.	×													S

Table 7-10   *(continued)*

Minerals and Location	Red	Pink	Orange	Yellow-Orange	Yellow	Yellowish-Green	Green	Blue	Pale Blue	Brown	Violet	Bluish-Violet	White	Wavelength— S (short) L (long)
Hydrozincite														
Utah									×					S
Keeler, Cal.									×					S
Lepidolite														
Keystone, S. Dak.							×							
Manganapatite														
Valyermo, Cal.			×											S
Portland, Conn.			×											S
Center Strafford, N.H.			×											S
Natrolite			×											
Nepheline														
Bancroft, Ontario			×											
Opal														
(numerous locations)						×								S,L
Pectolite														
Magnet Cove, Ark.			×	×	×	×								L
Paterson, N.J.			×	×	×	×								L
Lake Co., Cal.			×	×	×	×								
Powellite														
Michigan copper region					×									
Quartz (amethyst)														
N. Carolina							×							
Madagascar							×							
Scapolite														
Grenville, Quebec				×	×									L
Scheelite														
Mill City, Nev.									×					S
Malay States									×					S
Spodumene														
(kunzite) Portland, Conn.	×													
(kunzite) Pala, Cal.			×	×	×					×				
(hiddenite) N.C.		×												

Table 7-10 *(continued)*

Minerals and Location	Red	Pink	Orange	Yellow-Orange	Yellow	Yellowish-Green	Green	Blue	Pale Blue	Brown	Violet	Bluish-Violet	White	Wavelength— S (short) L (long)
Sphalerite														
Tsumeb, South West Africa			X											
Searls Lake, Cal.								X						
Bisbee, Ariz.			X											
Medford, Utah			X											
Spinel (red)														
Ceylon	X													
Sodalite (hackmannite)														
Bancroft, Ontario		X	X	X	X									L
Tourmaline (rose, rubellite)														
Pala, Cal.											X			L
Newry, Me.											X			L
Wavellite														
Mt. Holly, Pa.							X							
Willemite														
Franklin, N.J.							X							S,L
Witherite														
Cumberland, England								X						S
Wollastonite			X	X	X		X	X						
Zircon														
Minas Gerais, Brazil			X											
Ontario, Canada			X											
North Carolina			X											
California			X											
Idaho			X											
Oregon			X											
Montana			X											
Zoisite (thulite)														
Mitchell Co., N.C.					X									L

Table 7-10 *(continued)*

Thermoluminescent Minerals

Thermoluminescent minerals phosphoresce when heated at low temperatures (below red heat). The effect is best seen in total darkness.

MINERAL	COLOR OF LUMINESCENCE
Apatite	Bluish-white
Barite	Yellow-orange
Celestite	Yellowish-green
Diamond	
Fluorite (chlorophane)	Green
Lepidolite	Lilac
Spodumene (kunzite)	Red, reddish-yellow
Scheelite	Blue

Triboluminescent Minerals

Triboluminescent minerals will emit visible light when subjected to mechanical forces; as when they are ground, scratched, or struck with a hammer.

Corundum
Diamond
Fluorite (chlorophane)
Lepidolite
Sphalerite
Willemite

8.

Mineral Descriptions

NATIVE ELEMENTS

In nature, most elements combine to form compounds; but approximately twenty native elements may be found in their uncombined states as minerals. The NATIVE ELEMENTS may be divided into METALS, SEMIMETALS, NONMETALS, and one LIQUID METAL (mercury). Such elements as tin (Sn), lead (Pb), tellurium (Te), selenium (Se), tantalum (Ta), osmium (Os), palladium (Pd), and iridium (Ir) occur in the native state but are rare.

METALS

Gold Group
- Gold   Au
- Silver   Ag
- Copper   Cu

Platinum Group
- Platinum   Pt
- Mercury   Hg

Iron Group
- Iron   (Fe, Ni)
- Nickel-iron   (Ni, Fe)

SEMIMETALS

Arsenic Group
- Arsenic   As
- Antimony   Sb
- Bismuth   Bi

NONMETALS

Sulfur Group
- Sulfur   S

Carbon Group
- Diamond   C
- Graphite   C

Metals

The metals have similar physical properties. They are HEAVY and more or less MALLEABLE (ability to be hammered into a sheet without breaking). Gold, silver, and copper, which have similar habits, commonly occur in branching and wire-like shapes (Plates 1 and 10).

GOLD

Au   Isometric
Hardness $2\frac{1}{2}$–3   Sp. Gr. 19.3 (pure)

Summary

A soft, sectile, malleable metal that does not tarnish. *Color* and *streak:* golden-yellow when pure, silvery-yellow when alloyed with silver, reddish-orange when alloyed with copper. *Luster:* metallic.

A x 0.63

SMITHSONITE
$ZnCO_3$

Botryoidal aggregate colored yellow by greenockite (CdS): Marion County, Arkansas.

B x 0.60

SMITHSONITE

Bluish-green botryoidal aggregate: Magdalena, New Mexico.

C x 0.50

MAGNESITE
$MgCO_3$

Dull white compact porcelain-like mass with conchoidal fracture: Oberdorf, Styria, Austria.

D x 0.54

RHODOCHROSITE
$MnCO_3$

Concentric structure: Catamarca Province, Argentina.

E x 1.0

RHODOCHROSITE

Rhodochrosite crystal showing rhombohedral cleavage: Lake County, Colorado.

F x 0.44

RHODOCHROSITE

Small rhombohedral crystals with green fluorite: Saguache County, Colorado.

Cleavage: none. *Fracture:* hackly. Usually occurs in irregular masses (nuggets), scales, plates, grains, or dendritic shapes (Plate 10(*A, B*), facing p. 67). Occurs most frequently in placer deposits and in quartz veins. Pyrite (fool's gold), chalcopyrite, and scales of weathered biotite have at times been mistaken for gold. BRIGHT GOLDEN-YELLOW COLOR, METALLIC LUSTER, HIGH SPECIFIC GRAVITY, AND EXTREME MALLEABILITY AND SECTILITY ARE HIGHLY DISTINCTIVE.

Composition. Gold (Au). Commonly alloyed with silver. Natural gold with 20% or more of silver is called ELECTRUM; also alloyed with copper and iron, but less commonly with bismuth, platinum, lead, zinc, and tin.

Tests. Fuses easily (1026°C) to a bright metallic globule. Pure gold is insoluble in ordinary acids, but soluble in aqua regia (a mixture of 3 parts of concentrated hydrochloric acid and 1 part of concentrated nitric acid).

Crystallization. Isometric; hexoctahedral class, $\frac{4}{m}\,\overline{3}\,\frac{2}{m}$. Seldom in cubic, octahedral, or dodecahedral crystals. Usually occurs as nuggets, scales, or grains; also in dendritic, net-like, spongy, wire-like, and distorted crystal shapes.

Alteration. Resistant to change.

Associated Minerals. Vein gold is often associated with quartz and pyrite. Other associated minerals are pyrrhotite, arsenopyrite, sphalerite, galena, and molybdenite.

Minerals of Similar Appearance

Pyrite. Distinguished from gold by superior hardness (6–6½), brittleness, black streak, and moderate specific gravity (5.0±).

Chalcopyrite. Is brittle, has a black streak, a specific gravity of 4.1–4.3, and a hardness of 4.

Weathered Biotite. Golden flakes of weathered biotite may resemble scales of gold, but biotite can be easily crushed with a knife.

Occurrence. Widely distributed in small amounts; occurs most frequently in placer deposits and in quartz veins associated with silica-rich igneous rocks. The world's most important gold-producing district is the Witwatersrand near Johannesburg in the Transvaal, Union of South Africa. The gold of the Rand is scattered through quartz conglomerates. Vast placer deposits occur in Siberia on the eastern and western slopes of the Ural Mountains. Other important gold-producing localities include Canada, Australia, and India. In the United States, the principal gold-producing areas are California (the Mother Lode), Carlin, Nevada and South Dakota (the Homestake mine).

Use. As a monetary standard; in jewelry, in scientific equipment, and in dentistry.

SILVER

Ag   Isometric
Hardness 2½–3   Sp. Gr. 10.5 (pure)

Summary

A soft, sectile, malleable metal. *Color:* silver-white on fresh surface, but tarnishes readily to brown, gray, or black. *Luster:* metallic. *Streak:* shining silver. *Fracture:* hackly. *Cleavage:* none. Commonly occurs in dendritic and wire-like shapes (Plate 10(*C, E*), facing p. 67); also in irregular masses and scales. Resembles *platinum*. BRIGHT SILVER COLOR AND STREAK, HIGH SPECIFIC GRAVITY, AND EXTREME MALLEABILITY AND SECTILITY ARE DISTINCTIVE.

Composition. Silver (Ag). Commonly alloyed with mercury or gold, but less commonly with copper, antimony, bismuth, and arsenic.

Tests. Fuses (961°C) to a bright silver globule. Soluble in nitric acid—addition of a few drops of hydrochloric acid gives a white, curdy precipitate of silver chloride.

Crystallization. Isometric; hexoctahedral class, $\frac{4}{m}\,\overline{3}\,\frac{2}{m}$. Rarely in cubic or octahedral crystals; commonly in irregular masses, scales, plates, dendritic, and wire-like shapes; also as coatings.

Alteration. Alters to and from silver halides, sulfides, and sulfosalts.

Associated Minerals. Calcite, barite, quartz, fluorite, copper, uraninite; also sulfides and arsenides of silver, nickel, cobalt, copper, and lead minerals.

Minerals of Similar Appearance

Platinum. Resembles silver in color and malleability, but is harder (4–$4\frac{1}{2}$); does not tarnish, and is insoluble in nitric acid.

Bismuth. Pale reddish-silver on a fresh surface, and usually occurs in lamellar masses.

Occurrence. Found in hydrothermal veins with (1) sulfides and arsenides of nickel, cobalt, and silver minerals (Cobalt, Ontario, Canada); (2) uraninite (Great Bear Lake, Canada; and Czechoslovakia); (3) calcite and other silver minerals (Kongsberg, Norway; and Mexico). Also widely distributed in small amounts in the oxidized zone of certain ore deposits. In northern Michigan, native silver and copper occur together in aggregates known as "HALFBREEDS." Twisted aggregates of wire-silver are found in Batopilas, Mexico (Plate 10); and in Aspen, Colorado. Plate 10 also shows silver dendrites from Mexico. Good specimens of dendritic silver embedded in pinkish-white calcite are found in Cobalt, Ontario.

Use. Coinage, plating, jewelry, dentistry, and photography.

COPPER

Cu   Isometric
    Hardness $2\frac{1}{2}$–3   Sp. Gr. 8.95

Summary

A soft sectile malleable metal. *Color:* copper-red on a fresh surface but tarnishes readily to dull brown. *Streak:* shining copper. *Luster:* metallic. *Fracture:* hackly. *Cleavage:* none. Usually occurs in irregular masses; also in branching shapes (dendritic copper, Plate 1(*A*), facing p. 2), and in distorted groups of crystals (Plate 10(*F*), facing p. 67). COPPER-RED COLOR AND STREAK, MALLEABILITY, SECTILITY, AND HIGH SPECIFIC GRAVITY ARE DISTINCTIVE.

Composition. Copper (Cu). May contain small amounts of iron, silver, gold, mercury, arsenic, bismuth, and antimony.

Tests. Fuses at 1083°C. Dissolves readily in nitric acid, coloring the solution bluish-green (Cu).

Crystallization. Isometric; hexoctahedral class, $\frac{4}{m}\,\bar{3}\,\frac{2}{m}$. Rarely in cubic, octahedral, and dodecahedral crystals; crystals are usually rounded and distorted; occurs most frequently in masses and scales; also in branching and wire-like groups.

Associated Minerals. Malachite, cuprite, azurite, chalcopyrite, boronite, gold, silver, calcite, datolite, and zeolite minerals.

Alteration. Oxidizes to cuprite, malachite, and azurite.

Minerals of Similar Appearance

Niccolite. Resembles copper in color, but is brittle, has a brownish-black streak, and has a hardness of 5–$5\frac{1}{2}$.

Occurrence. Small amounts of copper are commonly associated with (red) cuprite, (green) malachite, and (blue) azurite in the oxidized zone above copper sulfide deposits. Huge masses of copper weighing many tons have been found in the Keweenaw Peninsula in northern Michigan. The copper occurs in veins intersecting basic lava flows, in small cavities in the lavas, and as the cementing material in associated conglomerates. Interesting copper specimens found in Michigan include halfbreed aggregates of silver and copper; branching tree-like aggregates (Plate 1, facing p. 2); and bright crystals of copper included in colorless calcite crystals.

Use. A minor ore of copper. Used primarily in alloys and electrical wires. Copper sulfides, such as chalcocite and chalcopyrite, are the major ores of copper.

PLATINUM

Pt   Isometric

   Hardness 4–4½   Sp. Gr. 14–19 (for native
platinum alloyed
with other metals)

Summary

A *silver-gray* malleable metal that does not tarnish. *Luster:* metallic. *Streak:* gray. *Fracture:* hackly. *Cleavage:* none; sectile. May be magnetic if rich in iron. Usually occurs in grains or scales; sometimes in nuggets (Plate 10(*D*), facing p. 67). Associated with ultrabasic igneous rocks. Resembles *silver*. SILVER-GRAY COLOR, METALLIC LUSTER, HIGH SPECIFIC GRAVITY, MALLEABILITY, AND INSOLUBILITY ARE DISTINCTIVE.

Composition. Platinum (Pt)—alloyed with iron, iridium, rhodium, palladium, and osmium.

Tests. Infusible in the blowpipe flame; dissolves in hot aqua regia, but not in other acids; may be weakly magnetic.

Crystallization. Isometric; hexoctahedral class, $\frac{4}{m}\,\bar{3}\,\frac{2}{m}$. Rarely in cubic crystals; usually in small grains or scales; less frequently in nuggets.

Associated Minerals. Olivine, pyroxene, magnetite, chromite.

Alteration. Resistant to change.

Minerals of Similar Appearance

Silver. Resembles platinum in color and malleability, but tarnishes readily, is easily fusible; soluble in nitric acid.

Occurrence. Associated with such basic or ultrabasic igneous rocks as olivine-gabbros, pyroxenites, peridotites, and dunites; also with placer deposits derived from these rocks. Large quantities of platinum have been found in placer deposits along the slopes of the Ural Mountains, U.S.S.R. Platinum occurs in dunite in the Bushveld igneous complex in the Transvaal, Union of South Africa. An important source is Sud-

bury, Ontario, Canada, where the platinum occurs as the mineral SPERRYLITE (PtAs₂) rather than in its native state. Sperrylite is associated with pyrrhotite-pentlandite ores.

Use. Jewelry, surgical instruments, chemical and electrical equipment, thermocouples.

MERCURY

Hg   Hexagonal—rhombohedral ($-39°$C)
Sp. Gr. 13.59

Summary

A bright *silver-white liquid* that solidifies at minus 39°C. Usually occurs as small drops or films associated with bright red cinnabar (HgS). Mercury is the only metal that is liquid at ordinary temperatures; thus, it does not resemble any other mineral. LIQUID NATURE AND ASSOCIATION WITH CINNABAR ARE DISTINCTIVE.

Composition. Mercury (Hg)—sometimes with small amounts of gold or silver.

Tests. Vaporizes under the blowpipe at $350° \pm 10°$; dissolves in nitric acid (HNO₃).

Occurrence. Native mercury is not a common mineral. It is usually of secondary origin associated with cinnabar in regions of volcanic activity or hot springs. Found in Almadén, Spain; at Idria, Gorizia, Italy; and at Mount Avala, Yugoslavia. In the United States, it is found in California, and in Terlingua, Texas.

Use. In the amalgamation process, used for the recovery of gold and silver; also in medicine, scientific apparatus, and in pigments. Cinnabar is the chief ore of mercury.

Metallic Iron

Metallic iron is rare in terrestrial rocks, but is a common constituent of meteorites (see p. 301). It is nickel-bearing, and may be of two types:

I. NICKEL-POOR IRON (Fe, Ni) contains minor amounts of nickel (about 2–7 percent) in a body-centered cubic structure.

II. NICKEL-RICH IRON (Ni, Fe) has a face-centered cubic structure that contains approximately 24–77 percent nickel.

LOW-NICKEL IRON

Fe, Ni   Isometric
    Hardness 4   Sp. Gr. 7.3–8.87

Type I is strongly magnetic, metallic, steel gray to iron black, and malleable; has poor cubic cleavage, and frequently occurs as thin plates in iron meteorites. It is also found as blebs and as larger masses in terrestrial rocks, but is rarely in crystals. The most important terrestrial occurrence is at Disko Island, Greenland, where it is found in large masses that occasionally weigh many tons; it is also found in small grains embedded in basalt. Terrestrial low-nickel-iron usually contains only about 2 percent nickel. Meteoric low-nickel-iron commonly occurs as the natural alloy KAMACITE which usually contains about 5.5 percent nickel.

HIGH-NICKEL IRON

Ni, Fe   Isometric
    Hardness 5   Sp. Gr. 7.8–2.22

Type II is strongly magnetic, metallic, silver to grayish-white, malleable and commonly occurs in meteorites, as, for example, the nickel-rich mineral TAENITE. The most important terrestrial occurrence is in Josephine County, Oregon, where large masses of nickel-iron may exceed 100 lb.

Semimetals

The semimetals are fairly brittle, and thus differ from the metals which can be hammered into thin sheets without breaking.

ARSENIC

As   Hexagonal
    Hardness $3\frac{1}{2}$ Sp. Gr. 5.63–5.78

Summary

Brittle. *Color:* tin white on a fresh surface, but readily tarnishes to dull grayish-black. *Luster:* Metallic. *Streak:* gray. *Cleavage:* perfect basal, but rarely seen. Commonly occurs in granular masses or botryoidal crusts with concentric layers (Plate 11(*C*), facing p. 82); rarely in crystals. Powdered mineral; has a garlic (As) odor. BOTRYOIDAL HABIT, TARNISH, AND TESTS FOR ARSENIC ARE DISTINCTIVE.

Composition. Arsenic (As). Frequently contains some antimony. The mineral allemontite contains approximately equal amounts of arsenic and antimony.

Tests. Volatilizes completely without melting, giving off a strong garlic odor and white fumes. Open tube—white fumes and minute brilliant crystals. Closed tube—arsenic mirror.

Crystallization. Hexagonal; scalenohedral class, $\bar{3}\frac{2}{m}$; commonly in botryoidal crusts or in granular masses; rarely in rhombohedral crystals that resemble cubes.

Alteration. Oxidizes to form a blackish crust of arsenolite, As_2O_3.

Occurrence and Associated Minerals. Occurs in hydrothermal veins with silver, nickel, and cobalt ores. Typical botryoidal crusts occur in the Harz Mountains, Germany (Plate 11, facing p. 82).

Use. A minor ore of arsenic.

ANTIMONY

Sb   Hexagonal
    Hardness $3–3\frac{1}{2}$ Sp. Gr. 6.61–6.72

Summary

Very brittle. *Color:* silver-white. *Luster:* bright metallic. *Streak:* gray. *Cleavage:* perfect basal. Commonly in radiating masses with a granular texture (Plate 11(*B*), facing p. 82); rarely in crystals. SILVER-WHITE COLOR, BRIGHT METALLIC LUSTER, BRITTLENESS, AND GRANULAR TEXTURE ARE DISTINCTIVE.

Composition. Antimony (Sb). May contain arsenic, iron, or silver.

Tests. Fuses easily at 630°C to a metallic globule which becomes coated with needles of artificial valentinite (Sb_2O_3). Open tube—dense heavy white fumes. Colors flame bluish-green, and gives off white fumes.

Crystallization. Hexagonal, scalenohedral class, $\bar{3}\frac{2}{m}$. Usually massive or lamellar; commonly radiating masses with a granular texture; rarely in pseudocubic crystals.

Alteration. Alters to valentinite (Sb_2O_3).

Occurrence and Associated Minerals. Occurs in veins with silver, antimony, and arsenic ores. Commonly associated with stibnite; also with galena, pyrite, sphalerite, and quartz. Found in Sala, Sweden; Andreasberg, Germany; Chile; New Brunswick, Canada; and in Kern and Riverside Counties, California.

Use. Minor ore of antimony. (Stibnite (Sb_2S_3) is the chief source.)

BISMUTH

Bi   Hexagonal
  Hardness 2–2½   Sp. Gr. 9.7–9.83

Summary

A soft, sectile, *pale reddish-silver* mineral that readily tarnishes iridescent. *Luster:* metallic. *Streak:* shining silver. *Cleavage:* perfect basal; brittle. Commonly in lamellar masses (Plate 11(*A*), facing p. 82); rarely in crystals. Distinguished from SILVER by reddish hue and lamellar structure. PALE REDDISH-SILVER COLOR, SECTILITY, AND LAMELLAR HABIT ARE DISTINCTIVE.

Composition. Bismuth (Bi). May contain minor amounts of sulfur, arsenic, antimony, tellurium.

Tests. Fuses at a comparatively low temperature (270°C) to a metallic globule. Soluble in nitric acid.

Crystallization. Hexagonal; scalenohedral class, $\bar{3}\frac{2}{m}$. Rarely in crystals; artificial crystals frequently occur in parallel groups of pseudocubic hopper-shaped crystals; usually massive; sometimes in reticulated or branching shapes.

Occurrence. Bismuth is a fairly uncommon mineral that occurs in hydrothermal veins with ores of nickel, cobalt, silver, tin, tungsten. The major deposits occur in Bolivia. Found with cobalt and silver minerals at Cobalt, Ontario, Canada; in the silver mines at Kongsberg, Norway; at Freiberg and Schneeberg, Saxony; and in Spain. Also found in Australia, in Mexico, and in Japan.

Use. Major ore of bismuth; used in medicine, cosmetics, and in low-melting-point alloys.

Nonmetals

SULFUR

S   Orthorhombic
  Hardness 1½–2½   Sp. Gr. 2.07

Summary

A soft, *yellow,* translucent mineral that can be ignited with a match; burns with a blue flame that yields strong fumes of sulfur dioxide. *Color:* yellow, but impure varieties may be orange-red, brown, green, yellowish-gray, or black. *Luster:* resinous. *Streak:* yellow. *Cleavage:* Imperfect basal, prismatic, and pyramidal. *Fracture:* conchoidal to uneven; brittle to subsectile. Bipyramidal or thick tabular crystals are common (Plate 5, facing p. 34). Commonly associated with celestite (Plate 11(*E*), facing p. 82), gypsum, calcite, and aragonite. YELLOW COLOR, EASE OF BURNING, LOW HARDNESS, AND LOW SPECIFIC GRAVITY ARE DISTINCTIVE. Resembles ORPIMENT.

Composition. Sulfur (S). May contain small amounts of selenium and tellurium; also clay and hydrocarbons.

Tests. Melts easily (113°C), burns at 270°C with a blue flame yielding sulfur dioxide (SO_2). Insoluble in water and most acids; soluble in carbon disulfide (CS_2).

Crystallization. Orthorhombic; dipyramidal class, $\dfrac{2}{m}\dfrac{2}{m}\dfrac{2}{m}$. Sulfur also occurs in two monoclinic forms but they rarely occur as minerals. Usually massive; also in crusts, reniform, and stalactitic shapes; crystals commonly bipyramidal.

Minerals of Similar Appearance

Orpiment. Resembles sulfur in color, luster, and hardness (Plate 16, facing p. 99), but has perfect cleavage in one direction, thus producing a foliated structure with flexible cleavage flakes.

Alteration. Alters from sulfides.

Occurrence. 1. As a volcanic sublimate (Japan; Chile). 2. In sedimentary deposits of gypsum and limestone (Sicily; U.S.S.R.). Large crystals occur with celestite at Girgenti, Sicily. 3. In the caprock of salt domes (Louisiana; Texas) (Figure 8-1). The salt dome sulfur is extracted by the ingenious Frasch process (Figure 8-2); this is feasible because of the low melting point of sulfur. A casing is driven through the overlying formations to the sulfur-bearing formation. Inside the casing is a pipe for heated brine, and inside the hot water pipe is a return sulfur pipe containing a narrow air pipe. Hot brine (300°F) is pumped under pressure into the sulfur deposit. The sulfur melts at 283°F, and the thin molten sulfur is forced up the return sulfur pipe by hot compressed air. The liquid sulfur is then piped to a stockpile where it solidifies. 4. Sulfur is also produced by the decomposition of metallic sulfides, such as pyrite. 5. Reduced from "sour gas" containing hydrogen sulfide in oil fields.

Use. In the manufacture of sulfuric acid, vulcanizing rubber, manufacture of wood pulp, matches, gunpowder, fertilizers, insecticides.

DIAMOND

C   Isometric
Hardness 10   Sp. Gr. 3.50–3.53

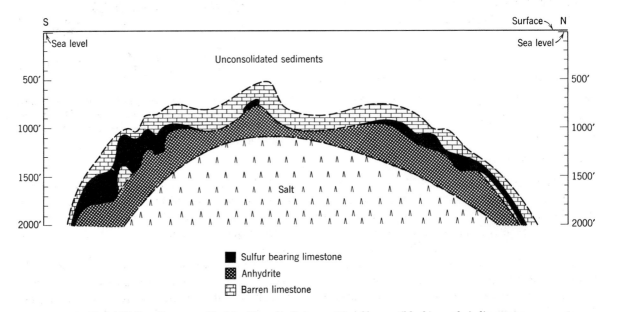

Sulfur bearing limestone
Anhydrite
Barren limestone

Fig. 8-1. N–S section across Hoskins Mound salt dome with sulfur cap (black) mostly in limestone above anhydrite.

Summary

Commonly occurs as rounded octahedral crystals with a *brilliant greasy luster.* *Color:* colorless to slightly tinted, rarely in deeper shades; may be black in impure varieties. *Cleavage:* perfect octahedral; brittle; sometimes luminescent in ultraviolet light; *harder than any other mineral.* Occurs in stream gravels, and in altered peridotite rock (kimberlite) (Plate 11(*D*), facing p. 82). DISTINGUISHED FROM OTHER COLORLESS MINERALS LIKE QUARTZ, TOPAZ, AND CORUNDUM BY SUPERIOR HARDNESS, OCTAHEDRAL CRYSTALS, BRILLIANT REFRACTION, DISPERSION WHEN CUT, AND PECULIAR BRILLIANT GREASY LUSTER WHEN UNCUT (RESEMBLES OILED GLASS).

Composition. Carbon (C).

Tests. Burns at about 1800°F to form carbon dioxide (CO_2); not attacked by acids or alkalis.

Crystallization. Isometric; hextetrahedral class, $\overline{4}\,3\,m$. Crystals usually octahedral, also dodecahedral; curved crystal faces are common; occurs sometimes in flat elongated crystals, or in spherical forms with a radial structure. Simple and multiple contact twins are common; also cyclic groups and penetration twins. Cubic crystals are rare (Figure 8-3). Figure 8-4 shows *synthetic* octahedral diamond crystals.

Varieties

Bort. Granular to microcrystalline aggregates; also badly flawed diamonds not suitable for gems.

Carbonado. Massive gray or black opaque bort; from Bahia, Brazil.

Alteration. Resistant to change.

Occurrence. Occurs most frequently in stream gravels (placer deposits) in the Congo, in Brazil, and in India. Also found in gravels along shorelines on the west coast of South Africa and on the ocean's floor. In South Africa, Tanganyika, Rhodesia, and at least one Arkansas locality, diamond crystals occur in a serpentinized porphyritic peridotite rock (kimberlite).

The South African diamonds occur in kimberlite "pipes" that are funnel-shaped bodies—cylindrical on the surface, but becoming elongated or fissure-like at

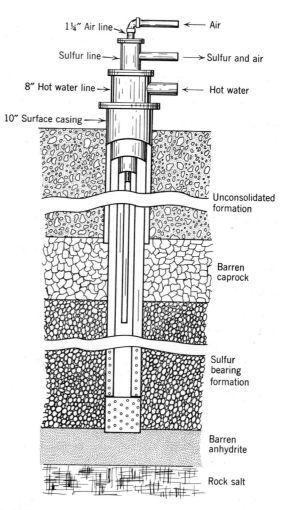

Fig. 8-2. Frasch Process of extracting sulfur (after W. T. Lundy, *Industrial Minerals and Rocks,* ed. by S. H. Dolbear and O. Bowles: Amer. Inst. Min. Met. Engrs; New York, 1949, Chapter 47).

depth. The Kimberly mine is 1600 ft. across, and was mined to a depth of 3500 ft. The giant Premier mine near Pretoria in the Transvaal measures 2800 ft. across. Figure 8-5 shows a vertical section of the De Beers mine.

The upper part of the "diamond pipes" is weathered to a soft yellowish material, known as "yellow ground"; at the lower part is the firmer bluish-green kimberlite or "blue ground." The olivine in the kimberlite has largely been altered to serpentine. Kimber-

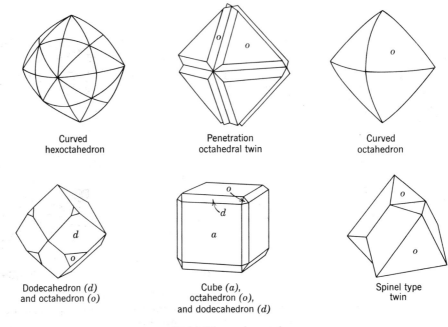

Curved
hexoctahedron

Penetration
octahedral twin

Curved
octahedron

Dodecahedron (d)
and octahedron (o)

Cube (a),
octahedron (o),
and dodecahedron (d)

Spinel type
twin

Fig. 8-3. Diamond crystals.

lite ranges in composition, and often contains a variety of inclusions from overlying and underlying rocks; some examples of these are garnet-bearing rocks that were carried up from depth, and fragments of fossiliferous shales that were dropped from above into the open cavity and then incorporated into the upwelling kimberlite magma which was not hot enough to appreciably alter them.

According to A. F. Williams,* kimberlite is a rock composed of two different ages. The matrix is a fine-grained mica-olivine rock, and the large olivine, enstatite, garnet, chrome-diopside, phlogopite, and ilmenite crystals that give the rock its porphyritic structure are transported minerals that crystallized out of the original magma at greater depths. Williams believes it was with these early-formed crystals that diamond crystallized before being eventually transported to its present resting place by the upwelling residual kimberlite magma. To support his view that the diamonds did not crystallize in situ, he supplies the following evidence: In the De Beers mine in South Africa,

*Genesis of the Diamond, Ernest Benn Limited, London, 1932.

two broken parts of a finger-shaped diamond were found at different and unconnected levels of the kimberlite pipe. The two pieces fitted exactly when placed together. Thus, the diamond obviously crystallized before reaching its present site.

The diamonds differ in size, color, and habit from mine to mine. For example, one mine may have a high percentage of large rounded octahedral crystals, whereas another mine will have few large diamonds, most of them being small dodecahedrons. The dodecahedral diamonds may possibly have formed as a result of preferred solution after growth.

Where it is believed that there has been a mixture of two types of kimberlite of different ages in the same mine, the crystal habits of the diamonds generally differ. In the manufacture of *synthetic* diamonds, temperature and chemical environment influence the color, and pressure the habit (see p. 26, Figure 2-34).

Use. For gems. Industrial diamonds as well as bort and carbonado are used for cutting tools, diamond drill bits; and as abrasives for grinding, cutting, and polishing wheels.

Plate 26. CARBONATES ▪ ARAGONITE GROUP: ARAGONITE, CERUSSITE, WITHERITE
DOLOMITE GROUP: DOLOMITE

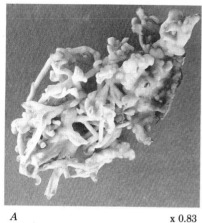

A   x 0.83

ARAGONITE
"Flos Ferri"

Curved twisted aggregate of aragonite (coralloidal structure): Styria, Austria.

B   x 0.70

ARAGONITE
CaCO₃

Twinned aragonite crystals. Intergrowth of 3 individuals produces a short pseudo-hexagonal prism: Molina de Aragon, Spain.

C   x 0.50

ARAGONITE

Slender pointed radiating crystals.

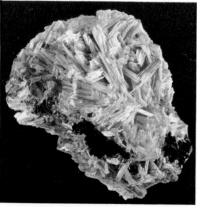

D   x 0.65

CERUSSITE
PbCO₃

Intergrowth of cerussite crystals forming a grid-like pattern (reticulated structure): Pima County, Arizona.

E   x 0.85

DOLOMITE
CaMg(CO₃)₂

Pale pink curved crystals ("Pearl Spar"): Joplin, Missouri.

F   x 0.70

WITHERITE
BaCO₃

Twinned witherite crystals resembling a series of hexagonal pyramids capping one another: Alston Moor, England.

Plate 27. CARBONATES CONTAINING HYDROXYL ■ MALACHITE, AZURITE, HYDROZINCITE, AURICHALCITE

A   x 0.72

MALACHITE
$Cu_2(CO_3)(OH)_2$

Botryoidal aggregate of malachite: Morenci, Arizona.

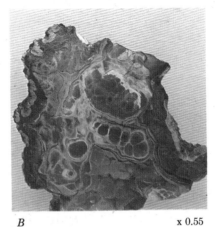

B   x 0.55

AZURITE

Botryoidal aggregate of azure blue azurite with green malachite.

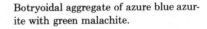

C   x 0.33

MALACHITE

Light and dark green layers of malachite.

D   x 0.60

AZURITE
$Cu_3(CO_3)_2(OH)_2$

Botryoidal aggregate of azurite showing characteristic azure blue color.

E   x 0.55

HYDROZINCITE
$Zn_5(OH)_6(CO_3)_2$

Dull white hydrozincite showing colloform structure: Kreuth, Carinthia.

F   x 0.66

AURICHALCITE
$(Zn, Cu)_5(OH)_6(CO_3)_2$

Crust of blue-green scales of aurichalcite: Zacatecas, Mexico.

GRAPHITE

C   Hexagonal
  Hardness 1–2   Sp. Gr. 2.09–2.23

Summary

Soft, sectile, greasy to the touch; stains fingers black. *Color:* black to steel gray. *Luster:* submetallic, sometimes dull. *Streak* (on smooth paper): gray-black. *Cleavage:* perfect basal (cleaves into thin flexible in-elastic plates). Occurs commonly in foliated masses (Plate 11(*F*), facing p. 82) or thin plates in such metamorphic rocks as crystalline limestones, schists, gneisses, and metamorphosed coal beds. Resembles *molybdenite*. GREASY FEEL, SOFTNESS, FOLIATED HABIT, MICACEOUS CLEAVAGE, AND LOW SPECIFIC GRAVITY ARE DISTINCTIVE.

Composition. Carbon (C). Often contains clay and iron oxides.

Tests. Infusible in the blow-pipe flame (melts at 3000°C); marks paper; insoluble in acids.

Fig. 8-4. Synthetic octahedral diamond crystals—about 35× magnification (courtesy of General Electric Research Laboratory).

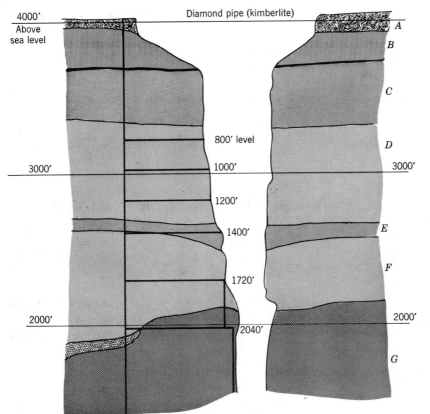

Fig. 8-5. The De Beers diamond-bearing pipe, South Africa. The mine-opening outlines the pipe. Here *A* represents dolerite; *B*, Dwyka shale; *C*, melaphyre; *D*, quartzite; *E*, melaphyre; *F*, quartz porphyry; *G*, granite (after A. F. Williams, *Genesis of the Diamond,* Ernest Benn Limited, London, 1932).

Crystallization. Hexagonal; dihexagonal dipyramidal class, $\frac{6}{m}\frac{2}{m}\frac{2}{m}$. Occasionally in thin tabular six-sided crystals; usually in foliated masses or scales.

Minerals of Similar Appearance

Molybdenite. Resembles graphite in softness, greasy feel, and foliated structure, but is *bluish* lead-gray with a green-gray streak on smooth paper.

Occurrence. Graphite usually occurs in metamorphic rocks. Large deposits are found in Ceylon, Madagascar, the U.S.S.R., Korea, Austria, and Mexico. Also occurs in Ticonderoga, New York; and as good crystals in marble at Sterling Hill, New Jersey.

Use. "Lead" pencils, high-temperature alloys, lubricants, refractory crucibles, protective paint, electrodes.

SULFIDES

The sulfides form a largely economic group that includes a number of important ore minerals. Metals like copper and lead, or semimetals like arsenic, antimony, and bismuth are combined with sulfur. Less common arsenides, antimonides, selenides, and tellurides are usually classified with the sulfides.

Where a semimetal and a metal are both present in a crystal structure, and where the role of the semimetal is similar to the metal, the mineral is referred to as a SULFOSALT (double sulfide). The following minerals are sulfosalts: PROUSTITE Ag_3AsS_3; PYRARGYRITE Ag_3SbS_3; POLYBASITE $(Ag,Cu)_{16}Sb_2S_{11}$; STEPHANITE Ag_5SbS_4; JAMESONITE $Pb_4FeSb_6S_{14}$; BOURNONITE $PbCuSbS_3$; ENARGITE Cu_3AsS_4; and TETRAHEDRITE $(Cu,Fe,Zn,Ag)_{12}(Sb,As)_4S_{13}$. These may be grouped with the sulfides according to important economic elements.

IRON SULFIDES

Pyrite, FeS_2
Marcasite, FeS_2
Pyrrhotite, $Fe_{1-x}S$

MERCURY SULFIDE

Cinnabar, HgS

NICKEL SULFIDES; ARSENIDE

Pentlandite, $(Fe,Ni)_9S_8$
Millerite, NiS
Niccolite, $NiAs$

COBALT SULFIDE; ARSENIDE

Cobaltite, $CoAsS$
Skutterudite, $(Co,Ni,Fe)As_3$

SILVER SULFIDES

Argentite, Ag_2S
Proustite, Ag_3AsS_3
Pyrargyrite, Ag_3SbS_3
Polybasite, $(Ag,Cu)_{16}Sb_2S_{11}$
Stephanite, Ag_5SbS_4

GOLD TELLURIDE

Calaverite, $AuTe_2$

GOLD-SILVER TELLURIDE

Sylvanite, $(Au,Ag)Te_2$

LEAD SULFIDES

Galena, PbS
Jamesonite, $Pb_4FeSb_6S_{14}$
Bournonite, $PbCuSbS_3$

ZINC SULFIDES

Sphalerite, ZnS

CADMIUM SULFIDE

Greenockite, CdS

COPPER SULFIDES

Chalcopyrite, $CuFeS_2$
Covellite, CuS
Chalcocite, Cu_2S
Bornite, Cu_5FeS_4
Enargite, Cu_3AsS_4
Tetrahedrite,
    $(Cu,Fe,Zn,Ag)_{12}(Sb,As)_4S_{13}$

ARSENIC SULFIDES

Orpiment, As_2S_3
Realgar, AsS
Arsenopyrite, $FeAsS$

ANTIMONY SULFIDE

Stibnite, Sb_2S_3

BISMUTH SULFIDE

Bismuthinite, Bi_2S_3

MOLYBDENUM SULFIDE

Molybdenite, MoS_2

PYRITE

FeS_2 Isometric
    Hardness 6–6½   Sp. Gr. 5.01

Summary

A hard, brittle, bright metallic, *brass-yellow* mineral (may tarnish darker or become iridescent). *Streak:* greenish-black. Often found in striated cubic crystals or pyritohedrons (Plate 12(A, B), facing p. 83); frequently massive. May resemble CHALCOPY-RITE, MARCASITE, PYRRHOTITE, PENTLANDITE, or GOLD.

DISTINCTIVE PROPERTIES ARE COLOR, LUSTER, HARDNESS, AND CRYSTAL FORM.

Composition. Iron disulfide, FeS_2 (46.6% Fe, 53.4% S). May contain cobalt and nickel in solid solution; may also contain as impurities small amounts of gold, copper, and arsenic.

Tests. Pyrite fuses easily, often with fumes of sulfur dioxide. The residue is strongly magnetic. Pyrite is essentially insoluble in hydrochloric acid, but the powdered form is soluble in concentrated nitric acid.

Crystallization. Isometric; diploidal class, $\frac{2}{m}\bar{3}$. Often occurs in striated cubes, pyritohedrons, octahedrons, or in combinations of these forms. Massive and granular aggregates are common. "Iron cross" penetration twins occur.

Associated Minerals. Pyrite is found alone or associated with many minerals. It frequently occurs with other sulfides such as sphalerite, chalcopyrite, and galena.

Minerals of Similar Appearance

Chalcopyrite. Deep brass-yellow, and inferior in hardness (H. = 4). Pyrite is paler brass-yellow, but tarnished surfaces are deceptive. An untarnished surface should be exposed.

Marcasite. Somewhat greenish, and differs in crystal habit. Twinned crystals may have a spearhead or cockscomb structure.

Pyrrhotite and **Pentlandite.** Bronze minerals with a hardness of 3¼–4. Pyrrhotite is moderately magnetic.

Gold. A malleable distinctively golden yellow mineral. Pyrite or "fool's gold" is hard and brittle.

Occurrence. Pyrite is one of the most ubiquitous of sulfides and occurs in veins, massive deposits, or as scattered crystals in numerous rock types. Great masses of gold-bearing pyrite associated with chalcopyrite are mined at the famous Rio Tinto deposit, near Huelva, in southwestern Spain. Pyrite masses in mines, as at Jerome, Arizona, and at Gilman, Colorado, have been known to constitute a fire hazard through spontaneous combustion. Cubes up to 6 in. have been found at Leadville, Colorado. Other occurrences of interest include: pyrite in quartz veins—Mother Lode, California; bright cubic crystals in chlorite schist—Chester, Vermont; disk-like radiating forms "pyrite suns" in shale—Sparta, Randolph County, Illinois; pyrite trilobite fossils in shale—northwest of Utica, New York (at Holland Patent); clusters of crystals in clay deposits—Long Island, New York; and iron-cross twins at Schoharie, New York.

Alteration. Limonite or goethite often form pseudomorphs after pyrite crystals.

Use. Pyrite is chiefly used to provide a source of sulfur for sulfuric acid and ferrous sulfate (copperas). Also yields impure iron when roasted. Some pyrite is mined for associated gold and copper.

MARCASITE (White Iron Pyrites)

FeS_2 Orthorhombic
   Hardness 6-6½   Sp. Gr. 4.88

Summary

A hard, brittle, metallic, *pale, brass-yellow* mineral *with a greenish cast,* deepening on exposure.

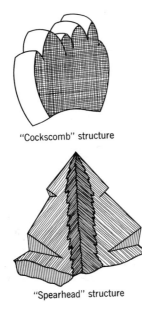

"Cockscomb" structure

"Spearhead" structure

Fig. 8-6. Marcasite.

Streak: greenish black; easily oxidized—may decompose on exposure to air. Commonly in acicular or radiating aggregates, or in crystals with a spearhead or cockscomb structure (Figure 8-6). Color is whiter than pyrite on a fresh surface. Often associated with galena and sphalerite. COLOR AND COCKSCOMB STRUCTURE ARE DISTINCTIVE (Plate 12(*D*), facing p. 83).

Composition. Iron disulfide, FeS_2, like pyrite, but decomposes more readily on exposure to air. Specimens tend to fall apart and give a brown color to paper trays and labels because of the release of sulfuric acid during oxidation.

Tests. Same as pyrite, except that marcasite powder when heated in concentrated nitric acid is decomposed, with the separation of sulfur producing a flocculent solution. Pyrite would be completely dissolved under the same conditions.

Crystallization. Orthorhombic; dipyramidal class, $\frac{2}{m}\frac{2}{m}\frac{2}{m}$. Marcasite is dimorphous with pyrite. Crystals are usually tabular parallel to a basal plane; also pyramidal. Frequently in twinned crystals with a spearhead structure (Figure 8-6). Massive, globular, and acicular radiating aggregates are common. Stalactitic aggregates may form with a radial inner structure covered with pyramidal crystals.

Associated Minerals. Galena, sphalerite, calcite, dolomite, and fluorite.

Minerals of Similar Appearance

Pyrite. Marcasite is distinguished from pyrite by color, crystallization, and ease of decomposition. Some specimens require x-ray identification or optical study on polished surfaces.

Millerite. Radiating fibrous aggregates of millerite commonly have dark velvety botryoidal surfaces, and may be associated with nickel and cobalt minerals.

Alteration. Crystals of marcasite may invert to pyrite—the more stable form of FeS_2—forming pyrite pseudomorphs after marcasite. Limonite pseudomorphs after marcasite are also common. Sulfuric acid and ferrous sulfate are formed when marcasite disintegrates.

Occurrence. Marcasite is less common than pyrite and crystallizes under a limited range of conditions. It is believed to be a low-temperature mineral formed from acid solutions. Marcasite frequently occurs as replacements in limestones and may be associated with sphalerite and galena in the Missouri, Kansas, Oklahoma (Tri-state) lead-zinc region.

Use. A minor source of sulfur.

PYRRHOTITE

$Fe_{1-x}S$ Hexagonal
    Hardness $3\frac{1}{2}$–$4\frac{1}{2}$ Sp. Gr. 4.6–4.7

Summary

A brittle, *brownish-bronze,* metallic mineral; usually moderately *magnetic* (but ranges in intensity from strong to weak). *Streak:* black. Ordinarily occurs in granular masses. Good crystals are rare (Plate 12(*C*), facing p. 83). May contain associated pentlandite and chalcopyrite. BROWNISH-BRONZE COLOR AND MAGNETISM ARE DISTINCTIVE.

Composition. Iron sulfide, $Fe_{1-x}S$. The x in the formula ranges from 0 to 0.2 owing to a deficiency of iron atoms in the pyrrhotite structure. Some of the least magnetic varieties are high in iron. The mineral *troilite,* which occurs as nodules in iron meteorites, is close to the FeS in composition, and is believed to be the end member of the pyrrhotite series.

Tests. Pyrrhotite fuses easily and becomes strongly magnetic when heated on charcoal or in the reducing flame; gives off fumes of hydrogen sulfide when dissolved in hydrochloric acid.

Crystallization. Hexagonal; dihexagonal-dipyramidal class, $\frac{6}{m}\frac{2}{m}\frac{2}{m}$. Crystals (rare) occur in a tabular hexagonal habit. Above 138°C the orthorhombic polymorph is stable.

Associated Minerals. Pentlandite, chalcopyrite, pyrite, and magnetite.

Minerals of Similar Appearance

Pentlandite. Exhibits octahedral parting, is a lighter bronze color, and is nonmagnetic. May be intimately intermixed with pyrrhotite.

Pyrite. Pale brass-yellow on a fresh surface; has a hardness of 6–$6\frac{1}{2}$, and is nonmagnetic.

Occurrence. At Sudbury, Ontario, pyrrhotite occurs in large masses intermixed with chalcopyrite and pentlandite. The rare platinum mineral sperrylite ($PtAs_2$) is also found there. Pyrrhotite is a common minor constituent of igneous rocks, and also occurs in pegmatites, high-temperature veins, and in contact deposits. Crystals of pyrrhotite have been found at the Morro Velho gold mine in Minas Gerais, Brazil, and at the Potosí mine in Chihuahua, Mexico. Massive pyrrhotite occurs with topaz and wolframite at Burnt Hill, New Brunswick, Canada. Troilite (FeS) has been found in serpentine in Del Norte County, California. It has also been found in meteorites.

Alteration. Alters to limonite, siderite, and iron sulfates.

Use. Concentrates of pyrrhotite and associated minerals at Sudbury serve as a source of iron ore, sulfur, nickel, copper, platinum, cobalt, and selenium.

CINNABAR

HgS   Hexagonal
    Hardness 2–$2\frac{1}{2}$ Sp. Gr. 8.09

Summary

A heavy, soft, *bright scarlet-red* mineral; brownish-red in impure varieties. *Streak:* scarlet to red brown. *Luster:* adamantine; dull in earthy varieties. *Cleavage:* perfect prismatic; translucent to transparent; opaque when earthy. Usually massive (Plate 12, facing p. 83). BRIGHT SCARLET COLOR AND STREAK, AND HIGH SPECIFIC GRAVITY FOR A NONMETALLIC MINERAL ARE DISTINCTIVE. May be confused with CUPRITE, REALGAR, and HEMATITE.

Composition. Mercury sulfide (86.2% Hg, 13.8% S). Often impure; may contain clay, iron oxides, bitumen.

Tests. Volatilizes on charcoal. Closed tube—minute droplets of metallic mercury are formed when cinnabar is mixed with sodium carbonate flux and heated. Open tube—a dark ring of metallic mercury forms when heated slowly.

Crystallization. Hexagonal; trigonal trapezohedral class: 32. Usually in granular masses; also earthy. Rarely in small rhombohedral crystals or in penetration twins.

Minerals of Similar Appearance

Realgar. Resembles cinnabar in color, but is usually more orange and is almost always associated with yellow orpiment. Cinnabar is an extremely heavy mineral (Sp. Gr. 8.09), whereas realgar has a low specific gravity (3.5).

Cuprite. Darker red than cinnabar, and is often associated with native copper and green or blue copper minerals.

Hematite. A cinnabar streak is vermilion in contrast to brick red hematite. Cinnabar also yields mercury tests.

Alteration. Alters to liquid mercury; at Terlingua, Texas, alters to such a rare mercury mineral as calomel $HgCl$ (white), which oxidizes to eglestonite Hg_4OCl_2 (yellow).

Associated Minerals. Mercury, pyrite, marcasite, stibnite, quartz, opal, calcite, dolomite.

Occurrence. Cinnabar is the important ore of mercury. It often occurs in veins or impregnations near recent volcanic rocks or hot spring deposits. The most important cinnabar deposits are in Almadén, Spain, and in Idria, Gorizia, Italy (formerly, in Carniola, Austria). Good crystals occur near Belgrade, Yugoslavia, and in Hunan, China. In the United States, deposits are found in California at New Almadén in Santa Clara County, and at New Idria in San Benito County. Also in Nevada, Oregon, Utah, and Texas. Good crystals also occur in Pike County, Arkansas.

Use. Major ore of mercury.

NICCOLITE

NiAs	Hexagonal
Hardness 5–5½	Sp. Gr. 7.78

Summary

A brittle, metallic, *pale copper-red* mineral with a dark tarnish; often coated with an alteration crust of pale apple-green annabergite. *Streak:* brownish-black. Usually massive (Plate 13, facing p. 90). Frequently occurs with silver and cobalt minerals. COPPER COLOR, APPLE-GREEN ALTERATION, AND MINERAL ASSOCIATIONS ARE DISTINCTIVE. Closely resembles the related mineral BREITHAUPTITE (NiSb).

Composition. Nickel arsenide (43.9% Ni, 56.1% As). Usually contains some antimony; may also contain small amounts of iron, cobalt, and sulfur.

Tests. Fuses easily on charcoal to a metallic globule that gives off white fumes of arsenious oxide with a garlic odor. With dimethylglyoxime, it gives a scarlet color in a nitric acid solution when neutralized with ammonia (test for nickel).

Crystallization. Hexagonal; dihexagonal dipyramidal class, $\frac{6}{m}\frac{2}{m}\frac{2}{m}$. Usually massive; sometimes botryoidal or columnar. Small pyramidal or tabular crystals are rare.

Associated Minerals. Silver, cobaltite, skutterudite, pentlandite, pyrrhotite, breithauptite, millerite, annabergite, erythrite, chalcopyrite, and arsenopyrite.

Alteration. Readily alters to pale green annabergite.

Minerals of Similar Appearance

Copper. Resembles niccolite in color, but is soft and sectile, and has a shining copper streak. Niccolite is brittle and has a brownish-black streak.

Breithauptite. A copper colored mineral with a strong violet cast. It resembles niccolite strongly but contains antimony.

Occurrence. At Cobalt, Ontario, Canada, niccolite occurs in large masses in veins with silver, nickel, and cobalt minerals. Also occurs in sulfide deposits associated with norite (a variety of the igneous rock gabbro). Alternating concentric shells of niccolite and arsenopyrite are found in the Natsume nickel deposits in Japan. The crystals, which are rare, occur in Richelsdorf, Germany.

Use. A minor ore of nickel.

PENTLANDITE

$(Fe,Ni)_9S_8$ Isometric
     Hardness $3\frac{1}{2}$–4   Sp. Gr. 4.6–5.0

Summary

A brittle, metallic, *yellowish-bronze* mineral that usually occurs in granular masses intermixed with pyrrhotite (Plate 13(*B*), facing p. 90). *Streak:* light brownish-bronze; nonmagnetic; octahedral parting. CLOSE TO PYRRHOTITE IN COLOR AND HARDNESS, BUT IS NONMAGNETIC.

Composition. Sulfide of iron and nickel $(Fe,Ni)_9S_8$. Commonly contains small amounts of cobalt.

Tests. Fuses easily to a magnetic steel globule. Gives nickel test (deep pink precipitate) with dimethylglyoxime.

Crystallization. Isometric; hexoctahedral class, $\frac{4}{m}\,\bar{3}\,\frac{2}{m}$. Usually in granular masses.

Occurrence and Associated Minerals. Pentlandite is intimately associated with pyrrhotite in ultrabasic igneous rocks. Major pyrrhotite-pentlandite deposits are located at Sudbury, Ontario, and in the Lynn Lake area in Manitoba, Canada. Both cobalt and platinum are associated with nickel at Sudbury. Important deposits of pyrrhotite-pentlandite are also found in Petsamo, U.S.S.R.

Use. The most important ore of nickel. Nickel is used primarily in such alloys as nickel steel; stainless steel; Monel metal (nickel and copper); Nichrome (nickel, chromium, iron); German silver (nickel, copper, and zinc); and nickel coins (25% Ni, 75% Cu). The chief attributes of nickel alloys are strength, resistance to corrosion and heat, and ductility.

MILLERITE

NiS   Hexagonal
     Hardness 3–$3\frac{1}{2}$ Sp. Gr. 5.5

Summary

A metallic, *brass-yellow* mineral with a *greenish tinge*. Commonly in needle-like radiating crystals (Plate 13(*C*), facing p. 90), and in matted hair-like tufts. *Streak:* greenish-black. *Cleavage:* 2-rhombohedral. Found in limestone cavities with calcite, dolomite, and fluorite; also in cavities in hematite and siderite. NEEDLE-LIKE AND HAIR-LIKE CRYSTALS, COLOR, AND METALLIC LUSTER ARE DISTINCTIVE.

Composition. Nickel sulfide, NiS (64.7% Ni, 35.3% S).

Tests. Fuses easily on charcoal to a magnetic globule. Gives a nickel test with dimethylglyoxime.

Crystallization. Hexagonal; scalenohedral class, $\bar{3}\,\frac{2}{m}$. Usually in needle- or hair-like crystals; also in radiating fibrous aggregates with dark velvety botryoidal surfaces. Felted hair-like masses may occur similarly to some jamesonite or chalcotrichite, but of different color.

Occurrence. Found in limestone cavities (Keokuk, Iowa; St. Louis, Missouri; Milwaukee, Wisconsin); in siderite cavities (Glamorgan, Wales); and in cavities in hematite (Sterling Mine, Antwerp, New York). Also found in carbonate veins; in meteorites; and as an alteration of other nickel minerals.

Use. A minor ore of nickel.

COBALTITE

CoAsS   Isometric
     Hardness $5\frac{1}{2}$ Sp. Gr. 6.33

Summary

A hard, brittle, *silver-white* metallic mineral with a slightly pinkish cast. Crystals are commonly cubes and pyritohedrons (Figure 8-7 and Plate 13(*D*), facing

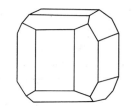

Fig. 8-7. Common cobaltite crystal (cube and pyritohedron).

p. 90); similar to pyrite crystals but differing in color. *Streak:* grayish-black. *Cleavage:* good cubic. Frequently associated with niccolite, skutterudite, and silver. Massive cobaltite is hardly distinguishable from massive silver-white cobalt, nickel, and iron sulfides and arsenides. SILVER COLOR AND CRYSTAL HABIT ARE DISTINCTIVE.

Composition. Cobalt sulfarsenide, CoAsS, usually contains iron and nickel.

Tests. Fusible. On charcoal, gives off arsenious oxide with an arsenic (garlic) odor. Dissolves in warm nitric acid and forms a rose colored (cobalt) solution.

Crystallization. Isometric; tetartoidal class, 23. Commonly occurs in cubes and pyritohedrons; also in granular masses.

Associated Minerals. Niccolite, skutterudite, rammelsbergite, silver, silver sulfosalts, bismuth, siderite, calcite, quartz.

Minerals of Similar Appearance

SKUTTERUDITE. Hard silver-white crystals of skutterudite commonly occur in cubes and octahedrons (Figure 8-8). Cobaltite crystals are frequently cubes and pyritohedrons.

Massive cobaltite greatly resembles massive hard silver-white iron, cobalt, and nickel sulfides and arsenides. Examples: LOELLINGITE, $FeAs_2$; SAFFLORITE, $(Co,Fe)As_2$; SKUTTERUDITE, $(Co,Ni)As_3$; RAMMELSBERGITE, $NiAs_2$; PARARAMMELSBERGITE, $NiAs_2$; ARSENOPYRITE, FeAsS; GLAUCODOT, $(Co,Fe)AsS$; ULLMANNITE, NiSbS; GERSDORFFITE, NiAsS.

Occurrence. A high-temperature mineral disseminated in metamorphic rocks (Skutterud, Norway); also in vein deposits with Co—Fe—Ni sulfides and arsenides (Cobalt, Ontario, Canada). Good pyritohedral crystals have been found in Tunaberg, Sweden.

Use. An ore of cobalt. Cobalt metal is used in high-speed tool steel, and in magnets. Cobalt oxide is used as a blue pigment.

SKUTTERUDITE

$(Co,Ni,Fe)As_3$ Isometric
Hardness $5\frac{1}{2}$–6   Sp. Gr. 6.5±

Summary

A hard, brittle, *silver-white* metallic mineral commonly in granular masses. Crystals are usually cubes or octahedrons or combinations of the two (Figure 8-8). *Streak:* black; alters to pink erythrite or, when rich in nickel (nickel-skutterudite), to pale green annabergite. *Cleavage:* none. Usually occurs in moderate temperature veins associated with cobalt, nickel, and silver minerals. When massive, difficult to distinguish from hard silver-white cobalt, nickel, and iron arsenides and sulfides. SILVER-WHITE CUBIC AND OCTAHEDRAL CRYSTALS ARE DISTINCTIVE.

Composition. Essentially a cobalt and nickel arsenide $(Co,Ni)As_3$, with iron (maximum to about 12%) substituting for some of the nickel or cobalt, $(Co,Ni,Fe)As_3$.

Tests. Fuses on charcoal to magnetic globule, and gives off arsenious oxide with a garlic (arsenic) odor. Soluble in nitric acid; yields a pink (Co) solution. A nickel-

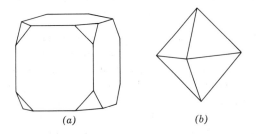

(a)   *(b)*

Fig. 8-8. Skutterudite crystals: *(a)* cube and octahedron; *(b)* octahedron.

skutterudite dissolved in nitric acid and neutralized with ammonia yields a violet solution with a pale green (Ni) precipitate.

Crystallization. Isometric; diploidal class, $\frac{2}{m}\bar{3}$. Usually in granular masses. Crystals are commonly cubes and octahedrons or combinations of the two. Pyritohedrons are rare.

Associated Minerals. Cobaltite, niccolite, arsenopyrite, silver, silver sulfosalts, bismuth, quartz, calcite, barite, and siderite.

Alteration. Alters to pink erythrite $(Co,Ni)_3(AsO_4)_2 \cdot 8H_2O$, or, when rich in nickel, to pale green annabergite $(Ni,Co)_3(AsO_4)_2 \cdot 8H_2O$.

Minerals of Similar Appearance

Cobaltite. Hard silver-white cobaltite crystals are usually cubes and pyritohedrons (Figure 8-7). Skutterudite crystals are usually cubes and octahedrons. Massive skutterudite is difficult to distinguish from massive silver-white iron, cobalt, and such nickel sulfides and arsenides as LOELLINGITE, $FeAs_2$; SAFFLORITE, $(Co,Fe)As_2$; RAMMELSBERGITE, $NiAs_2$; PARARAMMELSBERGITE, $NiAs_2$; ARSENOPYRITE, $FeAsS$; GLAUCODOT, $(CoFe)AsS$; COBALTITE, $CoAsS$; GERSDORFFITE, $NiAsS$; and ULLMANNITE, $NiSbS$.

Occurrence. Commonly occurs with cobalt, nickel, and silver minerals. Found in large amounts in the silver veins of Cobalt, Ontario; also occurs in Saxony at Freiberg, Schneeberg, and Annaberg.

Use. An ore of cobalt and nickel.

ARGENTITE (Silver Glance)

Ag_2S
    Hardness 2-2½

Isometric (above 179°C)
    Sp. Gr. 7.2-7.4

Summary

Soft, heavy, very sectile (can be cut with a knife like wax); *dark lead gray;* tarnishes readily to dull blackish gray. *Luster:* metallic. *Streak:* shining lead gray. Commonly occurs in groups of distorted crystals (Plate 13(*F*), facing p. 90); frequently with other silver minerals. EXTREME SECTILITY, DARK LEAD GRAY COLOR, HABIT, AND HIGH SPECIFIC GRAVITY ARE DISTINCTIVE. May resemble tarnished SILVER or CHALCOCITE.

Composition. Silver sulfide, Ag_2S, (Ag 87.1%, S 12.9%). Sometimes contains copper.

Tests. Fuses easily on charcoal in oxidizing flame; produces a silver globule and fumes of sulfur dioxide. Globule is soluble in nitric acid; when a few drops of HCl are added, a white, curdy precipitate forms, which darkens on exposure to air (test for silver).

Crystallization. Isometric; hexoctahedral class, $\frac{4}{m}\bar{3}\frac{2}{m}$ (at 179°C); below 179°C, the mineral ACANTHITE, which is the orthorhombic form of Ag_2S, is stable. Thus, specimens of Ag_2S are actually pseudomorphs of acanthite after argentite; usually massive, or in groups of branching and distorted crystals.

Associated Minerals. Silver minerals (silver, proustite, pyrargyrite); also galena, sphalerite, and nickel and cobalt minerals.

Alteration. Alters to silver and silver sulfosalts. Examples: proustite, pyrargyrite, polybasite, and stephanite.

Minerals of Similar Appearance

Silver. When tarnished, resembles argentite, but is bright silver-white on a fresh surface. Both silver and argentite are soft and sectile.

Chalcocite. Resembles massive argentite in color and hardness, but is less sectile, and is frequently associated with blue and green copper carbonates.

Occurrence. The most important primary ore of silver. Commonly found in low-temperature veins with other silver minerals; also found as microscopic inclusions in "argentiferous galena." Occurs in Kongsberg, Norway; Freiberg in Saxony; the silver mines of Mexico; in Peru; in Chile; in Bolivia; in Czechoslovakia; and in Nevada (Comstock Lode and Tonopah).

Use. An important ore of silver.

PROUSTITE (Ruby Silver)

Ag₃AsS₃   Hexagonal
    Hardness 2–2½   Sp. Gr. 5.57

Summary

A soft, *deep red* mineral with a brilliant luster and scarlet streak. *Cleavage:* rhombohedral; brittle; translucent to transparent. Usually massive. May occur in aggregates of distorted complex crystals; frequently with steep rhombohedral or scalenohedral terminations (Plate 13(*E*), facing p. 90). Usually found in low temperature veins with other silver minerals. Resembles *pyrargyrite* and *cuprite*. DEEP RED COLOR, SCARLET STREAK, BRILLIANT LUSTER AND SILVER ASSOCIATIONS ARE DISTINCTIVE.

Composition. Silver arsenic sulfide, Ag₃AsS₃ (65.4% Ag, 15.2% As, 19.4% S), with small amounts of antimony.

Tests. Fuses easily on charcoal; gives off white arsenious oxide fumes with a garlic odor; and forms a metallic silver globule. Globule dissolves in HNO₃ and gives silver test (white, curdy silver chloride precipitate upon the addition of HCl).

Crystallization. Hexagonal; ditrigonal pyramidal class, 3 *m*. Usually massive or in disseminated grains. Crystals are commonly prismatic with steep rhombohedral or scalenohedral terminations; frequently striated and distorted.

Associated Minerals. Other silver minerals like silver, argentite, pyrargyrite, polybasite, and stephanite. Also tetrahedrite, calcite, quartz.

Alteration. Alters to silver and argentite.

Minerals of Similar Appearance

Pyrargyrite. Crystals resemble proustite, but are usually darker in color (blackish-red) and have a purplish-red streak.

Cuprite. Resembles proustite in color, but is distinguished by crystal shape (commonly cube and octahedron) and association with other copper minerals.

Occurrence. Proustite is not a common mineral. It usually occurs as a low-temperature mineral with pyrargyrite and other silver minerals. Beautiful crystals, some 3 or 4 in. long, have come from Chañarcillo, Chile. Other notable locations are the Harz Mountains, Germany; Freiberg, Saxony; Guanajuato, Mexico; and Cobalt, Ontario. In the United States, it is found in small amounts in the silver deposits of Colorado, New Mexico, Idaho, Nevada, and California.

Use. An ore of silver.

PYRARGYRITE (Dark Ruby Silver)

Ag₃SbS₃   Hexagonal
    Hardness 2½   Sp. Gr. 5.85

Summary

A soft, *blackish-red* mineral with a brilliant almost metallic luster and a purplish-red streak. *Cleavage:* rhombohedral; brittle; translucent. Usually massive or in disseminated grains. Sometimes in prismatic crystals of hemimorphic development (upper crystal faces differ from lower faces). Occurs as a low temperature mineral with proustite and other silver minerals. Resembles *proustite* and *cuprite*. BLACKISH-RED COLOR, BRILLIANT LUSTER, CRYSTAL HABIT, AND SILVER ASSOCIATION ARE DISTINCTIVE.

Composition. Silver antimony sulfide, Ag₃SbS₃ (59.8% Ag, 22.4% Sb, 17.8% S), with small amounts of arsenic.

Tests. On charcoal, fuses easily to a metallic globule. A white ring of antimony trioxide forms around the globule. Malleable globule dissolves in HNO₃ and gives silver test (white, curdy silver chloride precipitate upon the addition of HCl). Powdered mineral is decomposed by a hot KOH solution; solution gives an orange precipitate on the addition of HCl (antimony test).

Crystallization. Hexagonal; ditrigonal pyramidal class, 3 *m*. Usually massive, or in disseminated grains. Crystals are commonly prismatic; often terminated with rhombohedrons or scalenohedrons; frequently of hemimorphic development.

Associated Minerals. Silver, argentite, proustite, polybasite, stephanite, tetrahedrite, calcite, quartz.

Alteration. Alters to argentite and silver.

Minerals of Similar Appearance

Proustite. Closely resembles pyrargyrite, but is usually lighter in color, is more translucent, and has a scarlet streak.

Cuprite. Resembles pyrargyrite in color, but is distinguished by crystal shape, tests for copper, and association with other copper minerals.

Occurrence. Pyrargyrite is more common than proustite. It forms as a late low-temperature silver mineral in the sequence of primary deposition. May also be derived as an alteration product of argentite, silver, and dyscrasite (Ag_3Sb). Occurs with other silver minerals in the Harz Mountains, Germany; Freiberg, Saxony; Příbram, Bohemia; Chañarcillo, Chile; Guanajuato, Mexico; Cobalt, Ontario, Canada. In the United States, it is found in small amounts in the silver deposits of Colorado, New Mexico, Idaho, Nevada, and California.

Use. An ore of silver.

POLYBASITE

$(Ag,Cu)_{16}Sb_2S_{11}$ Monoclinic
   Hardness 2–3   Sp. Gr. 6.0–6.2

Summary

A soft, *iron-black,* metallic mineral occurring in pseudohexagonal tabular crystals, often with triangular markings on the basal planes (Figure 8-9); also massive. *Streak:* black. *Cleavage:* imperfect basal. Usually associated with other silver minerals in low- to moderate-temperature veins. Resembles black tabular **HEMATITE** crystals with basal triangular patterns, but is considerably softer. **CRYSTALS ARE DISTINCTIVE.**

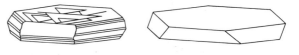

Fig. 8-9. Polybasite crystals.

Composition. Essentially a silver antimony sulfide in which copper substitutes for silver to about 30 atomic percent $(Ag,Cu)_{16}Sb_2S_{11}$.

Tests. Fuses easily. On charcoal, fuses to a metallic globule and forms a dense white coating of antimony trioxide (Sb_2O_3). Globule gives tests for silver.

Crystallization. Monoclinic; prismatic class, $\frac{2}{m}$. Occurs in tabular pseudohexagonal crystals, often with triangular patterns on the basal planes; also in granular masses.

Associated Minerals. Occurs with other silver minerals like pyrargyrite, proustite, stephanite, argentite, and silver; also with tetrahedrite, chalcopyrite, gold, quartz, calcite, dolomite, barite.

Alteration. Alters to silver and stephanite.

Occurrence. Occurs with silver ores in Mexico; Chile; Freiberg, Saxony; Příbram, Bohemia; and Andreasberg, in the Harz Mountains. In the United States it is found in small amounts in the silver mines of Colorado, Montana, Idaho, Arizona, and Nevada.

Use. An ore of silver.

STEPHANITE

Ag_5SbS_4 Orthorhombic
   Hardness 2–2½   Sp. Gr. 6.2

Summary

A soft, brittle, *iron-black,* metallic mineral; commonly in short prismatic crystals that are pseudohexagonal in shape (Figure 8-10); also massive. *Streak:* iron black. *Cleavage:* imperfect. A rare low-temperature silver mineral found in veins associated with other silver minerals. **CRYSTALS ARE DISTINCTIVE.**

Fig. 8-10. Stephanite crystal.

Composition. Silver antimony sulfide, Ag_5SbS_4 (68.5% Ag, 15.2% Sb, 16.3% S).

Tests. Fuses easily; fuses on charcoal to a metallic globule, and forms a dense white coating of antimony trioxide.

Crystallization. Orthorhombic; pyramidal class, $m\,m\,2$. Crystals are usually short prismatic or tabular; often pseudohexagonal in shape. Twinned crystals also produce pseudohexagonal shapes.

Associated Minerals. Other silver minerals like proustite, pyrargyrite, polybasite, argentite, and silver; also with tetrahedrite, galena, sphalerite, chalcopyrite, quartz, calcite, dolomite, and barite.

Alteration. Alters to silver, proustite, polybasite.

Occurrence. Stephanite is a rare low-temperature silver mineral which occurs with other silver minerals in veins. Good crystals have come from Příbram, Bohemia; Freiberg, Saxony; Andreasberg, Harz Mountains; Monte Narba, Sardinia; Cornwall, England; and Zacatecas, Mexico. In the United States, found in Nevada, Colorado, and California.

Use. An ore of silver.

CALAVERITE

$AuTe_2$ Monoclinic
    Hardness $2\frac{1}{2}$–3   Sp. Gr. 9.2

Summary

A soft, heavy, very brittle, *brass yellow* to *silver white* metallic mineral. *Streak:* yellowish and greenish gray. No *cleavage.* Usually in granular masses. Sometimes in small, deeply striated prismatic crystals.

Occurs with sylvanite and other gold-silver tellurides. Resembles *sylvanite.* DEEPLY STRIATED, ELONGATED CRYSTALS AND TESTS FOR GOLD AND TELLURIUM ARE DISTINCTIVE.

Composition. Gold ditelluride, $AuTe_2$ (44.03% Au, 55.97% Te). Small amounts of silver frequently replace gold.

Tests. Fuses easily on charcoal with a bluish-green flame, yielding a gold globule. Powdered mineral dissolved in concentrated hot sulfuric acid colors solution deep red (tellurium test) and separates a spongy gold mass.

Crystallization. Monoclinic; prismatic class, $\frac{2}{m}$. Usually in granular masses. Crystals are commonly prismatic and deeply striated; also twinned.

Associated Minerals. Sylvanite and other rare tellurides such as krennerite, $AuTe_2$; hessite, Ag_2Te; petzite, $(Ag,Au)_2Te$; altaite, $PbTe$; melonite, $NiTe_2$; tellurium, Te. Also gold, pyrite, fluorite, calcite, and quartz.

Minerals of Similar Appearance

Sylvanite. Striated prismatic crystals are similar to calaverite crystals, but sylvanite has perfect cleavage in one direction. Calaverite has no cleavage.

Alteration. Alters to spongy gold.

Occurrence. A low-temperature mineral; occurs with other tellurides at Cripple Creek, Colorado, and at Kalgoorlie, Western Australia.

Use. An ore of gold.

SYLVANITE (Graphic Tellurium)

$(Au,Ag)Te_2$ Monoclinic
    Hardness $1\frac{1}{2}$–2   Sp. Gr. 8.1

Summary

A soft, heavy, brittle, *silver-white,* metallic mineral with a slightly yellowish cast. *Cleavage:* perfect side pinacoid. *Streak:* light gray. Usually massive. Short prismatic crystals are rare. Sometimes in skeletal or arborescent forms that resemble writing. Resembles *calaverite, tellurium,* and *antimony.* "GRAPHIC" HABIT, CLEAVAGE, AND GOLD, SILVER, AND TELLURIUM TESTS ARE DISTINCTIVE.

Composition. Ditelluride of gold and silver, $(Au,Ag)Te_2$. The ratio of Au:Ag is close to 1:1 (Au 24.5%, Ag 13.4%, Te 62.1%). However, Au is sometimes present in excess of this ratio.

Tests. Fuses easily on charcoal with a blue flame; after continued blowing, yields a malleable gold-silver globule. Globule is decomposed in nitric acid, leaving a spongy mass of reddish gold. The addition of hydrochloric acid gives a curdy, white precipitate of silver chloride. Powdered mineral, heated in concentrated sulfuric acid, colors solution deep red (tellurium).

Crystallization. Monoclinic; prismatic class, $\frac{2}{m}$. Usually in granular or bladed masses. Sometimes in short prismatic, twinned, or skeleton crystals.

Minerals of Similar Appearance

Tellurium. Resembles soft, brittle, silver-white sylvanite but volatilizes completely on charcoal. Sylvanite leaves a metallic globule.

Calaverite. Has no cleavage. Sylvanite has perfect cleavage in one direction.

Antimony. Distinguished from sylvanite by tests for antimony.

Alteration. Alters to spongy masses of gold, also tellurite, TeO_2.

Occurrence and Associated Minerals. A low-temperature mineral, occurring with calaverite and other tellurides at Cripple Creek, Colorado, and at Kalgoorlie, Western Australia.

Use. An ore of gold and silver.

GALENA

PbS   Isometric
Hardness 2½   Sp. Gr. 7.58

Summary

A soft, heavy, *lead-gray* mineral, commonly occurring in cubic crystals (Plate 14, facing p. 91); *Luster:* brilliant metallic, but subject to dull gray tarnish. *Streak:* lead-gray. *Cleavage:* perfect cubic; brittle. CUBIC CRYSTALS AND CLEAVAGE, SOFTNESS, LEAD-GRAY COLOR, LUSTER, AND HIGH SPECIFIC GRAVITY ARE DISTINCTIVE.

Composition. Lead sulfide, PbS (86.6% Pb, 13.4% S). Commonly contains small amounts of silver, arsenic, antimony, and other impurities.

Tests. Fusible (M.P. 1115°C). Reduced on charcoal to a metallic lead globule surrounded by a yellow coating. Soluble in hot HCl. A few drops of dissolved mineral added to a potassium iodide solution produce a bright yellow precipitate.

Crystallization. Isometric; hexoctahedral class, $\frac{4}{m}\bar{3}\frac{2}{m}$. Commonly in cubic crystals; also cubes modified by hexoctahedrons. Frequently in cleavable and fine-grained masses.

Associated Minerals. Sphalerite, chalcopyrite, pyrite, tetrahedrite, silver minerals, calcite, barite, fluorite, quartz.

Alteration. Alters to cerussite, anglesite, pyromorphite, mimetite, and phosgenite.

Occurrence. Galena is a widely distributed mineral. It is found in hydrothermal sulfide veins; also in replacement deposits, usually in limestones and dolomites. In the United States, important deposits are located in southeastern Missouri and with zinc in the Joplin district of Missouri, Kansas, and Oklahoma. It is also found in the lead-silver deposits of Leadville, Colorado; in the Coeur d'Alene region of Idaho; and in the Park City and Tintic districts in Utah. Some important foreign localities are Freiberg, Saxony; Westphalia; Andreasberg, Harz Mountains; Cumberland, England; Broken Hill, Australia; and Eulalia, Mexico.

Use. The most important ore of lead. Galena frequently contains enough silver to be mined as a silver ore. Metallic lead is widely used in storage batteries and in various alloys such as solder (lead and tin), type metal (lead, antimony, and tin), low melting alloys (lead, bismuth, tin), and lead foil (lead, tin, and copper). Lead is also used in the manufacture of pigments.

wall, England. In the United States, found in Utah, Colorado, Idaho, Nevada, and California.

Use. A minor lead ore.

JAMESONITE (Brittle Feather Ore)

$Pb_4FeSb_6S_{14}$ Monoclinic
    Hardness $2\frac{1}{2}$ Sp. Gr. 5.63

Summary

A soft, brittle, *dark gray* metallic mineral; commonly in loosely matted hair-like crystals; also in fibrous feather-like masses. *Streak:* gray-black. Cleavage good across length. Associated with other lead sulfosalts, galena, sphalerite, tetrahedrite, pyrite, and stibnite in low- to moderate-temperature ore veins. Difficult to distinguish from BOULANGERITE and other related lead sulfosalts. HAIR-LIKE AND FEATHERY HABITS ARE DISTINCTIVE.

Composition. A lead iron antimony sulfide, $Pb_4FeSb_6S_{14}$, usually with some Cu and Zn.

Tests. Fuses easily on charcoal; forms yellow and white lead and antimony oxide coatings around the assay. Heated with a potassium iodide and sulfur flux on charcoal, it yields a chrome-yellow lead iodide coating.

Crystallization. Monoclinic; prismatic class, $\frac{2}{m}$. Commonly in fibrous feather-like masses; also in matted hair-like crystals.

Alteration. To bindheimite, $Pb_2Sb_2O_6(O,OH)$.

Minerals of Similar Appearance

Such "Feather Ore" minerals as BOULANGERITE ($Pb_5Sb_4S_{11}$) (Plate 14, facing p. 91); ZINKENITE ($Pb_6Sb_{14}S_{27}$); and MENEGHINITE ($Pb_{13}Sb_7S_{23}$) are similar to jamesonite in appearance and composition.

Occurrence. Associated with other lead sulfosalts, galena, sphalerite, tetrahedrite, pyrite, stibnite, quartz, and siderite in low- to moderate-temperature ore veins. Found at Příbram, Bohemia; Felsöbánya and Aranyidka, Roumania; Harz, Germany; and Corn-

BOURNONITE (Cog-wheel Ore)

$PbCuSbS_3$ Orthorhombic
    Hardness $2\frac{1}{2}$–3   Sp. Gr. 5.83

Summary

A soft, brittle, *steel gray* to *iron black,* metallic mineral; commonly in twinned crystals with a cogwheel shape (Figure 8-11, Plate 14(*F*), facing p. 91). *Streak:* steel gray to black. *Cleavage:* one good, two fair. Usually occurs in moderate-temperature hydrothermal veins associated with galena, tetrahedrite, sphalerite, chalcopyrite, pyrite, and quartz. WHEEL-LIKE TWINNED CRYSTALS ARE DISTINCTIVE.

Composition. A lead copper antimony sulfide, $PbCuSbS_3$; some arsenic can substitute for antimony.

Tests. Fuses easily on charcoal to a metallic globule with a white-and-yellow Pb and Sb oxide coating around the assay. It is decomposed by nitric acid, yielding a blue-green copper solution with a white Pb and Sb precipitate and yellowish sulfur residue.

Crystallization. Orthorhombic; dipyramidal class, $\frac{2}{m}\frac{2}{m}\frac{2}{m}$. Commonly in short prismatic or tabular crystals; also frequently in repeated twinned crystals, forming cogwheel shapes; and in granular masses.

Alteration. Alters to bindheimite, $Pb_2Sb_2O_6(O,OH)$; cerussite, malachite, azurite, and linarite.

Occurrence. Usually found in moderate temperature hydrothermal veins. Associated minerals include ga-

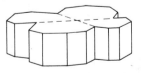

Fig. 8-11. Bournonite.

lena, sphalerite, tetrahedrite, chalcopyrite, pyrite, siderite, and quartz. Notable localities for crystals are the Harz Mountains, Germany; Liskeard, Cornwall, England; Bolivian tin-silver veins. Found in the United States in Park City, Utah; Yavapai County, Arizona; Austin, Nevada; and Inyo County, California.

Use. An ore of copper, lead, and antimony.

SPHALERITE (Zinc Blende, Black Jack)

ZnS   Isometric
   Hardness $3\frac{1}{2}$–4   Sp. Gr. 3.9–4.1

Summary

Sphalerite (Figure 8-12) is a difficult mineral to recognize because of its variable color and habit. Appropriately, the name comes from the Greek word meaning *treacherous*. *Color:* commonly ranges from amber yellow, to reddish-brown, to black (Black Jack); sometimes red (ruby zinc), green, and colorless when free of iron (rare). *Streak:* white, yellow, reddish-brown (always considerably lighter than the massive mineral). *Luster:* adamantine to resinous (Plate 14(*C, E*), facing p. 91); dull in fine-grained masses (Plate 14(*D*)). *Cleavage:* perfect dodecahedral. *Fracture:* conchoidal; brittle; transparent to opaque. Commonly in groups of rounded, distorted crystals; also coarse- to fine-grained cleavable masses. Frequently associated with galena, pyrite, and chalcopyrite. Some varieties resemble SIDERITE, GALENA, and CASSITERITE.

THE ADAMANTINE-LUSTER ON CLEAVAGES IS THE MOST DISTINCTIVE FEATURE.

Composition. Zinc sulfide, ZnS. Almost always contains iron (Zn,Fe)S; color deepens with increase in iron; manganese and cadmium are often present in small amounts. *Wurtzite,* the hexagonal form of ZnS, is stable above 1020°C.

Tests. Dissolves in HCl with evolution of H_2S gas (spoiled-egg odor). Occasionally luminescent and triboluminescent.

Crystallization. Isometric; hextetrahedral class, $\overline{4}\,3\,m$. Crystals are usually tetrahedral; sometimes dodecahedral; commonly twinned and in distorted aggregates of rounded crystals. Also in granular and fine grained masses.

Associated Minerals. Galena, pyrite, chalcopyrite, marcasite, greenockite, calcite, smithsonite, hemimorphite.

Alteration. Oxidizes to goslarite ($ZnSO_4 \cdot 7H_2O$); also alters to hemimorphite and smithsonite.

Minerals of Similar Appearance

Siderite. Coarse, cleavable masses of light brown siderite are distinguished from sphalerite by perfect rhombohedral cleavage and less brilliant luster. Sphalerite exhibits dodecahedral cleavage.

Galena. Massive galena may resemble black, almost metallic-looking varieties of sphalerite (Black Jack); but galena has cubic cleavage and a dark lead-gray streak. Black Jack sphalerite has a reddish-brown streak.

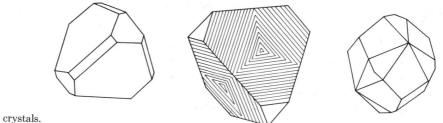

Fig. 8-12. Sphalerite crystals.

Cassiterite. Aggregates of brilliantly reflecting, rounded cassiterite crystals resemble somewhat the aggregates of distorted sphalerite crystals; but cassiterite is both heavier (Sp. Gr. 7.0) and harder (H. = 6–7).

Occurrence. Sphalerite is a common mineral in sulfide veins, in replacement deposits in limestones and other sedimentary rocks, and in contact metamorphic deposits. Important deposits in the United States occur in the Tri-State area of Missouri (Joplin), Kansas, and Oklahoma; also in Virginia and in Tennessee. Masses ranging in color from a beautiful transparent golden yellow to red occur in Picos de Europa, Santander, Spain.

Use. Most important ore of zinc. Also an important byproduct source of cadmium, gallium, indium, thallium, germanium. Metallic zinc is used for galvanizing iron; and for manufacturing brass (a copper-zinc alloy), wire and sheet zinc, and dry cell batteries. Zinc compounds are widely used in the manufacture of white paint.

GREENOCKITE

CdS	Hexagonal
Hardness 3–3½	Sp. Gr. 4.9

Summary

A soft, *yellow* to *orange* mineral; occasionally occurs as earthy coatings on sphalerite and other zinc minerals. *Luster:* adamantine, resinous, earthy. *Streak:* yellow to reddish. *Cleavage:* good prismatic. *Fracture:* conchoidal; brittle; translucent. YELLOW COLOR AND ASSOCIATION WITH ZINC MINERALS ARE DISTINCTIVE (Plates 14(*E*), facing p. 91, and 25(*A*), facing p. 154).

Composition. Cadmium sulfide, CdS (Cd 77.8%, S 22.2%). Complete solid solution exists between wurtzite (ZnS) and greenockite.

Tests. Infusible. Heated with a sodium carbonate flux on charcoal, it gives reddish-brown coating of cadmium oxide (CdO). Closed tube—turns carmine-red on heating; cools yellow. Soluble in concentrated HCl, giving off hydrogen sulfide gas.

Crystallization. Hexagonal; dihexagonal pyramidal class, 6 *m m*. Rarely occurs as small hemimorphic crystals; usually as earthy coatings.

Occurrence. Greenockite is an important ore of cadmium, but is not a common mineral. It usually occurs in small amounts as earthy coatings on sphalerite, smithsonite, and other zinc minerals. Found with zinc ores at Přibram, Bohemia; the Joplin district of Missouri; and as a bright yellow pigment in smithsonite in Marion County, Arkansas. Small crystals occur with prehnite, natrolite, and calcite at Renfrew, Scotland.

Use. Cadmium metal is used for electroplating other metals with a protective coating; also used in low-melting-point alloys, as control rods in nuclear reactors, and in nickel-cadmium batteries. Cadmium tungstate ($CdWO_4$) is used in fluorescent paint, and cadmium bromide ($CdBr_2$) is used in photography.

CHALCOPYRITE

$CuFeS_2$	Tetragonal
Hardness 3½–4	Sp. Gr. 4.1–4.3

Summary

A brittle, *deep golden,* metallic mineral; often subject to iridescent tarnish. *Streak:* greenish-black. Usually massive (Plate 15(*A*), facing p. 98). Crystals are sphenoidal, but suggest tetrahedrons (Figure 8-13). Resembles *pyrite* and *gold.* DEEP YELLOW COLOR, GREENISH-BLACK STREAK, SOFTNESS, AND CRYSTAL HABIT ARE DISTINCTIVE.

Fig. 8-13. Chalcopyrite crystal.

Plate 28. **SULFATES** ▪ BARITE, CELESTITE, ANGLESITE, CHALCANTHITE

A   x 0.48

BARITE
BaSO$_4$

Barite crystal with surface coating of red iron oxide: Cumberland, England.

B   x 0.72

BARITE

Aggregate of thin divergent barite plates ("crested barite"): Caribou Beach, Nova Scotia.

C   x 0.60

BARITE

Tabular barite crystals: Sterling, Colorado.

D   x 0.50

CELESTITE
SrSO$_4$

Celestite crystals with yellow sulfur: Val di Noto, Sicily.

E   x 0.46

ANGLESITE
PbSO$_4$

Fine-grained grayish-white anglesite formed by the alteration of galena (PbS), "woody anglesite." Dark gray areas are residual galena: Madonna Mine, Colorado.

F   x 0.60

CHALCANTHITE
CuSO$_4 \cdot$ 5H$_2$O

Azure blue chalcanthite crystal with minor light green alteration spots (upper). Chalcanthite crystal dehydrated to a powdery light green hydrous copper sulfate, CuSO$_4 \cdot$ H$_2$O (lower).

A   x 0.60

THENARDITE

Na_2SO_4

Twinned crystals of thenardite: Borax Lake, California.

B   x 0.70

LINARITE

$PbCu(SO_4)(OH)_2$

Deep azure blue botryoidal crusts of linarite: Inyo County, California.

C   x 0.53

ALUNITE

$KAl_3(SO_4)_2(OH)_6$

Massive flesh-colored alunite: Hyogoken, Japan.

D   x 0.62

COPIAPITE

$(Fe, Mg)Fe_4(SO_4)_6(OH)_2 \cdot 20H_2O$

Scaly aggregate of copiapite, pearly luster: Copiapó, Chile.

E   . x 0.60

POLYHALITE

$K_2Ca_2Mg(SO_4)_4 \cdot 2H_2O$

Polyhalite and halite. The red color of polyhalite is caused by impurities: western Texas.

F   x 0.70

BROCHANTITE

$Cu_4(SO_4)(OH)_6$

Fibrous veinlets of dark green brochantite: Chile.

Composition. Sulfide of copper and iron, $CuFeS_2$ (34.5% Cu, 30.5% Fe, 35% S). Commonly contains small amounts of silver and gold.

Tests. Fuses on charcoal to a magnetic globule. Soluble in concentrated HNO_3; the solution, neutralized with ammonia, yields a reddish-brown iron hydroxide precipitate in a deep blue solution.

Crystallization. Tetragonal; scalenohedral class, $\overline{4}\, 2\, m$. The characteristic crystals are sphenoidal but resemble tetrahedrons. Usually massive.

Associated Minerals. Pyrite, galena, sphalerite, chalcocite, malachite, covellite, quartz, calcite.

Alteration. Alters to chalcocite, covellite, malachite, and iron oxides.

Minerals of Similar Appearance

Pyrite. Harder than chalcopyrite (H. pyrite = $6-6\frac{1}{2}$, cannot be scratched by a knife). On a fresh surface, pyrite is a lighter brass yellow than chalcopyrite.

Gold. Resembles chalcopyrite in color, but is sectile, and has a pale, metallic gold streak. Chalcopyrite is brittle, and has a greenish-black streak.

Occurrence. Chalcopyrite is a common mineral of widespread occurrence in metallic sulfide vein deposits, disseminated copper deposits, contact metamorphic deposits, and magmatic segregation deposits. A few of the many important localities are Cornwall, England; Sudbury, Ontario, Canada; Butte, Montana; Bingham, Utah; and Jerome, Arizona.

Use. An important ore of copper.

COVELLITE

CuS   Hexagonal
    Hardness $1\frac{1}{2}-2$ Sp. Gr. 4.6–4.76

Summary

A soft, *dark indigo blue* mineral, subject to iridescent tarnish; usually in platy masses that split into thin flexible cleavage flakes (Plate 15(B), facing p. 98). *Luster:* submetallic. *Streak:* lead gray to black. *Cleavage:* perfect basal. Usually associated with copper minerals like chalcocite, bornite, and enargite, in zones of secondary sulfide enrichment. INDIGO BLUE COLOR AND PERFECT MICACEOUS CLEAVAGE ARE DISTINCTIVE.

Composition. Copper sulfide, CuS (66.4% Cu, 33.6% S). May contain small amounts of iron.

Tests. On charcoal, burns with a blue flame and fuses to a globule.

Crystallization. Hexagonal; dihexagonal dipyramidal class, $\frac{6}{m}\frac{2}{m}\frac{2}{m}$. Thin tabular hexagonal crystals are rare. Usually in platy masses; often intergrown with chalcopyrite or chalcocite.

Alteration. Covellite is derived from chalcopyrite, chalcocite, enargite, or bornite.

Occurrence. Covellite is a fairly uncommon mineral occasionally found with other copper sulfides in veins. Good crystals come from Alghero, Sardinia; and Butte, Montana.

Use. A minor ore of copper.

CHALCOCITE (Copper Glance)

Cu_2S Orthorhombic
    Hardness $2\frac{1}{2}-3$ Sp. Gr. 5.5–5.8

Summary

A soft, subsectile, *blackish, lead gray* mineral that tarnishes readily to dull black; frequently coated with a thin alteration film of green malachite or blue azurite. *Luster:* metallic. *Streak:* shining, dark lead gray. *Fracture:* conchoidal. Usually massive. Crystals are rare (Plate 15(C), facing p. 98). BLACKISH-GRAY COLOR, SHINING STREAK, SOFTNESS, SUBSECTILITY, AND ASSOCIATION WITH OTHER COPPER MINERALS ARE DISTINCTIVE.

Composition. Cuprous sulfide, Cu_2S (79.8% Cu, 20.2% S).

Tests. Fuses easily. After roasting with charcoal, gives a globule of copper in the reducing flame. Soluble in nitric acid, forming a green solution that becomes deep blue when neutralized with ammonium hydroxide.

Crystallization. Orthorhombic; dipyramidal class, $\frac{2}{m}\frac{2}{m}\frac{2}{m}$, below 105°C (above 105°C the hexagonal form of Cu_2S is stable). Commonly occurs in masses; sometimes soft and sooty. Rarely in small, striated tabular crystals. Twinned crystals usually look hexagonal. Twinning also produces geniculated (knee-like) twins and cross-like penetration twins.

Associated Minerals. Such copper minerals as azurite, malachite, cuprite, copper, enargite, chalcopyrite, tetrahedrite, covellite, and bornite.

Alteration. Alters to copper, malachite, azurite, covellite.

Minerals of Similar Appearance

Argentite. Resembles chalcocite in color and hardness, but is more sectile, and differs in habit and mineral associations.

Copper sulfides. Massive chalcocite can be distinguished by its sectility and shining streak from other common, soft, blackish-gray copper sulfides, such as enargite.

Digenite, $Cu_{2-x}S$. Isometric. Brittle, and usually deep blue in color; sometimes black. May be intergrown with chalcocite.

Occurrence. 1. Occurs frequently in the secondary enriched zone of copper-sulfide deposits (Rio Tinto, Spain; Morenci, Arizona; Ely, Nevada; Bingham, Utah; and elsewhere). 2. Also occurs in hydrothermal sulfide veins (Butte, Montana).

Exceptional crystals of chalcocite have been found at Bristol, Connecticut; and in Cornwall, England, and Messina, Transvaal, South Africa. Large chalcocite deposits occur at Tsumeb, South-West Africa; and in Chile, Peru, and Mexico.

Use. A very important ore of copper.

BORNITE

Cu_5FeS_4	Isometric
Hardness 3	Sp. Gr. 5.06–5.08

Summary

A brittle, metallic, *reddish-bronze* mineral that tarnishes quickly on exposure to iridescent blue and purple. *Streak:* gray-black. *Cleavage:* poor octahedral. Usually massive (Plate 15(*D*), facing p. 98). Commonly intermixed with chalcopyrite and chalcocite. REDDISH-BRONZE COLOR, ASSOCIATION WITH OTHER COPPER MINERALS, AND IRIDESCENT TARNISH ARE DISTINCTIVE. Because of its unusual tarnish, bornite is descriptively referred to as "peacock ore."

Composition. Sulfide of copper and iron, Cu_5FeS_4 (Cu 63.3%, Fe 11.2%, S 25.5%). May contain small amounts of lead, gold, and silver.

Tests. Fuses on charcoal to a brittle magnetic globule. Soluble in nitric acid. Gives tests for copper and iron.

Crystallization. Isometric; hexoctahedral class, $\frac{4}{m}\bar{3}\frac{2}{m}$. Aggregates of small, rough cubic crystals are rare. Usually massive.

Associated Minerals. Other copper minerals like chalcocite, chalcopyrite, enargite, covellite. Also pyrite, arsenopyrite, marcasite, and quartz.

Alteration. Alters to chalcocite, chalcopyrite, cuprite, covellite, chrysocolla, azurite, and malachite.

Occurrence. Bornite is widespread in copper deposits. It occurs frequently as a hypogene mineral in copper sulfide veins; less frequently as a supergene mineral in zones of secondary enrichment. Found in pegmatites, contact metamorphic deposits, replacement deposits. Also disseminated in basic rocks. Drusy crystals of bornite have been found at Bristol, Connecticut; at Butte, Montana; and at Cornwall, England. Large deposits occur in Mexico, Peru, Chile, Tasmania, Butte, and at the Magma Mine in Superior, Arizona.

Use. An ore of copper.

ENARGITE

Cu₃AsS₄   Orthorhombic
   Hardness 3   Sp. Gr. 4.4–4.5

Summary

A brittle, *iron black,* metallic mineral usually in bladed aggregates or in granular masses; sometimes in prismatic crystals (Plate 15(*E*), facing p. 98). *Streak:* black. *Cleavage:* perfect prismatic. Usually associated with other copper minerals (chalcocite, bornite, covellite, and tetrahedrite) in moderate-temperature vein and replacement deposits. BRITTLE, BLADED AGGREGATES WITH GOOD PRISMATIC CLEAVAGE ARE DISTINCTIVE.

Composition. A copper arsenic sulfide, Cu₃AsS₄; Sb substitutes for As to about 6 percent by weight; Fe and Zn are commonly present. The mineral FAMATINITE (Cu₃SbS₄) is the antimony analogue of enargite, but may differ in crystallization.

Tests. Fuses easily on charcoal, yielding white arsenious oxide fumes with a garlic (arsenic) odor. With borax fluxes, the roasted mineral yields a copper globule.

Crystallization. Orthorhombic; pyramidal class, 2 *m m*. Commonly in bladed or granular masses. Crystals usually prismatic with flat bases; also tabular.

Associated Minerals. Pyrite, sphalerite, galena, tetrahedrite, bornite, covellite, chalcocite, barite, quartz.

Alteration. May be replaced by tennantite (Cu₁₂As₄S₁₃).

Minerals of Similar Appearance

Chalcocite. Massive, grayish-black chalcocite is subsectile, and has a shining, dark gray streak. Enargite is brittle and cleavable.

Tetrahedrite. Has no cleavage. Enargite has perfect prismatic cleavage, and is blacker than tetrahedrite.

Stibnite. Striations across the cleavage faces of stibnite are distinctive. Bladed aggregates of black enargite resemble bladed stibnite masses, but stibnite is usu-ally lighter in color. Test for copper also distinguishes enargite from stibnite.

Occurrence. Usually found with other copper minerals in moderate-temperature vein and replacement deposits. Notable localities for enargite are Butte, Montana; Bingham and Tintic, Utah; Chuquicamata, Chile; Peru, and Mexico.

Use. A copper ore.

TETRAHEDRITE

(Cu,Fe,Zn,Ag)₁₂(Sb,As)₄S₁₃   Isometric
   Hardness 3–4½   Sp. Gr. 4.6–5.1

Summary

A brittle, *dark gray* to *iron-black,* metallic mineral; usually in granular masses (Plate 15(*F*), facing p. 98). *Streak:* grayish-black. *Cleavage:* none. Good crystals show a characteristic tetrahedral habit (Figure 8-14). The crystals are sometimes coated with yellow chalcopyrite. Usually in moderate- to low-temperature hydrothermal copper, lead, zinc, and silver veins. Frequently associated with chalcopyrite.

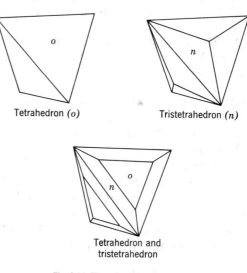

Tetrahedron (*o*)   Tristetrahedron (*n*)

Tetrahedron and tristetrahedron

Fig. 8-14. Tetrahedrite crystals.

pyrite, galena, sphalerite, and argentite. BRITTLENESS, GRAY COLOR, AND TETRAHEDRAL CRYSTALS ARE DISTINCTIVE.

Composition. Essentially a copper, iron, zinc, antimony, and arsenic sulfide $(Cu,Fe,Zn,Ag)_{12}(Sb,As)_4S_{13}$. Copper is the dominant metal. Small amounts of mercury may also be present. Arsenic can substitute for antimony in all proportions, thus forming a complete solid solution series with antimony (*tetrahedrite*) and arsenic (*tennantite*) end members. The dividing line between the two species is antimony:arsenic = 1:1. A variety rich in silver is called *freibergite*. Other varieties of tetrahedrite may contain mercury, lead, bismuth, nickel, or cobalt.

Tests. Fuses easily on charcoal to a brittle globule; evolves white antimony and arsenic fumes. Globule when dissolved in nitric acid and neutralized with ammonia yields a red-brown (iron) precipitate in a deep blue (copper) solution.

Crystallization. Isometric; hextetrahedral class, $\overline{4}\,3\,m$. Usually in granular masses. Crystals are commonly tetrahedral; sometimes in parallel groups.

Associated Minerals. Chalcopyrite, bornite, galena, pyrite, sphalerite, argentite, proustite, pyrargyrite, calcite, siderite, fluorite, barite, and quartz.

Alteration. Commonly alters to malachite, azurite, and bindheimite.

Occurrence. Tetrahedrite is a widespread and important economic mineral that is usually found in moderate- to low-temperature hydrothermal Cu, Pb, Zn, and Ag veins. It is frequently an important copper ore, and when rich in silver becomes a valuable silver ore. Notable localities are Freiberg, Saxony; Harz Mountains, Germany; Botés and Kapnik, Roumania; Příbram, Bohemia; Cornwall, England; and Bolivia; Chile; and Peru. Found in the United States in the silver, copper, and lead deposits of Idaho, Utah, Colorado, Montana, Nevada, New Mexico, Arizona, and California. Tennantite is rarer than tetrahedrite. At Tsumeb, South-West Africa, tennantite occurs intimately associated with germanite, Cu_3GeS_4.

Use. An ore of copper and silver.

ORPIMENT

As_2S_3 Monoclinic
Hardness $1\frac{1}{2}$–2   Sp. Gr. 3.49

Summary

A soft, sectile, *lemon yellow* mineral; commonly occurs in foliated masses that split into thin flexible cleavage flakes (Plate 16(*A*), facing p. 99). *Luster:* resinous to pearly. *Streak:* pale yellow. *Cleavage:* perfect in one direction; translucent. Commonly associated with realgar and stibnite in low-temperature deposits. LEMON YELLOW COLOR, PERFECT MICACEOUS CLEAVAGE, AND ASSOCIATION WITH ORANGE-RED REALGAR ARE DISTINCTIVE. Cleavage distinguishes orpiment from SULFUR.

Composition. Arsenic trisulfide, As_2S_3 (61.0% As, 39.0% S).

Tests. Fuses easily; when heated on charcoal, volatilizes completely, giving off white fumes of arsenious oxide with a garlic odor.

Crystallization. Monoclinic; prismatic class, $\frac{2}{m}$. Small prismatic crystals are rare. Occurs commonly in foliated, columnar, or fibrous masses.

Associated Minerals. Realgar, arsenic, stibnite, calcite, barite, gypsum, and ores of lead and silver.

Alteration. As an alteration product of realgar and arsenic.

Occurrence. Orpiment is a low-temperature mineral; found with realgar, and is formed under similar conditions. Occurs in hydrothermal veins; in hot spring deposits; as a volcanic sublimate; and as replacements in sedimentary rocks. Fine crystals occur at Mercur, Utah; and at Manhattan, Nevada.

REALGAR

AsS   Monoclinic
Hardness $1\frac{1}{2}$–2   Sp. Gr. 3.56

Summary

A soft, sectile, *orange-red* mineral that alters readily to yellow orpiment. *Streak:* orange-red. *Luster:* resinous; transparent to translucent. *Cleavage:* good. Crystals are rare. Usually in granular masses (Plate 16(*B*), facing p. 99). ORANGE-RED COLOR AND STREAK, RESINOUS LUSTER, GARLIC ODOR WHEN POWDERED, AND ASSOCIATION WITH ORPIMENT ARE DISTINCTIVE.

Composition. Arsenic sulfide, AsS (70.1% As, 29.9% S).

Tests. Fuses easily; when heated on charcoal, volatilizes completely, giving off white fumes of arsenious oxide with a garlic odor.

Crystallization. Monoclinic; prismatic class, $\frac{2}{m}$. Usually in granular masses; also earthy. Short, prismatic, vertically striated crystals are rare.

Associated Minerals. Orpiment, stibnite, and ores of lead, silver, and gold.

Alteration. On continued exposure to light, disintegrates to a reddish-yellow powder, which is a mixture of orpiment (As_2S_3) and arsenolite (As_2O_3).

Minerals of Similar Appearance

Cinnabar. Distinguishable from realgar by high specific gravity (8.09), and a moderate color difference.

Crocoite. An orange-red mineral with a brilliant luster and high specific gravity (5.9–6.1). Striated, elongated crystals are distinctive. Realgar is similar in color, but is usually massive, and can be scratched with a finger nail.

Occurrence. Occurs in hot spring deposits (Norris Basin, Yellowstone National Park, Wyoming); also as a volcanic sublimate. Found as a low-temperature mineral in hydrothermal veins (with lead and silver ores in Nagyág, Roumania). Occasionally in limestones and dolomites (occurs in good crystals at Binnental, Switzerland). Fine crystals occur in cavities with calcite at Mercur, Utah. Also at Manhattan, Nevada.

ARSENOPYRITE

FeAsS	Monoclinic
Hardness 5½–6	Sp. Gr. 6.07

Summary

A brittle, *silver gray,* metallic mineral; often in striated crystals with a rhombic cross section. Also frequently in twinned crystals, and in granular masses. *Streak:* black. *Cleavage:* prismatic. May yield a garlic odor when struck with a steel hammer or a pick. Some twinned crystals may resemble "spearhead" marcasite in shape (Figure 8-15). Resembles massive silver-white iron-cobalt-nickel arsenides and sulfides. SILVER-GRAY COLOR, CRYSTAL HABIT, AND GARLIC (ARSENIC) ODOR ARE DISTINCTIVE.

Composition. Essentially, an iron arsenide-sulfide, FeAsS, but may contain some cobalt. When the ratio of cobalt to iron is between 2:1 and 6:1, the mineral is referred to as *glaucodot,* (Co,Fe)AsS.

Tests. A brilliant gray "arsenic mirror" forms in a closed tube on the walls, with black arsenic sulfide nearby. On charcoal, gives off white fumes of arsenious oxide with a garlic odor. Decomposed in nitric acid with separation of sulfur. May yield a garlic odor when powdered.

Crystallization. Monoclinic; prismatic class, $\frac{2}{m}$. Commonly in striated pseudoorthorhombic crystals; also in twinned crystals that resemble marcasite; often in granular masses.

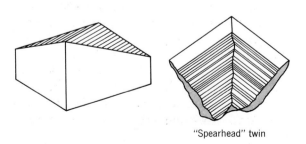

"Spearhead" twin

Fig. 8-15. Arsenopyrite crystals.

Associated Minerals. Wolframite, scheelite, molybdenite, cassiterite, gold, silver, galena, pyrite, and chalcopyrite.

Minerals of Similar Appearance

Marcasite. Twinned marcasite crystals may resemble arsenopyrite in shape. Arsenopyrite is silver-gray.

Cobaltite and **Skutterudite.** These two minerals closely resemble arsenopyrite when massive, but their isometric crystals are distinctive.

Safflorite, Rammelsbergite, Loellingite, and **Glaucodot.** Crystals of these hard, silver-colored Co-Ni-Fe arsenides and sulfides resemble arsenopyrite crystals, and are difficult to distinguish without x-ray patterns or chemical tests. A negative test for nickel helps to distinguish arsenopyrite from rammelsbergite ($NiAs_2$).

Alteration. Commonly alters to pale greenish scorodite, $Fe(AsO_4) \cdot H_2O$; the cobaltian varieties alter to pinkish erythrite, $(Co,Ni)_3(AsO_4)_2 \cdot 8H_2O$.

Occurrence. Arsenopyrite is an abundant arsenic mineral. It commonly occurs in high-temperature deposits, and is usually one of the first sulfide minerals to crystallize. Common in the tin deposits of Cornwall, England, and in those of Bolivia. Found with included gold at Boliden, Sweden. Frequently found in high-temperature gold-quartz veins; also found in contact metamorphic deposits. Good crystals have come from the silver-nickel mines of Freiberg, Saxony. Cobaltian arsenopyrite occurs in Sulitjelma, Norway, and in Nordmark, Sweden. In the United States, good crystals occur at Roxbury, Connecticut; at Franconia, New Hampshire; and at Franklin, New Jersey. Large spheroids consisting of alternating shells of niccolite and arsenopyrite occur at Natsume, Japan. Found in large masses in quartz at Deloro, Ontario, Canada.

STIBNITE

Sb_2S_3 Orthorhombic
    Hardness 2   Sp. Gr. 4.63

Summary

A soft, *steel gray,* metallic mineral, commonly occurring in long, bladed, striated crystals (Plate 16 (*E*), facing p. 99). Crystals are often bent, and subject to black iridescent tarnish (Figure 8-16). *Streak:* lead gray. *Cleavage:* parallel to length, cleavage faces have a brilliant metallic luster; subsectile. Resembles *bismuthinite.* **DISTINCTIVE CHARACTERISTICS ARE ELONGATED STRIATED CRYSTALS, STEEL-GRAY COLOR, SOFTNESS, AND PERFECT CLEAVAGE IN ONE DIRECTION WITH CROSS STRIATIONS.**

Composition. Antimony trisulfide, Sb_2S_3 (71.7% Sb, 28.3% S). Sometimes contains small amounts of Cu, Pb, Fe, Ag, Au, Co, Zn.

Tests. Fuses easily (M.P. 550°C). Volatilizes on charcoal, leaving a dense white coating. Powdered stibnite is soluble in concentrated HCl. A few drops of dissolved mineral, added to a solution of potassium iodide, produce a bright orange precipitate.

Crystallization. Orthorhombic; dipyramidal class, $\frac{2}{m}\frac{2}{m}\frac{2}{m}$. Commonly in large, vertically striated prismatic crystals; also in aggregates of slender, elongated crystals, and in granular masses.

Associated Minerals. Orpiment, realgar, galena, cinnabar, marcasite, pyrite, ankerite, calcite, barite, and chalcedony.

Alteration. Alters to cherry red kermesite; pale yellow stibiconite; and white valentinite. Pseudomorphs after stibnite may form.

Cleavage→

Cross striations

Fig. 8-16. Curved stibnite crystal.

Occurrence. Stibnite is a low-temperature mineral, occurs in hydrothermal veins, in hot spring deposits, and in replacement deposits. Beautiful groups of large crystals, some over a foot long, come from the mines of Ichinokawa, Iyo Province, Shikoku, Japan. Radiating aggregates of stibnite with barite occur at Felsö-bánya and Kapnik in Roumania. Important antimony ore deposits bearing stibnite are located in the provinces of Hunan and Kwantung in China. A common associate of mercury ores.

Use. The major ore of antimony. Antimony compounds are used in medicine, in the manufacture of matches and fireworks, in vulcanizing rubber, and in pigments. Antimony is alloyed with lead for storage battery plates; with lead and tin for type metal; and with tin and copper in antifriction metal.

BISMUTHINITE

Bi_2S_3 Orthorhombic
    Hardness 2   Sp. Gr. 6.8

Summary

A soft, *lead gray,* metallic mineral that resembles stibnite. Usually occurs in foliated or fibrous masses (Plate 16(*F*), facing p. 99). *Streak:* lead gray. *Cleavage:* perfect; slightly sectile. Commonly associated with bismuth and cassiterite.

Composition. Bismuth trisulfide, Bi_2S_3 (81.3% Bi, 18.7% S). May contain small amounts of Pb, Cu, Fe.

Tests. Fuses easily; when mixed with iodide flux and heated on charcoal, gives bright red coating. Dissolves in nitric acid; forms a white precipitate when diluted with water.

Crystallization. Orthorhombic; dipyramidal class, $\frac{2}{m}\frac{2}{m}\frac{2}{m}$. Usually in fibrous or foliated masses. Rarely in elongated, needle-like crystals.

Associated Minerals. Native bismuth, arsenopyrite, quartz, tourmaline, cassiterite, wolframite.

Minerals of Similar Appearance

Stibnite. Bismuthinite is distinguished from massive or acicular stibnite by tests for bismuth; bismuthinite has slightly darker color, and higher specific gravity.

Occurrence. Bismuthinite is an uncommon mineral found in high-temperature hydrothermal veins and granite pegmatites. Occurs with tin and tungsten in Bolivia.

Use. An ore of bismuth.

MOLYBDENITE

MoS_2 Hexagonal
    Hardness 1–1½   Sp. Gr. 4.62–4.73

Summary

A very soft, sectile, *bluish-gray, metallic* mineral with a greasy feel. Commonly occurs in foliated masses or scales (Plate 16(*D*), facing p. 99); sometimes in tabular six-sided crystals, or in fine granular masses. *Streak:* shining bluish-gray; greenish on glazed porcelain or glossy paper. *Cleavage:* perfect basal; easily separated into thin flexible plates. Closely resembles *graphite.* BLUISH-GRAY COLOR, SOFTNESS, METALLIC LUSTER, GREASY FEEL, AND PERFECT MICACEOUS CLEAVAGE ARE DISTINCTIVE.

Composition. Molybdenum sulfide, MoS_2 (60.0% Mo, 40.0% S).

Tests. Infusible (melting point 1185°C). When heated on charcoal in the oxidizing flame, gives a red coating near assay and, next to it, a yellow coating that turns white on cooling. When the white coating is touched with the reducing flame, it turns azure blue.

Crystallization. Commonly in foliated masses or in thin scales; also in tabular hexagonal crystals, and in fine granular masses.

Associated Minerals. Scheelite, wolframite, topaz, beryl, quartz, fluorite, arsenopyrite, chalcopyrite, and pyrite.

Minerals of Similar Appearance

Graphite. Grayish-black graphite closely resembles molybdenite, but molybdenite is bluish-gray and has a greenish streak on glazed porcelain. Graphite has a black streak.

Alteration. Alters to yellow ferrimolybdite; powellite (luminesces yellow); and ilsemannite (molybdenum blue).

Occurrence. Molybdenite is a high-temperature mineral, and an early metallic mineral to crystallize. It occurs as an accessory mineral in certain granites and in pegmatites. Also in high-temperature, deep-seated hydrothermal veins where it is associated with minerals like wolframite, scheelite, topaz, beryl, cassiterite, fluorite, and quartz; also in contact metamorphic deposits. The most important molybdenite deposit is at Climax, Colorado, where the molybdenite occurs with topaz and fluorite in disseminated particles, and as veinlets in granite. Often produced as a byproduct in copper mines. It is mined at Cuesta, New Mexico.

Use. Major ore of molybdenum.

OXIDES AND HYDROXIDES

Oxides such as cuprite (Cu_2O) or corundum (Al_2O_3) are compounds of a metallic element and oxygen. MULTIPLE OZIDES such as chromite ($FeCr_2O_4$), or columbite ($(Fe,Mn)Cb_2O_6$) have two metal ions coordinated with oxygen. The oxides are usually simple in composition, but range widely in physical properties.

HYDROXIDES contain the hydroxyl radical (OH). Some common hydroxides like goethite $FeO(OH)$ or psilomelane $BaMnMn_8O_{16}(OH)_4$ result from the weathering of primary minerals.

Simple Oxides

Cuprite, Cu_2O
Ice, H_2O
Zincite, ZnO
Hematite Group
    Corundum, Al_2O_3
    Hematite, Fe_2O_3

Ilmenite Series
    Ilmenite, $FeTiO_3$
    Geikielite, $MgTiO_3$
    Pyrophanite, $MnTiO_3$
Rutile Group
    Cassiterite, SnO_2
    Pyrolusite, MnO_2
    Rutile, TiO_2
Anatase (Octahedrite), TiO_2
Brookite, TiO_2
Uraninite, UO_2

Multiple Oxides

Spinel Group
  Spinel Series
    Spinel, $MgAl_2O_4$
    Gahnite, $ZnAl_2O_4$
  Magnetite Series
    Magnetite, Fe_3O_4
    Franklinite, $(Zn,Fe,Mn)(Fe,Mn)_2O_4$
  Chromite Series
    Chromite, $FeCr_2O_4$
Chrysoberyl, $BeAl_2O_4$
Columbite-Tantalite Series
    Columbite, $(Fe,Mn)(Cb,Ta)_2O_6$
    Tantalite, $(Fe,Mn)(Ta,Cb)_2O_6$

Hydroxides

Goethite Group
    Diaspore, $AlO(OH)$
    Goethite, $FeO(OH)$
    Limonite
Manganite, $MnO(OH)$
Psilomelane, $BaMn^{+2}Mn^{+4}_8O_{16}(OH)_4$
Brucite, $Mg(OH)_2$
Gibbsite, $Al(OH)_3$
    Bauxite
Heterogenite (a hydrous cobalt oxide)

CUPRITE (Ruby Copper; Red Copper Ore)

Cu_2O Isometric
    Hardness $3\frac{1}{2}$–4   Sp. Gr. 6.1

Plate 30. CHROMATES, MOLYBDATES, TUNGSTATES ▪ WULFENITE, CROCOITE, WOLFRAMITE, HUEBNERITE, SCHEELITE

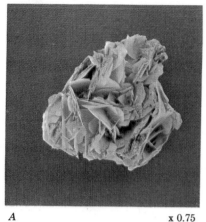

A   x 0.75

WULFENITE
PbMoO₄

Aggregate of very thin divergent wulfenite crystals: Mammoth Mine, Tiger Arizona.

B   x 0.68

WULFENITE

Thin tabular wulfenite crystals: Copperopolis, Arizona.

C   x 0.65

CROCOITE
PbCrO₄

Aggregate of prismatic crocoite crystals showing characteristic deep orange color and brilliant luster: Dundas, Tasmania.

D   x 0.4

WOLFRAMITE
(Fe, Mn)WO₄

Black tabular wolframite crystal: Zinnwald, Bohemia.

E   x 0.5

HUEBNERITE
MnWO₄

Brown thick tabular huebnerite crystal with quartz.

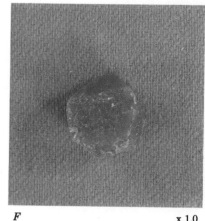

F   x 1.0

SCHEELITE
CaWO₄

Scheelite crystal (tetragonal bipyramid).

Plate 31. PHOSPHATES, ARSENATES, VANADATES ■ APATITE, COLLOPHANE, PYROMORPHITE, MIMETITE, VANADINITE

A   x 0.34

APATITE

Ca₅(F, Cl, OH)(PO₄)₂

Six-sided mottled brown-green apatite crystal showing typical "shattered" appearance: Rossie, New York.

B   x 0.5

COLLOPHANE

(massive cryptocrystalline apatite)

Collophane showing colloform structure: Dunnellon, Louisiana.

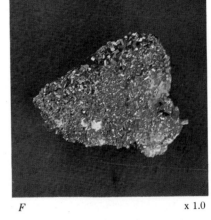

C   x 0.75

MIMETITE

Pb₅(AsO₄)₃Cl (campylite variety)

Aggregate of orange-brown melon-shaped crystals: Cumberland, England.

D   x 0.6

PYROMORPHITE

Pb₅(PO₄)₃Cl

Aggregate of hollow six-sided brown pyromorphite crystals: Oberlahnstein.

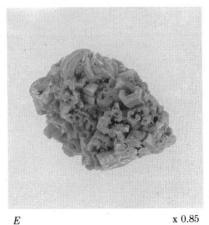

E   x 0.85

PYROMORPHITE

Cavernous crystals of pyromorphite showing typical hollow hexagonal prisms, Brisgourd, Germany.

F   x 1.0

VANADINITE

Pb₅(VO₄)₃Cl

Minute, bright red six-sided crystals of vanadinite with brilliant luster: Old Yuma Mine, Pima County, Arizona.

Summary

A *red* to *blackish-red* mineral, commonly in octahedral and cubic crystals (Figure 8-17); also massive, and in aggregates of capillary or hair-like crystals known as CHALCOTRICHITE (Plate 17, facing p. 114). Frequently alters to green malachite or copper. *Luster:* adamantine to submetallic to dull. *Streak:* bright brownish-red. *Cleavage:* poor octahedral; brittle. Found in the oxidized zone of copper deposits associated with limonite and secondary copper minerals; examples: malachite, azurite, native copper, chalcocite, and chrysocolla. DARK RED COLOR, BRIGHT LUSTER, CRYSTAL FORM, AND ASSOCIATED COPPER MINERALS ARE DISTINCTIVE. Resembles CINNABAR and HEMATITE.

Composition. Cuprous oxide, Cu_2O (88.8% Cu, 11.2% O).

Tests. Fuses on charcoal, and is reduced to a metallic copper globule. Soluble in concentrated HCl, and yields a white precipitate of cuprous chloride when solution is cooled and diluted with cold water.

Crystallization. Isometric; gyroidal class, 4 3 2. Commonly in octahedral crystals or in combinations of octahedral and cubic crystals; sometimes dodecahedral; also in cubes greatly elongated into needle-like or capillary shapes (chalcotrichite). Massive; in impure reddish-brown earthy masses, often mixed with hematite and limonite.

Alteration. Commonly alters to malachite and copper. Found as an alteration product of tetrahedrite and chalcopyrite.

Minerals of Similar Appearance

Cinnabar. Distinguishable from cuprite by high specific gravity (8.1) and scarlet color and streak. Impure cinnabar may resemble cuprite in color and streak.

Hematite. May resemble massive cuprite in color and red-brown streak; but cuprite is usually associated with other copper minerals, notably copper and blue and green copper carbonates (azurite, malachite).

Ruby Silvers (proustite; pyrargyrite). Distinguishable from cuprite by crystal shape and chemical tests.

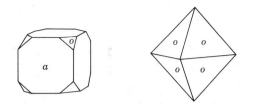

Fig. 8-17. Cuprite crystals (forms: *a*, cube; *o*, octahedron).

Occurrence and Associated Minerals. Cuprite is a widespread and important secondary ore of copper. It is found in the upper portions of copper sulfide deposits, formed by the oxidation of these copper minerals. Occurs with other secondary copper minerals like copper, azurite, malachite, chalcocite, chrysocolla, and tenorite (black copper oxide, CuO). Also occurs in clays and limonite. Pseudomorphs of malachite after octahedral cuprite crystals are common (Plate 4, facing p. 19). Cuprite crystals are sometimes found growing upon copper. Good crystals of cuprite have come from Cornwall, England; and Chessy, France. They are also found in Arizona at Bisbee, Morenci, Clifton, and Globe. It is an important ore of copper at Bisbee. Other notable localities are Cobar and Broken Hill, New South Wales, Australia; Coro-Coro, Bolivia; Chuquicamata, Chile; Congo, Africa; and the Ural Mountains, U.S.S.R.

Use. An ore of copper.

ICE

H_2O Hexagonal

Crystallization. Hexagonal; ditrigonal pyramidal class, 3 *m*. Occurs in masses, granular aggregates, and crystals. Snow crystals occur in a great variety of delicate, lacy, six-armed dendrites (skeleton crystals). Experiments indicate that the basic habits of ice crystals vary with the temperature of the air in which they form (see p. 23). At atmospheric pressure between 0° and −160°, three polymorphic forms of ice are known. High-pressure polymorphs also occur.

Physical Properties (Ice)

Hardness, $1\frac{1}{2}$
Specific Gravity, 0.916
Colorless to white; bluish in thick layers
Luster, vitreous
Fracture, conchoidal
Cleavage, none; but good basal gliding
Brittle, especially at low temperatures
Transparent

Occurrence. Snow, hail, and frost are precipitated from water vapor in the atmosphere when the temperature falls below the freezing point of water. Below the freezing point, ice also forms as coatings over bodies of water. Large masses of semipermanent ice are found in the ice caps of Greenland and Antarctica, and in mountain glaciers.

ZINCITE

ZnO   Hexagonal
Hardness 4   Sp. Gr. 5.68

Summary

A *deep red* to *orange-yellow* mineral, associated with white and pink calcite, pale green willemite, and black franklinite; occurs at Franklin and Sterling Hill, New Jersey, the important zincite localities. Usually massive (Plate 17(*D*), facing p. 114); crystals are rare. *Luster:* subadamantine. *Streak:* orange-yellow. *Cleavage:* perfect prismatic. *Fracture:* conchoidal; translucent; brittle. ORANGE-YELLOW STREAK, EVEN IN DEEP RED VARIETIES, AND ASSOCIATION WITH CALCITE, WILLEMITE, AND FRANKLINITE ARE DISTINCTIVE.

Composition. Zinc oxide, ZnO (80.3% Zn, 19.7% O). Contains small amounts of manganese.

Tests. Infusible. Powdered mineral when mixed with sodium carbonate and heated on charcoal gives a coating of zinc oxide that is yellow when hot and turns white on cooling. The white coating turns green when moistened with cobalt nitrate and heated in the oxidizing flame (test for zinc).

Crystallization. Hexagonal; dihexagonal pyramidal, $6\ m\ m$. Usually massive. Crystals are unique and rare. They are hemimorphic with a steep pyramid at one end and a flat base at the opposite end.

Occurrence. Zincite is a rare mineral except at the zinc deposits of Franklin and Sterling Hill, New Jersey, where it occurs with white calcite that luminesces red, yellowish green luminescent willemite and black octahedral crystals or masses of franklinite. Zincite has been reported in small amounts from Poland, Tuscany, Spain, and Tasmania.

Use. An ore of zinc.

CORUNDUM

Al_2O_3 Hexagonal
Hardness 9   Sp. Gr. 4.0–4.1

Summary

Next to diamond corundum is the hardest mineral. It occurs in a wide range of colors, with crystals frequently showing a zonal color distribution. Common corundum is usually *gray, gray-blue,* or *brown.* The black, abrasive EMERY is a mixture of fine-grained corundum and magnetite. RUBY is a deep red variety of corundum. SAPPHIRE is commonly blue. The less popular varieties of gem corundum are frequently named after other gems with the prefix *oriental.* Thus, yellow corundum is called *oriental topaz,* green corundum is called *oriental emerald,* and the violet variety is called *oriental amethyst.*

Some rare varieties of corundum have needle-like oriented inclusions of rutile that reflect light; thus, a spot source of light on a rounded polished surface reflects six rays in a star effect (star ruby, star sapphire; see p. 85).

Corundum commonly occurs in large six-sided crystals, frequently barrel shaped with rough, rounded surfaces (Plate 17(*E*), facing p. 114). Sets of fine striations are often caused by repeated twinning on the rhombohedron, and sometimes parallel to the base (Figure 8-18). *Cleavage:* none; but good rhombohedral

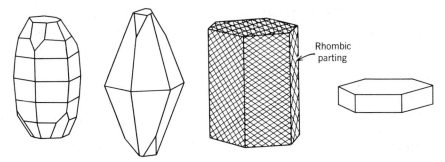

Rhombic parting

Fig. 8-18. Corundum crystals.

and basal parting exists. *Luster:* adamantine to vitreous; transparent to translucent; brittle. Sometimes luminesces red or orange-yellow under ultraviolet light. GREAT HARDNESS, SIX-SIDED CRYSTALS, AND SETS OF INTERSECTING PARALLEL STRIATIONS ON PARTING SURFACES ARE DISTINCTIVE.

Composition. Aluminum oxide, Al_2O_3 (52.9% Al, 47.1% O). Small amounts of Mn, Cr, Fe, Ti may replace Al. The red color of ruby is caused by small amounts of chromium oxide; and the blue of sapphire by iron and titanium oxides.

Tests. Infusible; insoluble in acids.

Crystallization. Hexagonal; scalenohedral class, $\overline{3}\frac{2}{m}$. Commonly in six-sided crystals with flat terminations; also in tapering hexagonal pyramids, frequently rounded into barrel shapes. Often in coarse- or fine-grained masses. Sometimes in thin, tabular hexagonal crystals.

Associated Minerals. Magnetite, hematite, kyanite, spinel, tourmaline, cordierite, micas, chlorites, garnet, oligoclase, anorthite, nepheline.

Alteration. Alters to such other aluminous minerals as kyanite, sillimanite, zoisite, and margarite.

Synthetic Corundum. Synthetic rubies and sapphires have been manufactured by the Verneuil process since 1902. The process consists in melting finely powdered aluminum oxide (Al_2O_3) in an oxy-hydrogen flame at temperatures of 2100° to 2200°C. The molten drops of alumina collect on a stage below the flame to form a carrot-shaped mass or "boule," which, despite its rounded shape, is a single crystal of corundum. The addition of small amounts of *chromium oxide* to the aluminum oxide produces a synthetic red corundum (synthetic ruby). Synthetic blue sapphire is produced by adding about 2% of *iron oxide* and 1% of *titanium oxide* to the powdered aluminum oxide. If about 3% of *vanadic oxide* is used as a coloring agent, then the synthetic corundum formed resembles the mineral alexandrite (a variety of chrysoberyl, $BeAl_2O_4$, that is usually green in daylight and violet-red in incandescent light). The synthetic corundum colored by vanadic oxide is gray-green in daylight and violet-red in incandescent light.

In 1947, the Linde Air Products Company produced star rubies and sapphires. The corundum boule formed at temperatures over 2000°C contains *titanium* in solid solution. At lower temperatures, the solid solution is not stable. If the boules are annealed in an oxidizing atmosphere at temperatures in the range of 1100° to 1500°C for a period of up to a week, then the corundum structure rejects the titanium. The titanium unmixes with, or separates from, the host structure in the form of microscopic oriented needles of rutile (TiO_2). The rutile inclusions arrange themselves in three intersecting sets of parallel needles in relation to the hexagonal symmetry of the host crystal. In comparison with natural stones, the synthetic star rubies and sapphires have more included rutile needles with a more uniform distribution. Thus, they have a slightly cloudy appearance, and their stars are more uniform and sharply defined.

Synthetic rubies and sapphires are also produced by hydrothermal methods, similar to those employed in the production of synthetic quartz.

Occurrence. Corundum occurs as an accessory mineral in igneous rocks that are poor in silica; namely, syenites, nepheline syenites, metamorphosed limestones, mica schists, and pegmatites. Rubies are found primarily in Burma, Siam, and Ceylon. The finest ruby crystals, some 2 in. long, come from near Mogok, upper Burma, where they occur in metamorphosed limestone and in the residual soil above the limestone. These mines have been operated since the fifteenth century. Important ruby and sapphire deposits occur near Bangkok, Siam. The rubies and sapphires occur with red spinel in clay. In Ceylon, rubies and sapphires occur in stream gravels derived from pegmatites. Attractive specimens of thin, tabular six-sided ruby crystals in green, chromium-bearing zoisite are found in Tanganyika. Tabular six-sided ruby crystals also occur in Goias, Brazil. Kashmir, in northern India, has long been an important location for sapphires. Other important sources of sapphires are the Ural Mountains, U.S.S.R.; Queensland, Australia; and Afghanistan. In the United States, sapphires have been mined near Helena, Montana; and rubies in North Carolina.

In the Transvaal, South Africa, large crystals of common corundum (Plate 17, facing p. 114) are found in pegmatites, or are loose in soil derived from the pegmatites. One giant crystal weighed 335 lb. In Madagascar, large crystals occur in mica schist. Corundum is abundant in Ontario, Canada, where it occurs as a primary constituent in syenites and nepheline syenites. Large emery deposits have long been mined on Cape Emeri on the Greek island of Naxos. In the United States, emery has been mined at Peekskill, New York, and at Chester, Massachusetts.

Use. As an abrasive; as gem material; and as bearings in watches; and in scientific instruments.

HEMATITE

Fe_2O_3 Hexagonal
     Hardness $5\frac{1}{2}$–$6\frac{1}{2}$ Sp. Gr. 4.9–5.3
     (soft in earthy varieties)

Summary

Hematite ranges from a soft, dull red, earthy material, to hard steel gray masses or grayish-black metallic crystals. It occurs in a wide variety of habits, frequently in botryoidal, micaceous, and oölitic aggregates (Plate 18, facing p. 115). Crystals are commonly thick to thin tabular (Figure 8-19). *Streak:* dark red to reddish-brown. *Cleavage:* none. *Parting:* rhombohedral, with nearly cubic angles; usually opaque but may be translucent and red in very thin scales. A BROWNISH-RED STREAK IS CHARACTERISTIC, EVEN IN BLACK VARIETIES. Red earthy hematite resembles CUPRITE and CINNABAR.

Composition. Ferric oxide, Fe_2O_3 (70% Fe, 30% O). May contain some titanium.

Tests. Infusible. Hematite heated in a reducing flame becomes strongly magnetic. A solution of hematite in concentrated hydrochloric acid plus potassium ferrocyanide gives a brilliant Prussian blue precipitate (test for ferric iron).

Crystallization. Hexagonal, scalenohedral class, $\bar{3}\,\frac{2}{m}$. Commonly in red, earthy masses; crystals are usually thick to thin tabular, often forming rosettes; also rhombohedral. Crystals are commonly striated; frequently in reniform or botryoidal aggregates with a radiating inner structure. Also foliated, micaceous, oölitic, and compact.

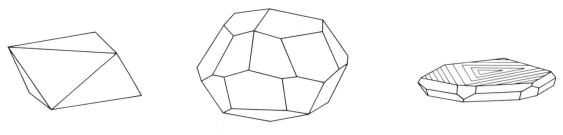

Fig. 8-19. Hematite crystals.

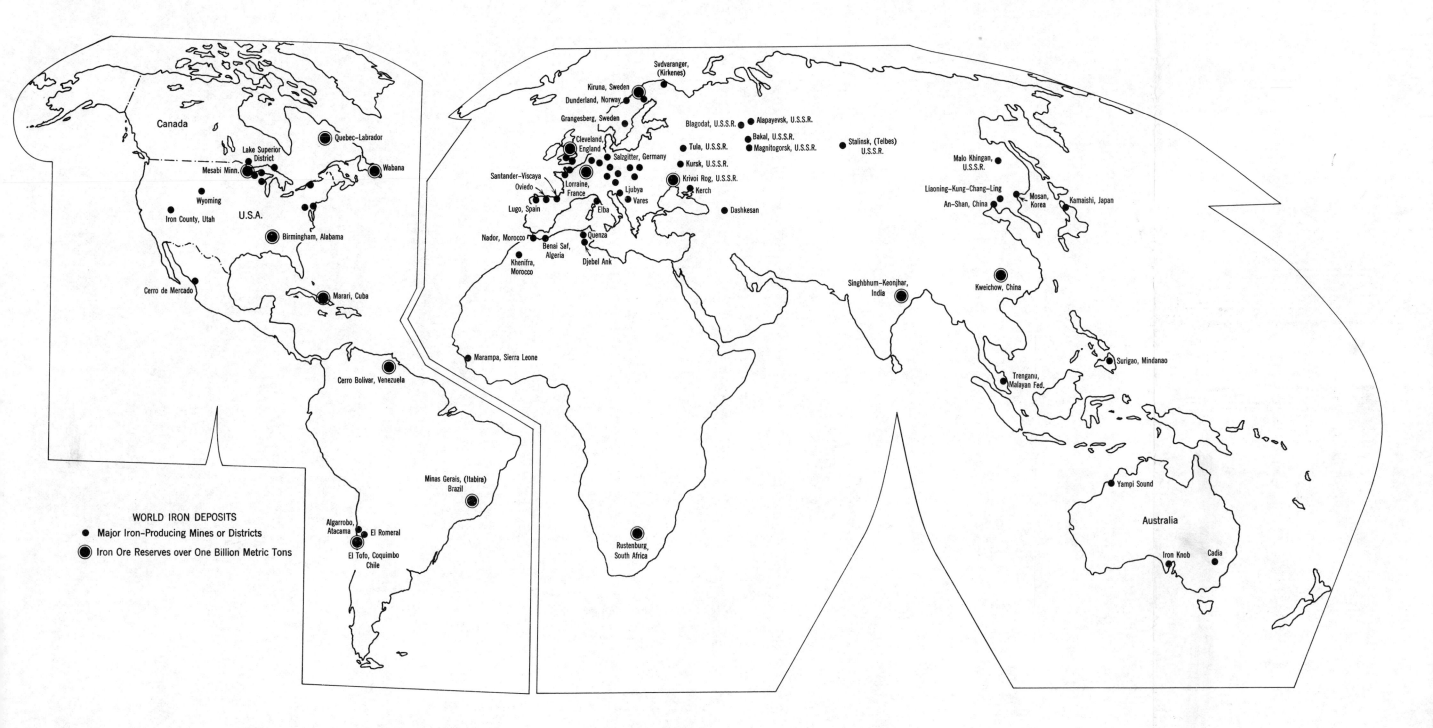

Fig. 8-20. The distribution of iron deposits.

Varieties

Red ocher	Red, earthy masses.
Kidney ore	Red or black reniform or botryoidal aggregates.
Micaceous hematite	Brilliant black foliated or micaceous aggregates.
Specular hematite	Splendent steel gray to black metallic crystals.
Iron roses	Thin, black plate-like crystals grouped in rosette fashion like the petals of a flower.
Martite	Octahedral pseudomorphs after magnetite.
Oölitic hematite	Aggregates of rounded red pellets that resemble fish roe; frequently contain fossil remains.

Associated Minerals. Magnetite, limonite, goethite, ilmenite, siderite, and chalcedony.

Minerals of Similar Appearance

Cuprite. Massive cuprite strongly resembles red, earthy hematite in color and streak; but is somewhat heavier (Sp. Gr. cuprite = 6.1) and is commonly associated with other copper minerals.

Cinnabar. Distinguishable from hematite by high specific gravity (8.1) and scarlet color and streak. Impure cinnabar may resemble hematite in color and streak.

Ilmenite. Black tabular crystals of hematite resemble ilmenite but have a distinctive red-brown streak.

Occurrence. An accessory mineral in many igneous rocks; also associated with metamorphic deposits and hydrothermal veins. Most large hematite deposits are of original sedimentary origin. Enormous deposits of hematite occur in the Lake Superior region of northern Michigan, Minnesota, Wisconsin, and Canada. Much of the *high-grade* hematite ore has already been mined in this region, but vast *taconite* deposits, which are low-grade silica-bearing iron ores, are being successfully mined, concentrated, and reduced. Large quantities of oölitic hematite occur in the Clinton Formation in New York, in eastern Tennessee, and in northern Alabama. The vast *ilmenite-hematite* deposit

at Allard Lake, Quebec, is of great economic value. Important deposits of hematite are found at Cerro Bolivar in Venezuela, and in Minas Gerais, Brazil. Good crystals of hematite occur in Cumberland, England; Bahia, Brazil; and on the island of Elba. "Iron roses" are found in the St. Gotthard area in Switzerland and in Minas Gerais, Brazil, where they are sometimes over 5 in. in diameter.

Alteration. Magnetite readily alters to hematite (martite pseudomorphs). Hematite is also found as pseudomorphs after, or alterations of limonite, siderite, and pyrite.

Use. Chief source of iron. Red ocher is used as a mineral pigment. Rouge is used as a polishing powder.

ILMENITE SERIES (Ilmenite, $FeTiO_3$; Geikielite, $MgTiO_3$; Pyrophanite, $MnTiO_3$)

	Hexagonal
Hardness 5–6	Sp. Gr. 4.0–4.7

Summary

Color: iron black (Fe-rich ilmenite) to brownish-black (Mg-rich geikielite) to deep red (Mn-rich pyrophanite). *Luster:* metallic to submetallic. *Streak:* black to brownish-red to yellow ochre. *Cleavage:* none for ilmenite, but rhombohedral for geikielite and pyrophanite; brittle. Ilmenite is sometimes weakly magnetic because of intermixed magnetite. Usually massive; also as a constituent of heavy black beach sands. Crystals are usually tabular (Figure 8-21). Often associated with magnetite, hematite, rutile, spinel, zircon, and monazite. TABULAR CRYSTALS AND TESTS FOR TITANIUM ARE DISTINCTIVE. Resembles HEMATITE, MAGNETITE, and COLUMBITE-TANTALITE.

Composition. Ilmenite—iron titanium oxide, $FeTiO_3$; related to hematite (Fe_2O_3), except that Ti replaces half

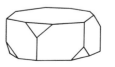

Fig. 8-21. Tabular ilmenite crystal.

the iron. Magnesium and manganese can replace the remaining iron, thus forming complete isomorphism between common ilmenite and the rare minerals geikielite and pyrophanite. Magnetite or hematite is sometimes intimately intermixed with ilmenite, caused in part by exsolution.

Tests. Infusible, becomes magnetic after heating. To test for titanium, fuse powdered mineral with sodium carbonate and dissolve in equal amounts of concentrated sulfuric acid and water. Very slowly, add acid with dissolved mineral to the water. Never add water to concentrated sulfuric acid. When solution is cool, the addition of a few drops of hydrogen peroxide (H_2O_2) will yield a yellow-amber color.

Crystallization. Hexagonal; rhombohedral class, $\overline{3}$. Usually massive. Crystals are usually tabular. Also as a major constituent in heavy and black beach sands.

Associated Minerals. Magnetite, hematite, rutile, spinel, zircon, monazite.

Minerals of Similar Appearance

Hematite. Iron black tabular hematite crystals resemble ilmenite crystals, but the former have a brownish-red streak, whereas Fe-rich ilmenite crystals have a black streak.

Columbite-tantalite. Thick, tabular columbite-tantalite crystals are much heavier than ilmenite, and are commonly associated with mica.

Magnetite. Differs in crystallization, and is strongly magnetic.

Alteration. Alters to leucoxene, a whitish, earthy material.

Occurrence. Ilmenite is found as a common accessory mineral in igneous rocks; as a magmatic segregation; in pegmatites; in high-temperature hydrothermal veins; in large masses associated with metamorphic rocks; and as a constituent of black beach sands. Large, thick, tabular crystals, some over 4 in. in diameter, occur in diorite veins in Kragerö, Norway. Some important locations for ilmenite are the black beach sands in India, Brazil, and Florida; the large ilmenite-

hematite masses in anorthosite near Allard Lake, Quebec; and the ilmenite-magnetite deposits at Kragerö and Ekersund in Norway. Ilmenite is also found in the Urals near Lake Ilmen, Miask, and in the Ilmen Mountains, U.S.S.R. In the United States, large titaniferous-magnetite deposits occur in the Adirondack region of New York. The rare magnesium-rich ilmenite (geikielite) has been found in the gem gravels of Ceylon. The manganese-rich ilmenite (pyrophanite) is also a rare mineral and occurs in Värmland, Sweden, and in Ouro Preto, Brazil.

Use. A source of titanium.

CASSITERITE (Tinstone)

SnO_2	Tetragonal
Hardness 6–7	Sp. Gr. about 7.0

(very heavy for a nonmetallic mineral)

Summary

A hard, heavy, *brown* to *brownish-black* mineral; occasionally yellow or white. *Luster:* adamantine to dull. *Streak:* white to pale brown. Frequently in brilliant twinned crystals with a characteristic re-entrant angle (Figure 8-22). Crystals often show an uneven distribution of color. Also commonly in dull brownish compact botryoidal masses with a concentric and radiating fibrous inner structure. This variety is known as wood tin (Plate 19(*A*), facing p. 122). Cassiterite is commonly associated with wolframite, topaz, tourma-

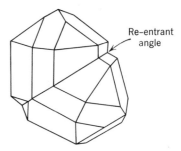

Fig. 8-22. Twinned cassiterite crystal.

line, fluorite, apatite, muscovite, and quartz. The high specific gravity and resistance to weathering frequently cause cassiterite to be concentrated in placer deposits, where it is often found as waterworn pebbles and grains (stream tin). HIGH SPECIFIC GRAVITY, ADAMANTINE LUSTER, PALE STREAK, TWINNED CRYSTALS, AND TESTS FOR TIN ARE DISTINCTIVE.

Composition. Tin dioxide, SnO_2 (78.6% Sn, 21.4% O). Some iron and small amounts of tantalum and columbium may substitute in part for tin.

Tests. Infusible. Fragments of cassiterite become slowly coated with a dull gray deposit of metallic tin when placed in dilute hydrochloric acid with a small piece of metallic zinc.

Crystallization. Tetragonal; ditetragonal dipyramidal class, $\frac{4}{m}\frac{2}{m}\frac{2}{m}$. Crystals are common, and are frequently in combinations of short prisms and bipyramids; frequently in twinned crystals; sometimes acicular. Commonly in compact concretionary masses or in botryoidal aggregates with a concentric radiating fibrous inner structure (wood tin). Also in granular masses.

Associated Minerals. Wolframite, molybdenite, topaz, tourmaline, apatite, muscovite, lepidolite, fluorite, bismuthinite, native bismuth, arsenopyrite, pyrrhotite, and quartz.

Occurrence. Usually found in high-temperature hydrothermal veins or in metamorphic deposits genetically associated with siliceous igneous rocks. Also found in pegmatites, greisen, and in rhyolite; and as waterworn pebbles in stream deposits. Botryoidal masses of wood tin occur in Durango, Mexico (Plate 19, facing p. 122);

they also occur in rhyolite at Lander County, Nevada, and Sierra County, New Mexico. Large brilliant crystals, some over $2\frac{1}{2}$ in. in diameter, are found in the bismuth-lead-silver sulfide deposits of Bolivia. Needle-like crystals ("needle tin") also occur in Oruro, Bolivia. Other notable localities for good crystals include Zinnwald, Bohemia (Plate 19, facing p. 122); Czechoslovakia; Saxony, Germany; and Cornwall, England. Most of the world's supply of cassiterite comes from the Malay States, Indonesia, Bolivia, Congo (Leopoldville), Thailand, and Nigeria. In the past, Cornwall, England, was an important source of cassiterite. No important tin deposits are known in the United States.

Use. Most important ore of tin.

RUTILE

TiO$_2$ Tetragonal
    Hardness 6–6$\frac{1}{2}$ Sp. Gr. 4.2

Summary

A *reddish-brown* to *almost black* mineral with an adamantine-submetallic luster. Commonly in long, vertically striated prismatic crystals; or in needle-like crystals included in such other minerals as rutilated quartz. Knee-like cyclic twins are common (Plate 8, facing p. 51). *Cleavage:* prismatic and basal; transparent in thin pieces; brittle. The minerals anatase (or octahedrite) and brookite have the same chemical composition as rutile, but differ in crystallization. RED-BROWN COLOR, ADAMANTINE LUSTER, AND LONG STRIATED CRYSTALS (Plate 19(*F*), facing p. 122) OR KNEE-LIKE TWINS (Figure 8-23) ARE DISTINCTIVE. Dark rutile crys-

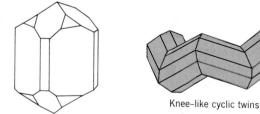

Fig. 8-23. Rutile crystals.

Knee-like cyclic twins

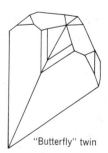

"Butterfly" twin

tals may resemble CASSITERITE (Sp. Gr. 7.0) but are much lighter in weight.

Composition. Titanium dioxide, TiO_2 (60% Ti, 40% O). Some iron, columbium, or tantalum may be present.

Tests. Infusible. Insoluble in acids. Can be dissolved in sulfuric acid after fusion with sodium carbonate and gives test for titanium (see ilmenite, p. 192).

Crystallization. Tetragonal; ditetragonal-dipyramidal class, $\frac{4}{m} \frac{2}{m} \frac{2}{m}$. Crystals are commonly long, vertically striated prisms; often in cyclic twins that are knee-shaped or ring-like; sometimes as "butterfly" twins (Figure 8-23). Also in granular masses.

Alteration. Alters from other titanium minerals like sphene and ilmenite. Found as paramorphs after anatase and brookite. Alters to leucoxene, a whitish, powdery material that is a mixture of titanium oxides.

Occurrence. Rutile is a high-temperature mineral, commonly found as a minor constituent in granites, pegmatites, gneisses, and schists. Also found in high-temperature quartz veins; metamorphosed limestones; and as a constituent of black beach sands, where it is associated with magnetite, zircon, and monazite. Needle-like or hair-like reddish- to yellowish-brown crystals of rutile are often included in colorless or smoky quartz (rutilated quartz, sometimes called Venus Hair stone). The best crystals of rutilated quartz occur in Minas Gerais and Bahia, Brazil. Plate 34, facing p. 210, shows pale reddish-brown hair-like inclusions of rutile in a quartz crystal. Golden-yellow rutile inclusions in smoky quartz are also common. Large cyclic and "butterfly" twins, some 2 in. long, are also common in Minas Gerais. Oriented microscopic inclusions of rutile that precipitated from solid solution may cause a star effect in rose quartz or corundum (see asterism, p. 85). The blue color of some quartz found in granites is believed to be caused by the selective scattering of blue wavelengths of light by microscopic oriented rutile inclusions. Large amounts of rutile occur in Australia and Norway; and in the Alpine regions of France, Italy, Switzerland, and Germany. In the United States, large crystals (some 4 in. long) and twinned crystals of rutile come from

Graves Mountain, Georgia. Also found in Magnet Cove, Arkansas; in Alexander County, North Carolina; and in the black sands of Florida.

Synthetic Rutile. In 1948, synthetic rutile was produced by the Verneuil process (see synthetic corundum, p. 189). By submitting them to the proper heat treatment in an oxidizing atmosphere, the boules can be made almost colorless, with a yellowish cast. The dark boules are deficient in oxygen, but as they are annealed in a stream of oxygen the ratio of titanium to oxygen is adjusted to correspond to the formula TiO_2, and the color changes from black to deep blue, light blue, green; and, finally, to a very pale yellow. This yellowish, almost colorless, synthetic equivalent of rutile looks very different from natural rutile, which is usually brownish-red to black. The synthetic rutile is cut into a very attractive gemstone called Titania. The indices of refraction and dispersion are higher than those of diamond. Dazzling flashes of dispersion colors are emitted by Titania, sometimes called Kenya Gem.

Use. Ore of titanium.

PYROLUSITE

MnO_2	Tetragonal
Hardness 1–6 (massive material) 6–6½ (crystals)	Sp. Gr. 4.4–5.0

Summary

Pyrolusite is one of the most common and important secondary manganese minerals. It is commonly formed by weathering of rocks and minerals containing manganese, and is usually associated with iron and manganese oxides like limonite, manganite, and psilomelane. Frequently found in soft, black fibrous masses that are pseudomorphs after manganite (Figure 8-24). Often occurs as dendritic coatings on rock surfaces (Figure 8-25); also as included dendrites in chalcedony (moss agate). Rarely in well-formed prismatic crystals. *Color:* steel gray to black. *Luster:* metallic. *Streak:* black (sooty and soft). *Cleavage:* prismatic; brittle;

A   x 0.75

LAZULITE
$MgAl_2(OH)_2(PO_4)_2$

Massive lazulite showing characteristic blue color (lower); steep pyramidal crystal with whitish surface alteration (upper): Graves Mountain, Georgia.

B   x 0.66

WAVELLITE
$Al_3(OH)_3(PO_4)_2 \cdot 5H_2O$

Radiating spherulitic aggregates of wavellite: Montgomery County, Arkansas.

C   x 0.85

VIVIANITE
$Fe_3(PO_4)_2 \cdot 8H_2O$

Deep blue tabular vivianite showing cracks parallel to cleavage direction: Ibex Mine, Leadville, Colorado.

D   x 0.62

ANNABERGITE
$Ni_3(AsO_4)_2 \cdot 8H_2O$

Pale green crusts of annabergite: Cobalt, Ontario.

E   x 0.88

ERYTHRITE
$Co_3(AsO_4)_2 \cdot 8H_2O$

Radiating crystals of pinkish-purple erythrite: Schneeberg, Saxony.

F   x 0.95

DESCLOIZITE
$ZnPb(VO_4)OH$

Drusy crusts of minute intergrown crystals: Grant County, New Mexico.

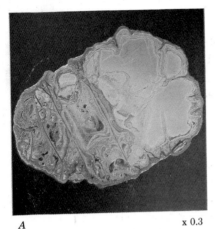

A   x 0.3

VARISCITE

$Al(PO_4) \cdot 2H_2O$

Polished specimen of pale green variscite partially altered to wardite and other phosphates: Fairfield, Utah.

B   x 1.0

TURQUOISE

$CuAl_6(PO_4)_4(OH)_8 \cdot 4H_2O$

Bluish-green turquoise showing dull porcelain-like luster: Mazapil, Zacatecas, Mexico.

C   x 0.75

CARNOTITE

$K_2(UO_2)_2(VO_4)_2 \cdot 3H_2O$

Loosely coherent aggregate of bright yellow carnotite: Rock Creek, Colorado.

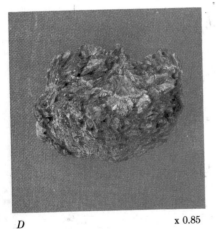

D   x 0.85

TORBERNITE

$Cu(UO_2)_2(PO_4)_2 \cdot 8\text{--}12H_2O$

Aggregate of green micaceous flakes: Shinkolobwe Mine, Katanga, Congo.

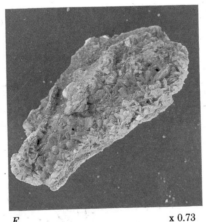

E   x 0.73

AUTUNITE

$Ca(UO_2)_2(PO_4)_2 \cdot 10\text{--}12H_2O$

Aggregate of yellow micaceous flakes: Johanngeorgenstadt, Saxony.

F   x 0.6

META-AUTUNITE

Aggregates of yellow-green thin tabular crystals standing on edge, forming serrated surfaces: Spokane Indian Reservation, Spokane, Washington.

Fig. 8-24. Fibrous pyrolusite: Ilmenau, Thuringia, Germany (x 1.0).

opaque. SOOTY BLACK STREAK AND SOFT FIBROUS AGGRE-GATES THAT SOIL FINGERS BLACK ARE DISTINCTIVE, but x-ray diffraction is often necessary to distinguish between massive black manganese oxides.

Composition. Manganese dioxide, MnO_2 (63.2% Mn, 36.8% O). Commonly contains some water and small amounts of heavy metals, phosphorus, and barium. The name WAD is used as a field term for obscure mixtures of hydrous manganese oxides. Pyrolusite and psilomelane are commonly the chief constituents. Wad is commonly bluish or brownish-black and frequently occurs in dull, loose, earthy masses that are light in weight.

Tests. Infusible on charcoal. Gives amethyst color to borax bead in oxidizing flame (test for Mn).

Crystallization. Tetragonal; ditetragonal dipyramidal class, $\frac{4}{m}\frac{2}{m}\frac{2}{m}$. Usually in fibrous masses that are pseudomorphs after manganite; also commonly occurs as dendritic coatings, concretionary forms, and as powdery and granular masses. Rarely in well-formed prismatic crystals.

Associated Minerals. Manganite, psilomelane, rhodonite, rhodochrosite, alabandite, hausmannite, braunite, limonite, hematite, siderite, and calcite.

Alteration. Alters from other manganese minerals like manganite, rhodonite, rhodochrosite, hausmannite, braunite, and alabandite.

Minerals of Similar Appearance

Manganite. Deeply striated prismatic crystals of manganite have a hardness of 4 and a dark, reddish-brown streak. Pyrolusite pseudomorphs after manganite are usually much softer and have a sooty black streak. Mixtures of pyrolusite and manganite are common.

Occurrence. Pyrolusite is one of the most common manganese minerals. It is of secondary origin and is formed under strong oxidizing conditions. Commonly occurs in bog, lake, and shallow marine deposits; as a weathering product of rocks and ores containing manganese; and in manganese deposits formed by circulating meteoric waters. Pyrolusite is frequently mixed with other manganese oxides (wad). The most extensive manganese deposits are in the U.S.S.R.; India; the Union of South Africa; Ghana, Africa; and Brazil. No large high-grade manganese deposits have been found in the United States. Well-formed, hard prismatic crystals (POLIANITE) have been found in Czechoslovakia.

Fig. 8-25. Pyrolusite dendrites on limestone (x 0.8).

Use. Chief ore of manganese. Manganese is one of the most essential elements in the production of steel. Approximately 12.5 lb. of manganese are used to deoxidize and desulfurize each short ton of steel produced. Manganese is also widely used in various alloys; as a decolorizer; and in dry-cell batteries.

In Kragerö, Norway, anatase occurs in apatite veins as an alteration product of sphene. Anatase is commonly found, partially altered to rutile, in the diamond placers of Minas Gerais and Bahia, Brazil. In the United States, fine deep blue crystals occur in Gunnison County, Colorado; also in Somerville, Massachusetts, with brookite and sphene.

ANATASE (Octahedrite)

TiO_2 Tetragonal
    Hardness $5\frac{1}{2}$–6   Sp. Gr. 3.9

Summary

A rare titanium mineral that has the same chemical composition as rutile and brookite, but differs in atomic arrangement. Usually in steep bipyramidal crystals. However, crystals sometimes look pseudo-octahedral—hence the name octahedrite. *Color:* yellowish to reddish-brown to blue-black. *Luster:* adamantine to submetallic. *Streak:* white to pale yellow. *Cleavage:* basal and pyramidal; transparent to translucent; brittle. Usually occurs with brookite and rutile in silicate veins and in crevices of gneisses and schists. Frequently alters to RUTILE, forming paramorphs of rutile after anatase. BIPYRAMIDAL CRYSTALS AND TESTS FOR TITANIUM ARE DISTINCTIVE.

Composition. Titanium dioxide, TiO_2 (60.0% Ti, 40.0% O)—polymorphous with rutile and brookite.

Tests. Same as rutile (see p. 194).

Crystallization. Tetragonal; ditetragonal dipyramidal class, $\frac{4}{m}\frac{2}{m}\frac{2}{m}$. Crystals are commonly steep pyramidal; also pseudo-octahedral, and tabular. The crystals are often complex.

Alteration. Frequently alters to rutile. Found as an alteration product of sphene and ilmenite.

Occurrence. Anatase is less common than rutile. It commonly occurs with brookite in silicate veins and in crevices of gneisses and schists. Also in placer deposits; and, occasionally, in pegmatites. Good crystals are found in crevices and veins in the Alpine Mountains.

BROOKITE

TiO_2 Orthorhombic
    Hardness $5\frac{1}{2}$–6   Sp. Gr. 4.1

Summary

Brookite is an uncommon titanium mineral that has the same chemical composition as rutile and anatase, but differs in atomic arrangement. Usually in long, flat, platy crystals. Crystals often have an uneven distribution of color, with dark corners caused by hourglass coloration. *Color:* various shades of brown to iron black. *Luster:* adamantine to submetallic. *Streak:* white to gray or yellowish; translucent; brittle. BROWN, PLATY CRYSTALS AND TESTS FOR TITANIUM ARE DISTINCTIVE.

Composition. Titanium dioxide, TiO_2, polymorphous with rutile and anatase.

Tests. Same as rutile (see p. 194).

Crystallization. Orthorhombic; dipyramidal class, $\frac{2}{m}\frac{2}{m}\frac{2}{m}$. Crystals are usually elongated and tabular, with prism faces vertically striated.

Alteration. Forms paramorphs of rutile after brookite.

Occurrence. Mineral associations and occurrences are similar to anatase. Occurs most frequently in veins and crevices of schists and gneisses; associated with anatase, rutile, hematite, sphene, quartz, adularia, albite, and chlorite. The titanium in these so-called Alpine-type deposits is derived from leaching of the gneisses and schists by hydrothermal solutions. Typical flat brown crystals are widespread in Alpine veins. Also commonly found in placer deposits in the Ural

Mountains, U.S.S.R., and in Bahia and Minas Gerais, Brazil. Seldom as a contact metamorphic mineral—as in Magnet Cove, Arkansas, where brookite occurs in untypical thick black crystals more or less altered to rutile. Also found in the United States in Ellenville, New York, and in Somerville, Massachusetts.

URANINITE

UO_2 Isometric

    Hardness 5–6   Sp. Gr. 10.9

    (Sp. Gr. decreases as U^4 oxidizes to U^6)

Summary

A hard, heavy, *black* radioactive mineral. Usually massive; often botryoidal with a pitch-like luster (pitchblende). Bright yellow, orange, and green alterations are common (Plate 19(*D, E*), facing p. 122). *Luster:* submetallic, pitch-like, or dull. *Streak:* brownish-black; brittle; opaque. **BLACK COLOR, HIGH SPECIFIC GRAVITY, RADIOACTIVITY, AND BRIGHT YELLOW, ORANGE, AND GREEN ALTERATIONS ARE DISTINCTIVE.**

Composition. Uranium dioxide, UO_2, but partial auto-oxidation produces UO_3; and the end products of radioactive decay—helium and lead—are always present. Small amounts of argon, nitrogen, radium, rare earths, and water may also be present. Thorium can substitute for uranium. A complete series from uraninite to blackish-gray thorianite (ThO_2) has been produced artificially.

Tests. Infusible. Radioactive. A bead made with sodium fluoride and powdered uraninite luminesces brightly under ultraviolet light.

Crystallization. Isometric; hexoctahedral, $\frac{4}{m} \bar{3} \frac{2}{m}$ (?). Usually massive; often botryoidal (pitchblende). Octahedral, dodecahedral, and cubic crystals are rare.

Alteration. Oxidizes to UO_3. Commonly alters to brightly colored secondary uranium minerals such as yellowish or orange gummite, reddish-orange curite, yellow autunite, and green torbernite.

Occurrence. Uraninite has various modes of occurrence that may be grouped as follows:

1. Uraninite occurs in pegmatites with zircon, tourmaline, monazite, and complex rare earth oxides (Bancroft, Ontario; Villeneuve, Quebec; Grafton Center, New Hampshire; Norway; and East Africa).

2. In high-temperature hydrothermal tin veins with cassiterite, arsenopyrite, pyrite, chalcopyrite, and galena (Cornwall, England).

3. In moderate-temperature hydrothermal veins with silver, cobalt, nickel, bismuth, and arsenic minerals (Joachimstal, Bohemia; Great Bear Lake, Canada). In veins and in replacement lenses of slates and dolomitic schists; associated with nickel, cobalt, and copper ores (Shinkolobwe, Katanga District, Congo).

4. In moderate-temperature hydrothermal veins without cobalt, and nickel minerals. Commonly associated with sphalerite, galena, and chalcopyrite (Gilpin County, Colorado).

5. In gold-bearing conglomerates (Witwatersrand, Union of South Africa).

6. In replacement deposits in sedimentary strata at many localities on the Colorado Plateau. The vein deposits of Canada, the Plateau deposits, and the Katanga deposits are extensive.

Use. Major ore of uranium. Used in the atomic bomb, in nuclear reactors, and as a source of radium. Small amounts of sodium uranate are used in ceramics to produce a yellow glaze.

SPINEL

$MgAl_2O_4$ Isometric

    Hardness 7.5–8   Sp. Gr. 3.58–4.6

    (gravity increases as Fe and Zn

                  substitution increases)

Summary

A hard, brittle, transparent to translucent mineral, usually occurring in octahedral crystals (Plate 20(*C, E, F*), facing p. 123). The *color* varies with composition from white to red, blue, violet, yellow, brown, blue-green, and black. *Luster:* vitreous to dull. *Fracture:* conchoidal. Deep red spinel, resembling ruby, turns dark brown or black on heating, but regains red color on cooling. Many historic "rubies" were probably red spinels. The famous Black Prince's Ruby in the crown jewels of England has proved to be a red spinel. HARD OCTAHEDRAL CRYSTALS ARE DISTINCTIVE.

Composition. Magnesium aluminum oxide, $MgAl_2O_4$, but iron, zinc, and manganese substitute for magnesium in all proportions. Gives rise to such related minerals as *gahnite* (zinc spinel, Plate 20, facing p. 123); *hercynite* (iron spinel); and *galaxite* (manganese spinel). The red magnesium spinels are known as *ruby spinel* or *ballas ruby* (Plate 20).

Tests. Infusible, insoluble. Red and violet spinels commonly luminesce under ultraviolet light. Co-test for aluminum. Soda fusion yields magnesium test.

Crystallization. Isometric; hexoctahedral class, $\frac{4}{m}\bar{3}\frac{2}{m}$. Usually in octahedral crystals. Twinned crystals (spinel twins) are common.

Alteration. A stable mineral that is usually very resistant to change, but may alter to talc, serpentine, or mica.

Associated Minerals. Chondrodite, scapolite, calcite, garnet, wollastonite, diopside, corundum, zircon, phlogopite, graphite, augite, magnetite, hematite, apatite, idocrase, and forsterite.

Occurrence. The spinels are high-temperature minerals, found as accessory minerals in basic igneous rocks; in contact metamorphic limestones; in highly aluminous schists; in pegmatites; and in placer deposits. The "ruby" spinels are found in contact-metamorphosed limestones and in stream placers associated with zircons, rubies, and sapphires in Ceylon and Burma. Gem spinels also occur in India, in Siam, and in Madagascar. In the United States, common spinels occur in the metamorphosed limestones of New York; also in Franklin, New Jersey, where octahedral gahnite crystals (zinc spinel) measuring 5 in. on edges have been found.

Synthetic Spinel is produced by the flame-fusion Verneuil process (see synthetic corundum, p. 189). In 1908, spinel boules were formed by accident; when, in an attempt to produce synthetic blue sapphire, magnesium oxide was added as a flux. The resulting boules were square in cross section rather than round, and proved to be isometric in crystallization instead of hexagonal. The square boules had the properties of spinel instead of corundum. Today, synthetic spinel is produced in many colors. The boules are colorless when no pigmenting agent is added to the mixture of magnesium oxide and aluminum oxide. Blue spinel is produced by the addition of cobalt oxide; green, by the addition of chromium oxide; and red, by annealing the green boules.

Use. As a gem stone when transparent.

MAGNETITE

$FeFe_2O_4$ Isometric
Hardness 6   Sp. Gr. 5.2

Summary

A brittle, *black,* strongly magnetic, metallic mineral; commonly massive or in octahedral crystals (Plate 20(*A*), facing p. 123). *Streak:* black. *Cleavage:* none. *Parting:* octahedral. Sometimes forms a natural magnet (lodestone). STRONG MAGNETISM AND BLACK OCTAHEDRAL CRYSTALS ARE DISTINCTIVE. May resemble ILMENITE, CHROMITE, FRANKLINITE, and MARTITE.

Composition. Ferrous and ferric iron, $Fe^{2+}Fe_2^{3+}O_4$, or sometimes written as Fe_3O_4. Magnesium, zinc, manganese, or nickel may substitute in part for ferrous iron (Fe^{2+}). Small amounts of aluminum, manganese, chromium, and vanadium may substitute in part for the ferric iron (Fe^{3+}). The substitutions give rise to a series of related minerals like MAGNESIOFERRITE, $MgFe_2O_4$; FRANKLINITE, $ZnFe_2O_4$; JACOBSITE, $MnFe_2O_4$; and TREVORITE, $NiFe_2O_4$.

Tests. Infusible. Strongly magnetic. Slowly dissolves in concentrated hydrochloric acid. Gives tests for both ferrous and ferric iron.

Crystallization. Isometric; hexoctahedral class, $\frac{4}{m}\bar{3}\frac{2}{m}$. Usually in granular masses. Crystals are commonly octahedral; less commonly dodecahedral.

Minerals of Similar Appearance

Chromite. Black chromite has a chestnut brown streak; is sometimes slightly magnetic; and is commonly associated with serpentine. Octahedral chromite crystals are rare.

Franklinite. Strongly resembles magnetite in hardness, color, and crystallization, but is low in magnetic intensity and is usually only slightly magnetic, if at all. Its streak is reddish-brown instead of black. It occurs largely at one important locality, Franklin-Ogdensburg, New Jersey.

Martite. Octahedral martite crystals, which are hematite pseudomorphs after magnetite, have a brick-red streak and are nonmagnetic.

Ilmenite. Lacks the strong magnetism of magnetite, and differs in crystallization.

Occurrence. Magnetite is an abundant ore of iron. The largest deposits, believed to be formed by magmatic differentiation, occur at Kiruna in northern Sweden. The mineral is also mined on a large scale in the Adirondack region of New York; in Norway; in the Ural Mountains of the U.S.S.R.; in Pennsylvania; and in the Bushveld Complex of South Africa. Magnetite is a common minor constituent in most igneous rocks, and also commonly occurs in metamorphic rocks and black beach sands. It may be closely associated with corundum and spinel in EMERY. It may form a natural magnet or lodestone which, among other localities, occurs at Magnet Cove, Arkansas. The outer portion of the fusion crust of iron meteorites is magnetite.

Alteration. To hematite (martite), and limonite.

Use. An ore of iron.

FRANKLINITE

$ZnFe_2O_4$ Isometric
   Hardness 6   Sp. Gr. 5.1–5.2

Summary

A hard, *black*, metallic mineral, commonly massive or in octahedral crystals (Plate 20, facing p. 123).

Streak: reddish-brown to dark brown; *slightly magnetic;* brittle. Occurs chiefly at Franklin-Ogdensburg, New Jersey, where it is associated with zincite and willemite in calcite. RESEMBLES MAGNETITE but is only slightly magnetic, and has a brown streak. MINERAL ASSOCIATIONS AND BROWN STREAK ARE DISTINCTIVE.

Composition. Essentially a zinc iron oxide, $ZnFe_2O_4$, but Mn^{2+} and Fe^{2+} substitute in part for zinc, and Mn^{3+} substitutes in part for ferric iron, $(Zn,Fe,Mn)(Fe,Mn)_2O_4$.

Tests. Infusible. Becomes strongly magnetic when heated in the reducing flame. Gives reddish color to borax bead in oxidizing flame.

Crystallization. Isometric; hexoctahedral class, $\frac{4}{m}\bar{3}\frac{2}{m}$. Commonly in octahedral crystals; also commonly in rounded grains, and granular masses.

Occurrence. Chiefly at Franklin-Ogdensburg, New Jersey, in crystalline limestone associated with zincite and willemite. Magnetite, rhodonite, and yellowish-brown garnet are also common associates. Octahedral franklinite crystals measuring over 6 in. on edge have been found in the New Jersey zinc deposit.

Use. A zinc and iron-manganese ore.

CHROMITE

$FeCr_2O_4$ Isometric
   Hardness 5.5   Sp. Gr. 4.5–4.8

Summary

A hard, brittle, *black* mineral with a *brown streak*. *Luster:* submetallic; sometimes weakly magnetic. Usually in granular masses that are frequently associated with serpentine (Plate 20(*D*), facing p. 123). Small octahedral crystals are rare. BROWN STREAK AND ASSOCIATION WITH SERPENTINE ARE DISTINCTIVE. Resembles MAGNETITE.

Composition. A ferrous chromic oxide, $FeCr_2O_4$, but usually with some magnesium substituting for iron and some aluminum substituting for chromium. *Magne-*

siochromite, $MgCr_2O_4$, is the magnesium member of the chromite series.

Tests. Infusible. Gives a green color to borax bead (chromium).

Crystallization. Isometric; hexoctahedral class, $\frac{4}{m}\,\bar{3}\,\frac{2}{m}$. Usually in granular masses. Small octahedral crystals are rare.

Associated Minerals. Olivine, pyroxene, sepentine, green chromium garnet (uvarovite), violet chromium chlorites, chromium spinel, pyrrhotite, niccolite, magnetite, corundum.

Alteration. Alters to goethite.

Occurrence. Occurs as an accessory mineral and as magmatic segregations in ultrabasic igneous rocks, such as peridotites, and in serpentine derived from them. Also occurs in stream placers. Meteorites frequently contain chromite. Commercial deposits occur in: the U.S.S.R.,; Southern Rhodesia; Turkey; the Union of South Africa; the Philippines; New Caledonia; Cuba; and India. In the United States, small amounts of chromite are mined in Montana, in California, and in Oregon. Rare octahedral crystals, some $\frac{1}{2}$ in. across, occur in West Africa.

Use. Major ore of chromium. Chromium is used in plating other metals because of its great hardness and nontarnishing properties. The addition of small amounts of chromium to steel produces a hard chrome steel. Chromium is a major constituent of hard corrosion-resistant alloys like stainless steel and nichrome. Chromium compounds are used in the dyeing and tanning of leather; and in the manufacture of yellow, red, orange, and green pigments. Chrome yellow consists largely of lead chromate. Chromite is also used in the manufacture of such refractories as furnace bricks. There are three grades of chromite ore: *metallurgical* grade with a Cr:Fe ratio of $2\frac{1}{2}$:1; *refractory;* and *chemical.*

CHRYSOBERYL

$BeAl_2O_4$	Orthorhombic
Hardness $8\frac{1}{2}$	Sp. Gr. 3.75

Summary

A hard, brittle, transparent to translucent, glassy mineral; commonly in twinned wedge-shaped or pseudohexagonal crystals (Figure 8-26). *Color:* various shades of green to yellow; also brown or gray. The gem variety, ALEXANDRITE, which is ordinarily green, is red by transmitted light. It also turns cherry red in incandescent light because of its strong absorption of yellow light and the difference in composition between sunlight and artificial light.

Chrysoberyl sometimes contains minute parallel needle-like inclusions or microscopic hollow tubes that reflect a narrow band of light at right angles to the inclusions; this usually occurs when the stone is cut and polished in a rounded form. The effect is known as CHATOYANCY. When the stone is turned, the position of the band of light across the curved surface changes, resembling the eye of a cat—hence the name *cat's eye.* YELLOW-GREEN COLOR, HARDNESS, AND TWINNED CRYSTALS ARE DISTINCTIVE.

Composition. An oxide of beryllium and aluminum, $BeAl_2O_4$ (19.8% BeO, 20.2% Al_2O_3). May contain small amounts of iron and chromium.

Tests. Infusible, insoluble.

Crystallization. Orthorhombic; dipyramidal class,

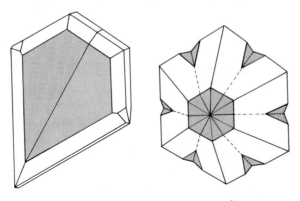

Fig. 8-26. Twinned chrysoberyl crystals.

$\frac{2}{m}\frac{2}{m}\frac{2}{m}$. Twinned crystals are common. Often cyclic, forming a wheel-like pseudohexagonal twin. Also flat, wedge-shaped twins with V-striations. Simple tabular crystals are rare.

Occurrence and Mineral Associations. Chrysoberyl is not a common mineral; usually occurs in granite pegmatites, aplites, and mica schists. Also as a placer mineral with cassiterite, corundum, diamond, and garnet. Pseudohexagonal twins of ALEXANDRITE, which are blue-green in daylight and cherry red under tungsten light, occur with emerald and phenakite in biotite schists in the Ural Mountains along the Takowaja River. The crystals range in size from about $\frac{1}{2}$ to 4 in. in diameter. Yellow-green to brownish alexandrites occur in Ceylon. CAT'S EYE chrysoberyl crystals are found primarily in Ceylon and Brazil. Good transparent yellow to olive green pseudohexagonal twins of chrysoberyl, which sometimes reach 3 in. in diameter, occur in Espírito Santo, Brazil. Chrysoberyl occurs in pegmatites in Madagascar; Germany; Italy; Norway; Orange Summit, New Hampshire; Haddam, Connecticut; and in various localities in Maine. Occurs in placer gravels in the Southern Urals; Ceylon; Japan; Brazil; Southern Rhodesia; and Katanga, Congo.

Use. A gem mineral.

COLUMBITE-TANTALITE SERIES

(Fe,Mn)(Cb,Ta)$_2$O$_6$	Orthorhombic
(Fe,Mn)(Ta,Cb)$_2$O$_6$	
Hardness 6–6$\frac{1}{2}$	Sp. Gr. 5.2 columbite;
	7.9 tantalite

Summary

A hard, *black,* submetallic mineral, often with an iridescent tarnish. *Streak:* very dark red to black. *Cleavage:* good side pinacoid; brittle. Commonly in short prismatic or tabular crystals often with attached plates of mica. Occurs in granite pegmatites, frequently associated with muscovite, lepidolite, spodumene, albite, beryl, apatite, samarskite, tourmaline,

cassiterite. Also in placer gravels derived from the pegmatites. TABULAR CRYSTAL HABIT, HIGH SPECIFIC GRAVITY OF TANTALUM-RICH MEMBERS, AND MINERAL ASSOCIATIONS ARE DISTINCTIVE. Resembles WOLFRAMITE, ILMENITE, and URANINITE.

Composition. An oxide of iron, manganese, columbium (niobium), and tantalum, ranging in composition from an almost pure columbite, (Fe,Mn)Cb$_2$O$_6$, to an almost pure tantalite, (Fe,Mn)Ta$_2$O$_6$. Small amounts of tin may substitute for iron and manganese, and small amounts of tungsten for columbium and tantalum.

Tests. Infusible and insoluble. Because of the difficulty in testing, the name tantalite is derived from the mythological Greek God *Tantalus* (father of Niobe) who was tantalized by water that receded when he tried to drink, and by fruit that vanished when he approached it.

Crystallization. Orthorhombic; dipyramidal class, $\frac{2}{m}\frac{2}{m}\frac{2}{m}$. Often in short rectangular prisms; also thick to thin tabular crystals; and parallel groups. Striated heart-shaped contact twins are common.

Alteration. Resistant to change.

Minerals of Similar Appearance

Uraninite. Distinguishable from columbite-tantalite by strong radioactivity.

Ilmenite. Considerably lighter in weight (Sp. Gr. 4.7).

Wolframite. Difficult to distinguish from columbite-tantalite which may contain small amounts of tungsten.

Occurrence. The columbite-tantalite minerals are found in granite pegmatites with such lithium silicates as spodumene, lepidolite, and Li-tourmaline; and with such lithium phosphates as amblygonite and triphylite. Also associated with albite, microcline, beryl, muscovite, samarskite, apatite, cassiterite, and microlite. Columbite-tantalite is commonly found as a detrital mineral because of its resistance to chemical alteration and high specific gravity.

Notable localities are Western Australia; Katanga, Congo; Brazil; Ilmen Mountains, U.S.S.R.;

Norway; South-West Africa; Canada; and Madagascar. In the United States, it is found in the New England pegmatites; Pike's Peak, Colorado; Keystone, South Dakota; North Carolina; and San Diego, California.

Use. Ore of columbium (niobium) and tantalum. Tantalum is used for making surgical instruments and laboratory apparatus because of its high resistance to acid corrosion and its malleability. Columbium (niobium) is used in high-temperature-resistant alloys, stainless steel, and in tin-niobium alloys for superconducting electromagnets.

DIASPORE

AlO(OH)	Orthorhombic
Hardness $6\frac{1}{2}$–7	Sp. Gr. 3.3–3.5

Summary

A very brittle, transparent to translucent, bright, glassy mineral (pearly on cleavage faces); commonly in thin, bladed masses (Plate 21(*F*), facing p. 130). *Color:* often very pale violet; also white, gray, yellowish, greenish. *Streak:* white. *Cleavage:* perfect side pinacoid. Commonly associated with corundum. Occurs as a constituent of bauxite and aluminous clays. CHARACTERIZED BY THIN, BLADED HABIT; CLEAVAGE, AND HARDNESS.

Composition. Hydrogen aluminum oxide, AlO(OH) (85% Al_2O_3, 15% H_2O). Small amounts of Mn and Fe may substitute for Al.

Tests. Infusible, insoluble. Decrepitates (separates into scales) and gives off water when heated in the closed tube.

Crystallization. Orthorhombic; dipyramidal class, $\frac{2}{m}\frac{2}{m}\frac{2}{m}$. Commonly in thin, platy crystals; also in bladed and foliated masses; finely disseminated.

Associated Minerals. Corundum, dolomite, chlorite, margarite, magnetite, spinel; and as a constituent of bauxite.

Alteration. Alters from corundum and may alter to kaolin.

Occurrence. Diaspore is found as microscopic crystals disseminated in bauxite with boehmite and gibbsite (Hungary; France; and Arkansas and Missouri in the United States). Found as a constituent of emery rock with corundum, and as bladed crystals in cavities of emery, in Chester, Massachusetts. Also in emery-bearing chlorite schists near Maramorskoi in the Ural Mountains, U.S.S.R. Occurs as small, colorless to pale yellow bladed crystals in cavities of crystalline limestones with corundum, tourmaline, and pyrite (in Campolongo, Switzerland).

GOETHITE

FeO(OH)	Orthorhombic
Hardness 5–$5\frac{1}{2}$	Sp. Gr. 3.3–4.3

(apparent H. = 1 for earthy porous varieties)

Summary

Color ranges from brilliant black to brown or brownish-yellow. *Luster:* black, varnish-like; dull or silky. *Streak:* yellowish-brown. *Cleavage:* perfect side pinacoid; translucent in thin splinters; brittle. Commonly in botryoidal or stalactitic aggregates with a fibrous radial inner structure (Plate 21(*A, B, C*), facing p. 130); also in soft, earthy, and compact masses. Usually formed as a weathering product of iron-bearing minerals. Much of the material formerly classified as limonite is now believed to be goethite. YELLOWISH-BROWN STREAK IS CHARACTERISTIC, EVEN IN BLACK VARIETIES. May resemble LIMONITE or black botryoidal and stalactitic varieties of HEMATITE and PSILOMELANE.

Composition. Hydrogen iron oxide, FeO(OH) (62.9% Fe, 27.0% O, 10.1% H_2O). Frequently contains small amounts of manganese and adsorbed capillary water.

Tests. Goethite fuses with great difficulty on charcoal, but becomes strongly magnetic. Gives off water in the closed tube and is converted to ferric oxide, Fe_2O_3. Soluble in HCl.

Crystallization. Orthorhombic; dipyramidal class, $\frac{2}{m}\frac{2}{m}\frac{2}{m}$. Occasionally in vertically striated prismatic crystals; thin, tabular, acicular, or velvet-like aggregates of capillary crystals. Usually massive; commonly in botryoidal or stalactitic aggregates with a fibrous radial or concentric inner structure. Brown and yellow color bands are often developed perpendicular to the fibers; also earthy, loose, porous, and compact masses.

Associated Minerals. Hematite, pyrite, magnetite, siderite, limonite, glauconite, psilomelane, pyrolusite, and calcite.

Minerals of Similar Appearance

Hematite. Black botryoidal hematite aggregates have a *red-brown* streak. Goethite has a *yellowish-brown* streak.

Psilomelane. Black botryoidal or stalactitic aggregates of psilomelane have a *black* streak.

Limonite. Goethite is distinguishable from amorphous limonite by a crystalline structure that is indicated by cleavage, fibrous radial growth, or crystal form.

Occurrence. Goethite is usually formed under oxidizing conditions as a weathering product of iron minerals like pyrite, magnetite, siderite, or glauconite. Also commonly found in old bogs formed by direct precipitation from water (bog iron ore). Rarely in low-temperature hydrothermal veins. Goethite and limonite frequently form a gossan which is a residual, weathered iron capping that is formed on deposits rich in iron-bearing sulfide minerals. Large laterite deposits of goethite occur in Cuba as a result of the weathering of serpentine. Extensive goethite deposits are found in Alsace-Lorraine, and in the Quebec-Labrador district of Canada. Goethite often forms pseudomorphs after pyrite. The mineral LEPIDOCROCITE, $FeO(OH)$, is dimorphous with goethite, and frequently occurs with it.

Alteration. Alters from pyrite, magnetite, siderite. Many of the pseudomorphs of "limonite" after pyrite are probably goethite.

Use. An iron ore and pigment.

LIMONITE

$FeO(OH) \cdot nH_2O$

Hardness 1–5½   Sp. Gr. 2.7–4.3

Summary

Limonite is essentially **an amorphous** equivalent of goethite with a variable **water** content. Much of the material formerly classified as limonite is now believed to be cryptocrystalline goethite with adsorbed or capillary water. The name LIMONITE is often used as a field term for hydrous iron oxides whose identity has not been determined.

Limonite closely resembles goethite, but is commonly vitreous and does not show fibrous internal structure, cleavage or any other evidence of crystallinity. Commonly occurs in stalactitic, colloform, earthy, or loose porous masses (Plate 21(*D, E*), facing p. 130). *Color:* yellow-ocher, brown, black. *Streak:* brownish yellow. *Luster:* vitreous; dull in earthy varieties; brittle.

Limonite is a secondary mineral that frequently occurs with goethite in gossans ("iron hats" of iron-bearing deposits) and in laterites as the decomposition product of iron-bearing minerals.

MANGANITE

$MnO(OH)$ Monoclinic
                                  (pseudo-orthorhombic)
Hardness 4   Sp. Gr. 4.33

Summary

A *black,* submetallic mineral, commonly in deeply striated prismatic crystals that are often grouped in bundles (Figure 8-27; Plate 22(*F*), facing p. 131). *Streak:* dark reddish-brown. *Cleavage:* perfect side; fair prismatic and basal; brittle; translucent in thin splinters. Frequently alters to pyrolusite. BUNDLES OF DEEPLY STRIATED PRISMATIC CRYSTALS WITH GROOVED, ROUNDED CROSS SECTIONS, BROWN STREAK, AND HARDNESS OF 4 ARE DISTINCTIVE.

Fig. 8-27. Manganite crystals: Ilfeld, Harz, Germany (x 0.71).

Composition. A basic oxide of manganese, MnO(OH) (62.4% Mn, 27.3% O, 10.3% H_2O).

Tests. Infusible on charcoal. Gives amethyst color to borax bead in oxidizing flame (test for Mn).

Crystallization. Monoclinic; prismatic class, $\frac{2}{m}$ (pseudo-orthorhombic). Commonly in short to long vertically striated prismatic crystals; often terminated by a flat base, and grouped in bundles. Also fibrous and columnar.

Associated Minerals. Pyrolusite, psilomelane, braunite, hausmannite, goethite, barite, calcite, and siderite.

Alteration. Alters to pyrolusite, psilomelane, braunite, and hausmannite. Pseudomorphs of pyrolusite after manganite are common.

Minerals of Similar Appearance

Pyrolusite. Pseudomorphs of pyrolusite after manganite have a black streak, and are usually much softer than manganite, often soiling fingers black.

Enargite. Black prismatic crystals and fibrous masses of enargite resemble manganite, but have a black streak

and are further distinguished by fusibility and tests for copper.

Occurrence. Manganite occurs in low-temperature hydrothermal veins. In Ilfeld, Harz, Germany, groups of striated crystals are abundant in veins associated with calcite and barite. Manganite also occurs in secondary deposits with pyrolusite and psilomelane. In the United States, good specimens have come from the iron district in Negaunee, Michigan.

Use. A minor ore of manganese.

PSILOMELANE

$BaMnMn_8O_{16}(OH)_4$ Orthorhombic
    Hardness 5–6   Sp. Gr. 4.7
    (soft in earthy varieties)

Summary

Psilomelane is a barium-bearing manganese oxide of secondary origin, commonly formed from weathering of other manganese minerals. It usually occurs in hard, black botryoidal or stalactitic aggregates; also in earthy masses; never in crystals. *Color:* iron black. *Streak:* black to brownish-black. *Luster:* submetallic to dull; opaque. Commonly associated with pyrolusite and limonite. SMOOTH, HARD, BLACK BOTRYOIDAL MASSES ARE DISTINCTIVE (Figure 8-28). Resembles GOETHITE and LIMONITE, but has a black streak.

Composition. A basic oxide of barium with bivalent and quadrivalent manganese, $BaMn^2Mn_8^4O_{16}(OH)_4$. Small amounts of copper, cobalt, nickel, magnesium, calcium, and tungsten are often present.

Tests. Infusible. When fused with borax, gives amethyst bead in oxidizing flame.

Crystallization. Orthorhombic. Commonly in hard, botryoidal or stalactitic aggregates; also massive and earthy. Crystals do not occur.

Associated Minerals. Pyrolusite, goethite, limonite, braunite, rhodonite, rhodochrosite, and calcite.

Fig. 8-28. Botryoidal aggregate of psilomelane: Langeberg, Saxony (x 0.83).

Alteration. Alters from other manganese minerals like manganite, rhodonite, and rhodochrosite.

Occurrence. Psilomelane is a secondary manganese mineral that commonly occurs with pyrolusite and is formed under similar conditions (see pyrolusite, p. 194).

Use. An ore of manganese.

HETEROGENITE

Hydrous cobalt oxide	Amorphous(?)
Hardness 3–4	Sp. Gr. 3.44

Heterogenite occurs as black to brownish-black, glassy, botryoidal masses with a conchoidal fracture. Resembles obsidian, but is considerably softer. *Streak:* blackish-brown. Occurs with malachite and chrysocolla in Katanga, Congo, where it is an important ore of cobalt. Also found at Schneeberg, Saxony; and in New Caledonia. Resembles stainierite, CoO(OH).

BRUCITE

$Mg(OH)_2$	Hexagonal
Hardness $2\frac{1}{2}$	Sp. Gr. 2.4

Summary

A soft, sectile, transparent to translucent mineral, commonly in aggregates of platy crystals (Plate 22(*E*), facing p. 131); fibrous masses, or foliated aggregates. *Color:* white, pale greenish, bluish, yellowish. *Luster:* waxy; pearly on cleavage faces. *Cleavage:* perfect micaceous, yielding cleavage plates that are flexible but not elastic. Commonly associated with serpentine, chlorite, calcite, talc, magnesite, and aragonite. MICACEOUS CLEAVAGE WITH PEARLY LUSTER ON CLEAVAGE FACES, SOFTNESS, AND FOLIATED HABIT ARE DISTINCTIVE. Resembles TALC and GYPSUM.

Composition. Magnesium hydroxide, $Mg(OH)_2$ (69.0% MgO, 31.0% H_2O). Small amounts of Mn, Fe, and Zu may substitute for magnesium.

Tests. Infusible, but glows in flame. Easily soluble in acids.

Crystallization. Hexagonal; scalenohedral class, $\bar{3}\frac{2}{m}$. Commonly in crystal plates; foliated masses; fibrous aggregates; and sometimes in fine-grained masses.

Alteration. Alters from periclase, MgO; and from serpentine.

Minerals of Similar Appearance

Talc. Foliated aggregates of talc have a greasy feel; are softer than brucite (talc H. = 1); insoluble in acids.

Gypsum. Fibrous aggregates of gypsum have a silky luster and are less soluble in hydrochloric acid than is brucite.

Occurrence. As a low-temperature hydrothermal vein mineral; commonly found in serpentine, or in chlorite and dolomite schists. Also found in crystalline limestone as an alteration product of periclase (MgO). Associated minerals include calcite, aragonite, dolomite, talc, magnesite, serpentine, and periclase.

Long fibers of brucite are found with serpentine at Asbestos, Quebec. Large crystals have been found at the Tilly Foster Mine, Brewster, New York. Broad plates, over 7 in. long, have been found at Wood's Mine, Lancaster County, Pennsylvania. Large deposits of brucite occur near Gabbs in Nye County, Nevada.

Use. As a refractory material.

BAUXITE

Hardness 1–3   Sp. Gr. 2–2.55

Summary

Bauxite is a group term for mixtures of hydrous aluminum oxides, composed primarily of microscopic crystals of the minerals GIBBSITE, $Al(OH)_3$, DIASPORE, $AlO(OH)$, BOEHMITE, $AlO(OH)$, and "amorphous" cliachite, $Al_2O_3nH_2O$. Since bauxite is a mixture of minerals, it is a rock name. It is also a commercial name, commonly applied to aluminum ore.
Color: white, yellowish, grayish, reddish-brown. *Luster:* dull to earthy. Commonly in fine-grained, clay-like masses. Also in pisolitic aggregates of rounded particles (Plate 22, facing p. 131). PISOLITIC AGGREGATES ARE CHARACTERISTIC. Resembles CLAYS.

Occurrence. Bauxite is the major ore of aluminum. It is of secondary origin; formed by the leaching of silica from low-silica aluminum-bearing rocks under tropical or subtropical weathering conditions. Residual soils composed primarily of hydrous aluminum and iron oxides are called LATERITES. Important producers of bauxite are Surinam (Dutch Guiana); British Guiana; the United States; France; Hungary; Jamaica; Brazil; the U.S.S.R. In the United States, large bauxite deposits occur in Arkansas, Georgia, and Alabama.

GIBBSITE

$Al(OH)_3$ Monoclinic
          Hardness $2\frac{1}{2}$–$3\frac{1}{2}$ Sp. Gr. 2.4

Summary

Gibbsite occurs in stalactitic aggregates, often with a smooth enamel-like surface and a radiating fibrous inner structure (Plate 22(*B, C*), facing p. 131). Also in compact earthy masses; rarely in pseudohexagonal plates. *Color:* white, but often tinted by impurities. *Luster:* vitreous to dull; pearly on cleavage faces. *Cleavage:* perfect basal, but rarely observed. Gibbsite gives off a strong, earthy clay-like odor when breath is blown on it. Gibbsite is usually of secondary origin, formed as a result of the alteration of aluminum-bearing minerals. An important fine-grained constituent of bauxite. MINUTE VEINLETS AND SMALL CRYSTALS IN PISOLITIC BAUXITE ARE APT TO BE GIBBSITE.

Composition. Aluminum hydroxide, $Al(OH)_3$ (65.35% Al_2O_3, 34.65% H_2O).

Tests. Infusible, but glows and gets harder and whiter. Not readily soluble in acids. Gives off water in closed tube.

Alteration. From aluminous minerals like corundum and feldspars.

Occurrence. Gibbsite is a common fine-grained constituent of bauxite along with diaspore cliachite and boehmite. Usually of secondary origin, formed by the weathering of aluminous minerals under tropical or subtropical climatic conditions. Sometimes as a low-temperature hydrothermal mineral.

Large crystals of gibbsite occur in talc schists near Slatoust, Urals, U.S.S.R. Stalactitic aggregates with radiating, fibrous inner structures are found near Ouro Preto; small transparent crystals at Pocos de Caldas in Minas Gerais (Brazil), and in Richmond, Berkshire County, Massachusetts (Plate 22, facing p. 131). Gibbsite occurs in bauxite in the Guianas; France; Germany; Hungary; the United States; India; and Ghana, Africa.

Cliachite, the "amorphous," hydrous aluminum oxide, which occurs as pisolitic masses in bauxite, frequently contains fine-grained gibbsite between the pisolites.

Use. Aluminum ore.

Halides and Some of Their Properties

MINERAL	HARDNESS	STREAK	SP. GR.	COMMON HABIT	REMARKS
Halite, NaCl	$2\frac{1}{2}$	White	2.16	Cubes	Salty taste
Sylvite, KCl	2	White	2.2	Cubo-octahedrons	Bitter salty taste
Cerargyrite, AgCl	$2\frac{1}{2}$	Shining white	5.56	Sectile masses	Bright wax-like luster; greenish gray
Cryolite, Na_3AlF_6	$2\frac{1}{2}$	White	2.97	Brittle masses	Resembles white paraffin
Fluorite, CaF_2	4	White	3.18	Cubes	Octahedral cleavage

HALIDES

The halides include those minerals whose principal anions are the halogen elements (fluorine, bromine, chlorine, and iodine). Halides are relatively soft, and are often light in color; some are water-soluble with a salty or bitter taste. Crystals are often cubes.

HALITE (Common Salt; Rock Salt)

NaCl   Isometric
    Hardness $2\frac{1}{2}$ Sp. Gr. 2.16

Summary

A soft, water-soluble, transparent to translucent mineral with a salty taste; commonly in cubic crystals (Plate 23(*A*), facing p. 146), or in cubes with step-like depressions in each face (hopper crystals, Figure 2-16, p. 16). Also granular to compact masses. *Color:* white or colorless; rarely with deep blue to violet color localized in irregular patches (Plate 23(*C*); also yellow and red. *Cleavage:* perfect cubic. *Fracture:* conchoidal. *Streak:* white. *Luster:* vitreous; fairly brittle; sometimes luminescent. Commonly associated with anhydrite, gypsum, polyhalite, and sylvite in sedimentary beds. CUBIC CRYSTALS, CLEAVAGE, AND SALTY TASTE ARE CHARACTERISTIC. Resembles SYLVITE.

Composition. Sodium chloride, NaCl (39.3% Na, 60.7% Cl). Often mechanically mixed with calcium and mag-

nesium sulfates and chlorides. Halite containing calcium or magnesium chlorides often absorbs moisture from the atmosphere and gradually dissolves (deliquesces).

Coloration. Irregular patches of deep blue to violet color are believed to be caused by certain defects in the crystal structure that absorb light (color centers). The cause of the structural defects has been attributed to an excess of sodium ions; to colloidal particles of sodium metal; to exposure to light after being subjected to pressure; or to irradiation with gamma rays.

Crystallization. Isometric; hexoctahedral class, $\frac{4}{m}\,\bar{3}\,\frac{2}{m}$. Commonly in cubes, often with hopper-like depressions; also coarse, granular, to compact masses.

Tests. Readily dissolves in water. Tastes salty.

Associated Minerals. Gypsum, anhydrite, sylvite, polyhalite ($K_2SO_4 \cdot MgSO_4 \cdot 2CaSO_4 \cdot 2H_2O$), carnallite ($KMgCl_3 \cdot 6H_2O$), kainite ($MgSO_4 \cdot KCl \cdot 3H_2O$); other salts, and clay minerals.

Minerals of Similar Appearance

Sylvite. Has a bitter taste, and commonly occurs in cubes modified by octahedrons. Halite does not ordinarily exhibit octahedral faces.

Occurrence. Halite is a common mineral, occurring in large beds formed by the evaporation of enclosed bodies of sea water. As the sea water evaporates, the concentration of salts increases; and, upon supersaturation, halite (NaCl), gypsum ($CaSO_4 \cdot 2H_2O$), anhydrite ($CaSO_4$), sylvite (KCl), and other salts are precipitated. The salt beds are often interstratified with shale and covered by other sedimentary rocks. The deformation of salt beds at depth sometimes causes the salt to flow under pressure and squeeze upward along lines of weakness in the overlying strata to form salt domes or plugs (see Figure 8-1, p. 160). Gypsum, anhydrite, sulfur, and oil are often associated with salt domes.

Large salt deposits occur in Stassfurt, Germany; Poland; Austria; the U.S.A.; England; France; the U.S.S.R.; and Czechoslovakia. In the United States, salt is recovered from salt domes in Texas and Louisiana. Extensive beds of halite occur in New York State, Michigan, Ohio, and New Mexico.

Small amounts of halite are formed as a volcanic sublimate.

Use. Most halite is widely used in the chemical industry as a source of sodium and chlorine. Used in winter to reduce ice on highways. Halite is most familiar as common table salt.

SYLVITE

KCl   Isometric
    Hardness 2   Sp. Gr. 1.98

Summary

Soft, water-soluble, transparent to translucent with a *bitter* salty taste. Commonly in cubes or cubes modified by octahedrons (Plate 23(*B*), facing p. 146); also in cleavable and fine grained masses. *Color:* white or colorless; frequently tinted red by hematite inclusions; also blue, violet, yellow, gray. *Streak:* white. *Cleavage:* perfect cubic. *Luster:* vitreous; not as brittle as halite. HABIT, CUBIC CLEAVAGE, AND BITTER SALTY TASTE ARE CHARACTERISTIC. Resembles HALITE but has a bitter taste, is often colored red, and commonly occurs in cubo-octahedral crystals.

Composition. Potassium chloride, KCl (52.4% K, 47.6% Cl). May be mixed with NaCl (halite).

Tests. Soluble in water. Bitter salty taste.

Crystallization. Isometric; hexoctahedral class, $\frac{4}{m}\bar{3}\frac{2}{m}$. Cubic crystals are common; they are also cubes modified by octahedrons; granular to compact masses.

Associated Minerals. Other potassium salts like carnallite ($KMgCl_3 \cdot 6H_2O$), polyhalite ($K_2SO_4 \cdot MgSO_4 \cdot 2CaSO_4 \cdot 2H_2O$), and kainite ($MgSO_4 \cdot KCl \cdot 3H_2O$). Also commonly associated with halite, gypsum, anhydrite, and clay minerals.

Occurrence. Sylvite occurs in salt beds with halite, but is less abundant. Also as a volcanic sublimate. Important sylvite deposits occur in Stassfurt, Germany; Saskatchewan, Canada; Utah; New Mexico; and Texas.

Use. Source of potassium compounds for fertilizers.

CERARGYRITE (Horn Silver)

AgCl   Isometric
    Hardness $2\frac{1}{2}$ Sp. Gr. 5.56

Summary

Soft, sectile (can be cut with a knife), translucent; commonly in masses or crusts with a bright, *waxy* or horn-like luster. Cubic crystals are rare. *Color:* usually gray or greenish-gray; sometimes colorless; rapidly darkens to violet-brown or purple on exposure to light. *Streak:* shining white. COLOR, WAX-LIKE APPEARANCE, AND EXTREME SECTILITY ARE CHARACTERISTIC.

Composition. Silver chloride, AgCl. A complete solid solution series exists between *cerargyrite* (AgCl) and *bromyrite* (AgBr). Small amounts of iodine may substitute for either chlorine or bromine.

Tests. Fuses easily on charcoal to a malleable globule of metallic silver. Globule dissolves in nitric acid and gives a white curdy precipitate of silver chloride on the addition of hydrochloric acid.

Crystallization. Isometric; hexoctahedral class, $\frac{4}{m}\bar{3}\frac{2}{m}$. Usually in wax-like masses or crusts; rarely in cubic crystals.

Alteration. Commonly alters from silver, argentite, and silver sulfosalts.

Occurrence. Cerargyrite and bromyrite are secondary minerals, formed from the oxidation of silver minerals; usually in arid regions. Found associated with silver, argentite, proustite, pyrargyrite, limonite, cerussite, galena, barite, wulfenite, malachite, and pyromorphite.

Cerargyrite occurs in notable amounts at Broken Hill, New South Wales; Atacama, Chile; Potosí, Bolivia; and in Peru; and Mexico. In the United States, important deposits are located at Leadville, Colorado; Treasure Hill, Nevada; in the silver districts of Utah and New Mexico; and in Arizona. Well-formed cubic and cubo-octahedral crystals have been found at the Poorman Mine, Owyhee County, Idaho.

Use. An ore of silver.

CRYOLITE

Na₃AlF₆   Monoclinic
Hardness 2½   Sp. Gr. 2.97

Summary

A soft, translucent, *snow white* to colorless mineral that looks somewhat like white paraffin. Usually in coarse granular masses; commonly associated with brown siderite (Plate 23(*F*), facing p. 146). Pseudocubic crystals are rare. *Luster:* wax-like. *Cleavage:* none. *Parting:* pseudocubic. *Streak:* white; brittle. Fragments of cryolite in water are almost invisible because the refractive indices of cryolite are very close to the index of water. The main cryolite deposit is in a unique pegmatite in a granite stock at Ivigtut, Greenland, where cryolite is associated with siderite, topaz, fluorite, microcline, quartz, and minor amounts of sulfide minerals. WHITE PARAFFIN-LIKE APPEARANCE AND ASSOCIATION WITH BROWN SIDERITE ARE CHARACTERISTIC.

Composition. Sodium aluminum fluoride, Na₃AlF₆ (32.8% Na, 12.8% Al, 54.4% F). Traces of iron, calcium, manganese, or organic material may color cryolite reddish or brownish.

Tests. Fuses easily on charcoal to form a colorless bead, which becomes opaque white on cooling. Colors flame bright yellow (sodium).

Crystallization. Monoclinic; prismatic class, $\frac{2}{m}$. Usually in coarse, granular masses. Pseudocubic or pseudo-cubo-octahedral crystals are rare.

Alteration. Commonly alters to rare aluminum fluoride minerals.

Associated Minerals. Siderite, topaz, microcline, quartz, fluorite, sphalerite, galena, chalcopyrite, pyrite, and molybdenite.

Occurrence. Cryolite is an uncommon mineral that occurs in large amounts in only one location—Ivigtut, Greenland. Masses of cryolite occur with the associated minerals in the Ivigtut pegmatite.

Use. Used as a flux in the electrolytic process of refining bauxite ore to obtain aluminum. At present, much cryolite used for this process is manufactured artificially from fluorite.

FLUORITE

CaF₂   Isometric
Hardness 4   Sp. Gr. 3.18

Summary

Usually in glassy transparent to translucent cubes (Plate 23(*D*), facing p. 146). Penetration twins, the corners of one penetrating the cube faces of the other, are common (Figure 8-29); also in coarse- to fine-grained masses. Fluorite occurs in many *colors,* most frequently violet, yellow, blue-green, light green, pink, white, and colorless. Fluorite associated with radioactive minerals tends to become black or very dark purple. Crystals often exhibit an uneven distribution of color, frequently in bands parallel to the crystal faces. The massive material also shows color

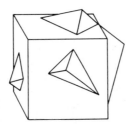

Fig. 8-29. Fluorite—penetration twin.

zoning, sometimes zigzag bands of different colors. *Cleavage:* perfect octahedral (Plate 23(*E*), facing p. 146). *Streak:* white; brittle. Frequently *luminesces violet-blue* under ultraviolet light. The word "fluorescence" is derived from the mineral fluorite. Chlorophane is a thermoluminescent variety of fluorite, usually rose-pink, that phosphoresces bright green in the dark after being heated to about 150°C. PERFECT OCTAHEDRAL CLEAVAGE, CUBIC CRYSTALS, AND HARDNESS ARE CHARACTERISTIC. May resemble CALCITE, or APOPHYLLITE.

Composition. Calcium fluoride, CaF_2 (51.3% Ca, 48.7% F). Yttrium and cerium may substitute partially for calcium. Other rare earth elements may be present in minute amounts.

Tests. In the closed tube at low heat (about 150°C) fluorite sometimes becomes phosphorescent. The phosphorescence may be green, blue, purple, or reddish, but must be observed in the dark. Fuses on charcoal. May lose color on heating. Frequently luminesces under ultraviolet light. Powdered mineral in hot concentrated sulfuric acid liberates hydrofluoric acid, which etches glass.

Crystallization. Isometric; hexoctahedral class, $\frac{4}{m}\,\bar{3}\,\frac{2}{m}$. Crystals are usually cubic, less often octahedral or dodecahedral. Cube faces are usually shiny and smooth; octahedral faces dull and rough. Octahedral crystals are believed to have formed at high temperatures, while cubic crystals are an indication of low temperature deposition. Pink or green are the common colors of the high-temperature octahedral crystals. Cubic crystals are found in a great variety of colors. Cubic penetration twins are common. Also coarse- to fine-grained masses. Color banding is common in crystals and in massive material.

Minerals of Similar Appearance

Quartz. Purple fluorite is often mistaken for amethyst quartz but can easily be distinguished by hardness, habit, and cleavage. Fluorite (H. = 4) can be scratched by a knife, but quartz (H. = 7) cannot. Fluorite commonly occurs in cubes with perfect octahedral cleavage—quartz in six-sided crystals, cleavage rarely seen.

Calcite. Effervesces with a drop of HCl.

Apophyllite. Tetragonal but crystals may look cubic. However, the faces of the pseudocube differ in luster; the two basal faces are commonly pearly, whereas the other four have a glassy luster.

Occurrence. Fluorite is a common mineral. It occurs most often as a vein mineral associated with calcite, dolomite, barite, quartz, galena, and sphalerite. In high-temperature veins with cassiterite, wolframite, topaz, tourmaline, molybdenite, apatite, and quartz. Also in dolomites and limestones; and as a minor accessory mineral in igneous rocks.

In the United States large fluorite deposits occur in veins near Rosiclare, Illinois; and in Kentucky. Associated minerals include calcite, barite, dolomite, quartz, galena, and sphalerite. Other important fluorite producers are Germany, Mexico, the U.S.S.R., Canada, Italy, England, and France. Beautiful groups of large cubic crystals come from Cumberland and Derbyshire, England. Some of the crystals are enclosed in crusts of quartz. Rose-colored octahedral crystals occur in St. Gotthard, Switzerland, and elsewhere in the Alps. Green octahedral crystals occur in Cera, Brazil; Kongsberg, Norway; Bancroft, Ontario; and South-West Africa. CHLOROPHANE, a thermoluminescent variety of fluorite, phosphoresces bright green when heated. It is found in small amounts at the Burnt Hill Tungsten Mine, New Brunswick, Canada; and at Trumbull, Connecticut.

Use. Fluorite is used as a flux in the manufacture of steel. The name is derived from the Latin word *fluere*

A   x 0.61

RUTILATED QUARTZ

Reddish-brown hair-like inclusions of rutile in quartz. The random orientation of the rutile inclusions indicates that they were formed first and later enveloped by the quartz crystal: Minas Gerais, Brazil.

B   x 0.62

ROSE QUARTZ

Aggregate of small rose quartz crystals: Brazil.

C   x 0.45

SMOKY QUARTZ

Dark smoky brown quartz crystal: Minas Gerais, Brazil.

D   x 0.70

QUARTZ (AMETHYST)

Pale amethyst quartz crystals showing uneven color distribution: Minas Gerais, Brazil.

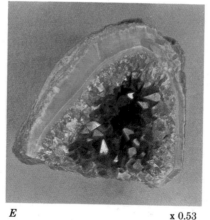

E   x 0.53

QUARTZ (AMETHYST GEODE)

Cavity lined first with gray chalcedony and then with amethyst crystals: Hungary.

F   x 0.30

PEGMATITIC QUARTZ

Clusters of small rose quartz crystals in subparallel arrangement on large slightly smoky quartz crystals. Mica and platy albite are also present: Brazil.

A   x 0.50

**SILICIFIED WOOD
(CHALCEDONY)**

Wood replaced by red jasper and gray chalcedony (petrified wood): Petrified Forest, Arizona.

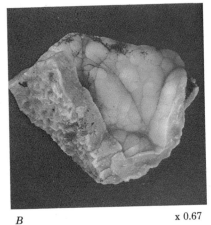

B   x 0.67

CHALCEDONY

Botryoidal aggregate of bluish-gray chalcedony: Cornwall, England.

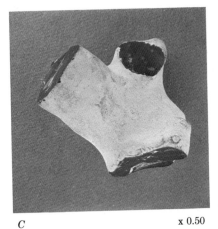

C   x 0.50

FLINT (CHALCEDONY)

Grayish-black flint occurring in white chalk: Dover, England.

D   x 0.81

CHRYSOPRASE (CHALCEDONY)

Translucent apple green chalcedony: Tulare County, California.

E   x 0.65

AGATE (CHALCEDONY)

Chalcedony showing concentric banding.

F   x 0.51

JASPER (CHALCEDONY)

Banded jasper: Germany.

(to flow), because fluorite melts easily. Also used in the preparation of hydrofluoric acid; in the manufacture of opal glass; and in fluorocarbon plastics and gases. Colorless transparent crystals are used for making lenses; also spectrographic prisms that transmit ultraviolet light.

CARBONATES

The carbonates include a number of common minerals. *Calcite, smithsonite, aragonite, witherite, strontianite, malachite,* and *azurite* will effervesce in cold dilute HCl, giving off bubbles of CO_2. Others require hot HCl or HNO_3. The carbonates are relatively soft, and can be scratched by a knife blade.

The members of the CALCITE GROUP belong to the hexagonal-scalenohedral class of the hexagonal system. They all have *perfect rhombohedral cleavage,* but the cleavage is not often seen in smithsonite because of its botryoidal habit.

The members of the ARAGONITE GROUP belong to the dipyramidal class of the orthorhombic system. The crystals are often characterized by twinned intergrowths that are pseudohexagonal in appearance (Plate 26, facing p. 162).

Carbonate Minerals

CALCITE GROUP

Calcite, $CaCO_3$
Magnesite, $MgCO_3$
Siderite, $FeCO_3$
Rhodochrosite, $MnCO_3$
Smithsonite, $ZnCO_3$

DOLOMITE GROUP

Dolomite, $CaMg(CO_3)_2$

ARAGONITE GROUP

Aragonite, $CaCO_3$
Witherite, $BaCO_3$
Strontianite, $SrCO_3$
Cerussite, $PbCO_3$

HYDROUS CARBONATES

Malachite, $Cu_2(CO_3)(OH)_2$
Azurite, $Cu_3(CO_3)_2(OH)_2$
Aurichalcite, $(Zn,Cu)_5(OH)_6(CO_3)_2$
Hydrozincite, $Zn_5(OH)_6(CO_3)_2$

CALCITE

$CaCO_3$　　　　　　　　　　　　　Hexagonal
　Hardness 3　　　　　　　　Sp. Gr. 2.71
　(on cleavage face)

Summary

A soft, vitreous, transparent to translucent mineral with *perfect rhombohedral cleavage* (Plate 24, facing p. 147). *Color:* usually white or colorless, but may assume a variety of tints. *Streak:* white; brittle. The habit of calcite is extremely varied, and often complex. Over 300 form-combinations have been recorded, but the most common are the rhombohedron and the scalenohedron (Plate 6(*E*), facing p. 35). Repeated lamellar twinning and contact twins are common (Plate 8(*B*), facing p. 51). Sometimes luminesces red, yellow, pink, or blue in ultraviolet light (Plate 9, facing p. 66). HARDNESS, RHOMBOHEDRAL CLEAVAGE, AND EFFERVESCENCE IN COLD DILUTE HCl ARE CHARACTERISTIC. Resembles ARAGONITE and DOLOMITE.

Composition. Calcium carbonate, $CaCO_3$ (56.0% CaO, 44.0% CO_2). Mn, Zn, Fe, and Co may substitute in part for calcium. A complete substitutional series extends to rhodochrosite ($MnCO_3$), and a partial series extends towards smithsonite and siderite.

Tests. Dissolves with effervescence in cold dilute HCl. Infusible. Sometimes luminesces under ultraviolet light. Often thermoluminescent.

Crystallization. Hexagonal; hexagonal scalenohedral class, $\bar{3}\frac{2}{m}$. Crystals are common but their habits are extremely varied (Figure 8-30). The most important habits are (**1**) thick to thin rhombohedra, (**2**) scalenohedral, and (**3**) short to long prisms. Crystals are fre-

Some Properties of the Carbonates

CALCITE GROUP	COLOR	HARDNESS	EFFERVESCENCE	REMARKS
Calcite	White; numerous tints	3	Cold dilute HCl	Perfect rhombohedral cleavage
Magnesite	White	$3\frac{1}{2}$–5	Hot dilute HCl	Usually in dull compact masses
Siderite	Brown; gray	3.7–4.2	Hot dilute HCl	Perfect rhombohedral cleavage; brown color
Rhodochrosite	Pink; rose red	$3\frac{1}{2}$–$4\frac{1}{2}$	Hot dilute HCl	Pink color; rhombohedral cleavage
Smithsonite	White; green; blue; yellow; pink; gray	4–$4\frac{1}{2}$	Cold dilute HCl	Botryoidal aggregates
ARAGONITE GROUP				
Aragonite	White; various tints	$3\frac{1}{2}$–4	Cold dilute HCl	Pseudohexagonal prisms
Witherite	White	$3\frac{1}{2}$	Cold dilute HCl	Pseudohexagonal bipyramids; Sp. Gr. 4.29
Strontianite	White; gray; yellow; pink; greenish	$3\frac{1}{2}$–4	Cold dilute HCl	Fibrous aggregates; Sp. Gr. 3.72
Cerussite	White	3–$3\frac{1}{2}$	Warm dilute HNO_3	Brilliant luster; heavy; grid-like aggregates
DOLOMITE GROUP				
Dolomite	White; pale tints	$3\frac{1}{2}$–4	Hot HCl; powder in cold HCl	Curved crystal aggregates
HYDROUS CARBONATES				
Malachite	Green	$3\frac{1}{2}$–4	Cold dilute HCl	Green color
Azurite	Azure blue	$3\frac{1}{2}$–4	Cold dilute HCl	Azure blue color
Aurichalcite	Greenish-blue	2	Hot dilute HCl	Blue-green pearly scales
Hydrozincite	White	2–$2\frac{1}{2}$	Hot dilute HCl	Dull white masses; usually luminesces pale blue to violet

quently complex. Polysynthetic twinning is common; so are simple contact twins.

Varieties

Iceland Spar. Colorless, transparent calcite showing strong double refraction.

Dog-tooth Spar. Acute scalenohedral crystals.

Satin Spar. Fibrous calcite with a silky luster.

Sand Calcite. Crystals of calcite containing up to 60% quartz sand (Figure 2-28, p. 21).

Mexican Onyx. Banded calcite or aragonite (Plate 24(*A*), facing p. 147).

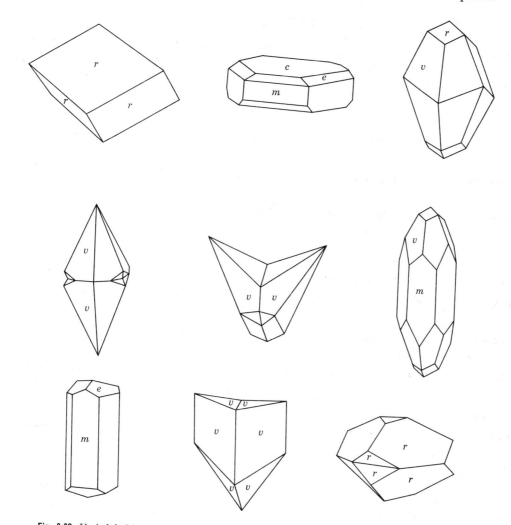

Fig. 8-30. Varied habits of calcite crystals. Forms: (*r*) positive rhombohedron; (*e*) negative rhombohedron; (*c*) pinacoid; (*m*) prism; (*v*) scalenohedron.

Cave Deposits. *Stalactites*—icicle-shaped masses suspended from the roofs of caverns, formed by dripping mineral-charged waters. *Stalagmites*—cone-shaped masses built up on the floor of caverns by dripping water (Figure 8-31).

Travertine or **Tufa.** Deposits of calcite left by hot or cold calcareous springs. The masses are often porous.

Limestone; Chalk. Dull compact rocks whose chief or only constituent may be calcite.

Marble. Coarse- to fine-grained metamorphic equivalent of limestone.

Minerals of Similar Appearance

Aragonite. Lacks rhombohedral cleavage and differs in crystallization.

Dolomite. Only slightly soluble in cold HCl; curved, saddle-shaped aggregates are characteristic.

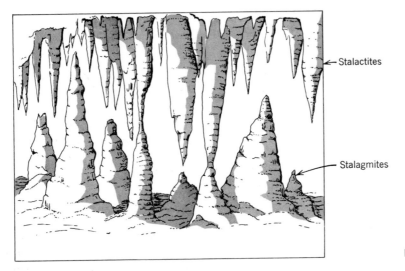

Fig. 8-31. Stalactites and stalagmites.

Occurrence. Calcite is a common and widely distributed mineral. Huge sedimentary limestone deposits have been formed primarily by the very thick deposition and compaction of calcareous skeletons and shells of sea animals on the ocean bed. Limestones recrystallized by metamorphism may be known as marbles (or crystalline limestones). Calcite is the predominant mineral and sometimes the only mineral in many limestones.

Calcite commonly occurs as a hydrothermal vein mineral with sulfide ores; also as travertine deposits around hot and cold springs and streams. An important spring deposit occurs at Mammoth Hot Springs in Yellowstone National Park.

Cave Deposits. Huge limestone caverns like the Mammoth Cave in Kentucky, the Luray Caverns in Virginia, or the Carlsbad Caverns in New Mexico are the result of ground water carrying carbon dioxide (carbonic acid, H_2CO_3) that slowly dissolves masses of limestone. Calcium carbonate is only slightly soluble in water; however, it dissolves readily in water charged with carbon dioxide, and forms soluble calcium bicarbonate ($Ca(HCO_3)_2$). The reaction is:

$$CaCO_3 + CO_2 + H_2O \rightleftharpoons Ca^{2+} + 2HCO_3^{-}.$$

When the water containing calcium bicarbonate in solution enters a cavern, it looses some of its carbon dioxide and the above reaction is reversed. The calcium carbonate is deposited on the roof of the caverns by the dripping water, and slowly builds up to form STALACTITES. Corresponding STALAGMITES rise from the floor of the cavern where the water strikes as it drips from the roof.

The important occurrences of calcite are too numerous to list. A few of the outstanding locations are the Tri-State district of Missouri, Kansas, and Oklahoma where very large golden yellow scalenohedral crystals of calcite are found (Plate 24(*B*), facing p. 147), some a few feet in length. Fine crystals of colorless to pale yellow transparent calcite, some over 5 in. long, enclose bright copper in the copper deposits of the Keeweenaw Peninsula in northern Michigan. Good crystals also occur in the Harz Mountains of Germany, in Iceland (Iceland spar), and in England. Banded calcite and aragonite, known as Mexican onyx or onyx marble, occurs chiefly in Mexico, California, and Utah. The important limestone districts in this country are in Illinois, Michigan, Ohio, New York, Pennsylvania, Indiana, and Wisconsin.

Use. Calcite is used in the production of Portland cement, quicklime (CaO), fertilizers, fluxes. Also widely used as a building and ornamental material.

SIDERITE

FeCO$_3$ Hexagonal
     Hardness 3.7–4.2   Sp. Gr. 3.9±

Summary

A soft, light to dark *brown* to gray vitreous mineral with perfect *rhombohedral cleavage*. *Streak:* white; translucent. Rhombohedral crystals are frequently curved; resemble saddle-shaped dolomite. Cleavable masses (Plate 24, facing p. 147) are common. Effervesces in hot dilute HCl. Alters readily to goethite. BROWNISH COLOR, CURVED CRYSTALS, RHOMBOHEDRAL CLEAVAGE, AND MAGNETISM AFTER HEATING ARE DISTINCTIVE. May resemble massive SPHALERITE.

Composition. Ferrous carbonate, FeCO$_3$ (62.1% FeO, 37.9% CO$_2$). Manganese and magnesium substitute for iron. A complete substitutional series extends to rhodochrosite (MnCO$_3$) through the substitution of Mn^{2+} for Fe^{2+}, and to magnesite (MgCO$_3$) by the substitution of Mg^{2+} for Fe^{2+}. Limited amounts of calcium can replace part of the ferrous iron.

Tests. Siderite dissolves in hot HCl. Turns black and magnetic when heated on charcoal or in the reducing flame.

Crystallization. Hexagonal; hexagonal-scalenohedral class, $\overline{3}\,\frac{2}{m}$. Commonly in coarse, cleavable or fine-grained masses. Crystals are usually rhombohedral, frequently with acute angles. Faces are often curved or composite; sometimes in tabular or scalenohedral crystals. Also botryoidal, concretionary, oölitic, compact, and in earthy masses.

Minerals of Similar Appearance

Dolomite and **Ankerite.** Curved siderite crystals strongly resemble dolomite, but brownish color and high specific gravity are distinctive (Sp. Gr. dolomite 2.85). Usually, brownish rhombohedral ankerite crystals (Ca(Mg,Fe)(CO$_3$)$_2$) cannot readily be distinguished from siderite crystals.

Sphalerite. The perfect rhombohedral cleavage of siderite distinguishes it from massive sphalerite, which has dodecahedral cleavage.

Occurrence. Siderite frequently occurs in sedimentary deposits of clay, coal seams, or shale. It is usually massive or concretionary. When mixed with clay, siderite is known as clay-ironstone; it is known as black-banded ore when mixed with carbonaceous material. These sedimentary deposits are extensive in England, Scotland, and western Pennsylvania. Siderite also occurs with sulfide minerals in hydrothermal veins containing ores of silver, galena, pyrite, chalcopyrite, and tetrahedrite. Extensive deposits of siderite may be formed by hydrothermal replacement of limestone by iron-bearing solutions. Deposits of this type occur in Spain, Algeria, and Austria. Siderite is also found with cryolite in the pegmatites of Ivigtut, Greenland.

Alteration. To goethite.

Use. An ore of iron.

SMITHSONITE

ZnCO$_3$ Hexagonal
     Hardness 4–4½   Sp. Gr. 4.4

Summary

Usually in botryoidal or stalactitic masses (Plate 25(*A, B*), facing p. 154), or in dull honeycombed masses that resemble dry bone (dry-bone ore). *Color:* varies—white, gray-white, pale brown, green to blue-green (from Cu), bright yellow (from geenockite CdS), and pink (from Co). *Luster:* vitreous to dull. *Streak:* white. *Cleavage:* rhombohedral but not often seen because of a common botryoidal or bone-like habit; brittle; translucent. Effervesces in cold HCl. Smithsonite is found in the weathered zone of zinc ore deposits, and is usually formed by the alteration of a primary zinc mineral such as sphalerite. Commonly associated with sphalerite, hemimorphite, galena, cerussite, anglesite, pyromorphite, malachite, azurite, and calcite. BOTRYOIDAL OR STALACTITIC HABIT, EFFER-

VESCENCE IN COLD HCl, AND HIGH SPECIFIC GRAVITY FOR A CARBONATE ARE CHARACTERISTIC. Resembles PREHNITE and HEMIMORPHITE.

Composition. Zinc carbonate, $ZnCO_3$ (64.8% ZnO, 35.2% CO_2). Ferrous iron commonly substitutes for part of the zinc. Smaller amonts of Mn, Mg, Cu, Ca, Cd, Co, and Pb may also substitute in part for zinc.

Tests. Infusible. Soluble with effervescence in cold HCl. When heated on charcoal, gives a white sublimate of zinc oxide; yellow when hot. Sublimate turns green when moistened with cobalt nitrate and heated again (test for zinc, but difficult to obtain).

Crystallization. Hexagonal; hexagonal scalenohedral class, $\bar{3}\frac{2}{m}$. Commonly in botryoidal or stalactitic masses, and in porous bone-like masses known as *drybone ore*. Also in granular to compact masses. Rarely in rhombohedral crystals with curved faces, or in rounded scalenohedrons.

Minerals of Similar Appearance

Hemimorphite. Botryoidal masses of hemimorphite may resemble smithsonite but do not effervesce in acid, and are lighter (Sp. Gr. = 3.4–3.5).

Prehnite. Green botryoidal masses of prehnite cannot be scratched by a knife blade (H. = 6–6½); are much lighter in weight than smithsonite (Sp. Gr. prehnite = 2.8–2.9); and do not dissolve with effervescence in acid.

Alteration. From sphalerite.

Associated Minerals. Sphalerite, hemimorphite, galena, cerussite, anglesite, pyromorphite, mimetite, aurichalcite, azurite, malachite, and hydrozincite.

Occurrence. Smithsonite is a secondary mineral formed by the alteration of such primary zinc minerals as sphalerite, and occurs in the oxidized zone of zinc ore deposits.

Beautiful blue-green botryoidal masses come from the Kelly Mine, Magdalena, New Mexico (Plate 25(*B*), facing p. 154). Stalactitic and botryoidal aggregates colored bright yellow by greenockite, CdS (turkey-fat ore), occur in Marion County, Arkansas, and also in the lead-zinc deposits of Sardinia. Smithsonite

is mined as a zinc ore at Leadville, Colorado. Botryoidal crusts occur abundantly in a variety of colors at Laurium, Greece. Small, curved rhombohedral crystals occur in Broken Hill, New South Wales; and in Tsumeb, South-West Africa.

Name. Smithsonite was named after James Smithson, founder of the Smithsonian Institution in Washington. Smithsonite is often called CALAMINE in Europe. In the United States, the zinc silicate hemimorphite was called calamine, but hemimorphite is now the preferred name.

Use. A minor zinc ore. Also used as a decorative material.

RHODOCHROSITE

$MnCO_3$	Hexagonal
Hardness 3½–4	Sp. Gr. 3.7

Summary

A *pale pink* to *rose-red* mineral (Plate 25(*D, E, F*), facing p. 154), sometimes grayish or brown. Weathers readily to blackish manganese oxides. *Cleavage:* perfect rhombohedral. *Streak:* white. *Luster:* vitreous to pearly; transparent to translucent; brittle. Effervesces in hot HCl. Usually in coarse, cleavable masses. Less frequently in rhombohedral crystals or in botryoidal crusts. PINK COLOR, RHOMBOHEDRAL CLEAVAGE, AND EFFERVESCENCE IN HOT HCl ARE CHARACTERISTIC. Resembles RHODONITE.

Composition. Manganese carbonate, $MnCO_3$ (61.7% MnO, 38.3% CO_2). Ferrous iron and calcium substitute for manganese, and a complete substitutional series extends to siderite $FeCO_3$, and calcite $CaCO_3$. Limited amounts of Mg, Zn, and Co may substitute for Mn.

Tests. Infusible. Soluble with effervescence in hot HCl. Powdered mineral colors borax bead violet in oxidizing flame (Mn test).

Crystallization. Hexagonal; hexagonal-scalenohedral class, $\bar{3}\frac{2}{m}$. Usually in cleavable masses; also compact, and botryoidal. Rhombohedral crystals with curved faces are rare.

Minerals of Similar Appearance

Rhodonite. Resembles massive rhodochrosite in color, but is harder (H. = $5\frac{1}{2}$–6); lacks rhombohedral cleavage; and does not dissolve in HCl.

Alteration. To pyrolusite (MnO_2) or manganite ($MnO(OH)$) on exposure to the atmosphere.

Occurrence. Usually occurs as a primary mineral in hydrothermal lead, copper, or silver ore veins. Less frequently in high-temperature metamorphic deposits with garnet, rhodonite, tephroite, and manganese oxides. Sometimes in pegmatites. Also as a secondary mineral in residual iron and manganese oxide deposits.

Beautiful, deep rose-red, transparent rhombohedral crystals of rhodochrosite, ranging in size from $\frac{1}{2}$ to 2 in., occur near Leadville, Lake County, Colorado. Massive, pink rhodochrosite is mined as a manganese ore mineral at Butte, Montana. Found in the silver veins of Roumania, and in Freiberg, Germany. Aggregates with a concentrically banded inner structure of light and dark pink rhodochrosite (Plate 25, facing p. 154) occur in the Catamarca Province in Argentina.

Use. A minor ore of manganese.

MAGNESITE

$MgCO_3$ Hexagonal
   Hardness 4   Sp. Gr. 3.0±

Summary

Usually in dull white, compact porcelain-like masses with a conchoidal fracture (Plate 25(*C*), facing p. 154). Also in coarse, cleavable, marble-like masses. Rarely in rhombohedral crystals. *Color:* white, sometimes tinted gray, yellowish, or brown. *Luster:* dull to glassy. *Streak:* white. *Cleavage:* rhombohedral. *Fracture:* conchoidal; transparent to translucent; brittle. Effervesces in hot HCl. Sometimes luminesces blue or green under ultraviolet light. COMPACT, DULL WHITE MASSES THAT RESEMBLE UNGLAZED PORCELAIN AND EFFERVESCENCE IN HOT HCl ARE OFTEN CHARACTERISTIC. May resemble calcite, dolomite, kaolin, microcrystalline datolite, or chert.

Composition. Magnesium carbonate, $MgCO_3$ (47.8% MgO, 52.2% CO_2). A complete substitutional series extends to siderite, $FeCO_3$, as a result of the substitution of ferrous iron for magnesium. Manganese, calcium, and cobalt may substitute for magnesium.

Tests. Infusible. Dissolves with effervescence in hot HCl. Sometimes shows bluish or greenish luminescence under ultraviolet light. May be triboluminescent.

Crystallization. Hexagonal; hexagonal-scalenohedral class, $\bar{3}\frac{2}{m}$. Usually in dull white microcrystalline masses; less frequently in coarse, cleavable masses. Rarely in crystals.

Minerals of Similar Appearance

Calcite. Dissolves readily in cold HCl; magnesite dissolves in hot HCl.

Dolomite. Cleavable masses of dolomite strongly resemble coarse, cleavable masses of magnesite and usually require x-ray tests to differentiate. Specific gravity in liquids may be used.

Kaolin. May resemble microcrystalline magnesite, but is softer and does not effervesce in HCl.

Microcrystalline datolite. Does not dissolve in HCl.

Chert. Resembles porcelain-like magnesite but cannot be scratched by a knife blade (H. = 7), and does not dissolve in HCl.

Alteration. Alters from rocks that are rich in magnesium. Frequently replaces calcite or dolomite.

Occurrence. 1. Magnesite forms by the alteration of such magnesium-rich rocks as serpentine, dunite (olivine), and peridotite; the alteration is caused by waters carrying carbon dioxide. 2. Through the replacement of limestones and dolomites by magnesium-bearing solutions. 3. Rarely found in hydrothermal ore veins, in cavities in lavas, or as a primary mineral in igneous rocks. Coarse, cleavable masses of magnesite are of metamorphic origin, and are commonly associated with talc, chlorite, and mica schists.

Huge deposits of magnesite occur in Styria, Austria; in Manchuria; the island of Euboea, Greece; and in California. Large, colorless, and transparent rhombohedral crystals and cleavage rhombs (2″ to 3″ across) occur in a pegmatite at Bom Jesus dos Meiras, Bahia, Brazil.

Use. In the manufacture of magnesium oxide for refractory bricks; in the production of oxychloride cement; as a source of carbon dioxide; magnesium compounds for medicines; and as an ore of magnesium metal. The latter has been extracted from sea water.

ARAGONITE

$CaCO_3$ Orthorhombic
    Hardness $3\frac{1}{2}$–4   Sp. Gr. 2.94

Summary

Aragonite is dimorphous with calcite ($CaCO_3$), but is much less common and less stable. Paramorphs of calcite after aragonite are common. Three common habits are illustrated in Plate 26(*A*, *B*, *C*), facing p. 162: (**1**) short pseudohexagonal prisms formed by the intergrowth of three individuals; (**2**) slender, pointed crystals, often in radiating groups; and (**3**) curved twisted coral-like aggregates known as "flos ferri."

Color: usually white, colorless, or pale tints. *Luster:* vitreous. *Streak:* white. *Cleavage:* poor parallel to side pinacoid and prism; transparent to translucent; brittle; readily effervesces in cold HCl. Sometimes luminesces under ultraviolet light. PSEUDOHEXAGONAL PRISMS AND EFFERVESCENCE IN COLD ACID ARE CHARACTERISTIC. Resembles CALCITE but lacks rhombohedral cleavage.

Composition. Calcium carbonate, $CaCO_3$, like calcite (56.0% CaO, 44.0% CO_2). Strontium, lead, and zinc may substitute in part for calcium. Strontium variety (mossottite); lead (tarnowitzite); zinc (nicholsonite).

Tests. Infusible. Effervesces in cold dilute HCl. Decrepitates as it transforms to calcite when heated at low red heat in the closed tube.

Crystallization. Orthorhombic; dipyramidal class, $\frac{2}{m}\frac{2}{m}\frac{2}{m}$. Commonly in pseudohexagonal twinned crystals; single crystals are frequently long, slender and pointed; less frequently as tabular crystals. Also in botryoidal, columnar, stalactitic, fibrous, coralloidal, and pisolitic aggregates.

Alteration. To calcite.

Occurrence. Aragonite is less common than calcite. It is always a low-temperature mineral formed under near surface conditions. It occurs in hot spring deposits; in recent marine sediments; in gypsum or clay beds; as a constituent of the "mother of pearl" on the inside of some shells; as the inorganic constituent of pearls; with sedimentary limonite and siderite deposits; and in the weathered zone of ore deposits associated with secondary minerals like limonite, malachite, azurite, smithsonite, and cerussite.

Good pseudohexagonal twinned crystals are found in Aragon, Spain; in the sulfur mines of Girgenti, Sicily, and at Bastennes, France. Flos ferri (iron flowers) occur in the Austrian iron mines; at Bisbee, Arizona; in New Mexico; and also in Mexico. Long pointed crystals are common in Alston Moor and Cumberland, England. "Mexican onyx," containing aragonite, is found in California and Mexico.

WITHERITE

$BaCO_3$ Orthorhombic
    Hardness 3–$3\frac{1}{2}$ Sp. Gr. 4.29

Summary

Crystals of witherite are twinned, forming pseudohexagonal pyramids; usually with deep horizontal striations, and often resembling a stack of pyramids capping one another (Plate 26(*F*), facing p. 162). Commonly in fibrous, botryoidal, or granular masses. *Color:* white, colorless, or pale tints. *Luster:* vitreous. *Streak:* white. *Cleavage:* one distinct, two imperfect;

Plate 36. **SILICATES** ■ OPAL, $SiO_2 \cdot nH_2O$, CHALCEDONY, SiO_2

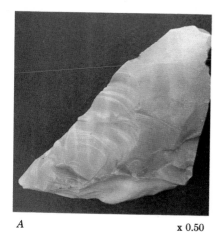

A   x 0.50

OPAL

$SiO_2 \cdot nH_2O$

Common opal showing conchoidal fracture: Virgin Valley, Nevada.

B   x 0.60

OPAL (HYALITE)

Clear colorless opal (hyalite) with a globular structure: San Luis Potosi, Mexico.

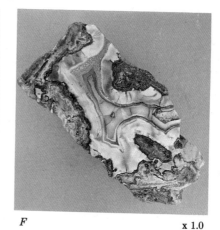

C   x 0.95

OPAL

Precious opal showing a play of colors: Queensland, Australia.

D   x 0.65

OPAL (GEYSERITE)

Porous opal deposited by thermal waters: Yellowstone Park, Wyoming.

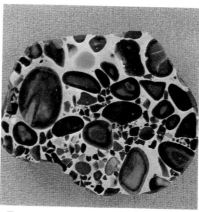

E   x 1.2

QUARTZ (PUDDINGSTONE)

A conglomerate of rounded nodules of chalcedony cemented by a grayish-white chalcedony groundmass: France.

F   x 1.0

CHALCEDONY

(an encrusted cavity) SiO_2

Polished specimen of banded chalcedony: Brazil.

A   x 0.45

MICROCLINE (AMAZON STONE)
K(AlSi₃O₈)

Green microcline (Amazon stone) showing perthite structure, an intergrowth of microcline with wavy white bands of albite: Hybla, Ontario, Canada.

B   x 0.62

MICROCLINE

Reddish microcline showing perthite structure and blocky cleavage in two directions at almost 90°: Chester County, Pennsylvania.

C   x 0.55

ADULARIA
K(AlSi₃O₈)

Twinned glassy white crystal: St. Gotthard, Switzerland.

D   x 0.63

OLIGOCLASE (SUNSTONE)

Oligoclase containing minute "parallel" hematite inclusions that reflect light (sunstone), and showing polysynthetic twinning striations characteristic of the plagioclase feldspars: Tvedestrand, Norway.

E   x 0.50

LABRADORITE

Polished section of labradorite showing a brilliant iridescent play of blue and green colors: Labrador.

F   x 0.60

ALBITE (CLEAVELANDITE)

Aggregate of thin plate-like albite crystals (Cleavelandite): Amelia Court House, Virginia.

transparent to translucent; brittle; effervesces in cold dilute HCl; often luminesces blue under ultraviolet light.

Witherite is not a common mineral. It usually occurs in low-temperature hydrothermal veins associated with galena and barite. HIGH SPECIFIC GRAVITY, EFFERVESCENCE IN COLD ACID, AND PSEUDOHEXAGONAL PYRAMIDS ARE DISTINCTIVE.

Composition. Barium carbonate, $BaCO_3$ (77.7% BaO, 22.3% CO_2). Small amounts of strontium may substitute for barium.

Tests. Fuses and gives yellow-green flame test for barium. Dissolves with effervescence in cold dilute HCl. The addition of sulfuric acid yields a white precipitate of barium sulfate, even in very dilute solutions. Often luminesces blue under ultraviolet light.

Crystallization. Orthorhombic; dipyramidal class, $\frac{2}{m}\frac{2}{m}\frac{2}{m}$. Inverts at 811°C to a hexagonal form, and to an isometric form at 982°C. Crystals are twinned forming pseudohexagonal pyramids, usually deeply horizontally striated. Commonly in botryoidal, fibrous, columnar, and granular masses.

Alteration. To barite.

Occurrence. In low-temperature hydrothermal veins associated with galena and barite. Found in good crystals at Cumberland and Northumberland, England. In the United States, large crystals occur in Rosiclare, Illinois, associated with fluorite. Abundant in massive varieties at El Portal, Yosemite National Park, California.

Use. A minor ore of barium.

STRONTIANITE

$SrCO_3$ Orthorhombic
    Hardness $3\frac{1}{2}$–4   Sp. Gr. 3.72

Summary

Usually in white, radiating fibrous masses. *Color:* white to colorless; also gray, greenish, yellowish, and pink. *Luster:* vitreous. *Streak:* white. *Cleavage:* good

prismatic; brittle; transparent to translucent. Soluble with effervescence in cold dilute HCl. May luminesce under ultraviolet light; sometimes thermoluminescent. Occurs as a low-temperature hydrothermal mineral commonly associated with barite, calcite, and celestite. HIGH SPECIFIC GRAVITY, FIBROUS AGGREGATES, FLAME TEST FOR STRONTIUM, AND EFFERVESCENCE IN COLD ACID ARE DISTINCTIVE. Resembles fibrous aggregates of ARAGONITE (Sp. Gr. 2.9) but strontianite is heavier.

Composition. Strontium carbonate, $SrCO_3$ (70.2% SrO, 29.8% CO_2). Some calcium usually substitutes for strontium.

Tests. When heated, it swells, sprouts, and colors flame crimson (test for Sr). Dissolves with effervescence in cold dilute HCl. Precipitates strontium sulfate when sulfuric acid is added to a medium strength solution.

Crystallization. Orthorhombic; dipyramidal class, $\frac{2}{m}\frac{2}{m}\frac{2}{m}$. Inverts to a hexagonal modification at 929°C. Usually in fibrous masses. Crystals are commonly acicular, or spear-shaped; also pseudohexagonal twins.

Alteration. To celestite, $SrSO_4$.

Occurrence. 1. In low-temperature hydrothermal veins in limestones and marls, usually associated with barite, celestite, and calcite. 2. In low-temperature sulfide veins. 3. In geodes, or as concretionary masses in clay or limestone beds.

The largest commercial strontianite deposit occurs in Westphalia, Germany. In the United States, large deposits occur in the Strontium Hills, San Bernardino County, California.

Use. As an ore of strontium. Strontium is used in the refining of sugar, and in fireworks, rockets, and medicines.

CERUSSITE

$PbCO_3$ Orthorhombic
    Hardness 3–$3\frac{1}{2}$ Sp. Gr. 6.55

Summary

A soft, heavy, brittle, *colorless* to *white* mineral with a *brilliant luster*. The habit is extremely varied, but intergrowths of slender twinned crystals forming a grid-like pattern are the most characteristic (Plate 26(*D*) facing p. 162). *Streak:* white. *Cleavage:* prismatic. *Fracture:* conchoidal; transparent to translucent. Cerussite is a secondary mineral found in the upper zone of lead deposits, where it alters from galena by the action of carbonated waters. Specimens of cerussite commonly contain cores of galena. VERY HIGH SPECIFIC GRAVITY FOR A NONMETALLIC MINERAL, BRILLIANT LUSTER, GRID-LIKE AGGREGATES, AND EFFERVESCENCE IN NITRIC ACID ARE DISTINCTIVE. Distinguishable from ANGLESITE ($PbSO_4$) by effervescence in acid.

Composition. Lead carbonate, $PbCO_3$ (83.5% PbO, 16.5% CO_2). May contain small amounts of silver that were present in the galena from which the cerussite altered.

Tests. On charcoal, fuses easily to a yellow bead and finally to a metallic lead globule. When heated in the closed tube, turns yellow, then reddish brown; turns yellow again on cooling. Dissolves with effervescence in dilute nitric acid.

Crystallization. Orthorhombic; dipyramidal class, $\frac{2}{m}\frac{2}{m}\frac{2}{m}$. Crystal habit is varied; small, tabular crystals are common. Also reticulated or grid-like groups, formed by twinned crystals crossing each other at 60° angles. Twinning also frequently produces star-like pseudohexagonal shapes and pseudohexagonal bipyramids. Also found in granular to compact masses.

Alteration. From galena, PbS; and from anglesite, $PbSO_4$.

Associated Minerals. Galena, sphalerite, anglesite, pyromorphite, limonite, smithsonite, malachite, azurite, phosgenite, and wulfenite.

Occurrence. Cerussite is a common and important secondary ore of lead. It is found in the oxidized zone of lead deposits, where it alters from galena by the action of carbonated waters. A few of the many important locations are Broken Hill, New South Wales, where cerussite commonly occurs as twinned grid-like aggregates. Fine, large crystals associated with anglesite, malachite, azurite, and smithsonite occur at Tsumeb, South-West Africa. In the United States, found as an abundant ore at Leadville, Colorado. Occurs in part as large heart-shaped twins in Dona Ana County, New Mexico, where cerussite is associated with bright yellow and orange wulfenite, anglesite, and vanadinite.

Use. An important ore of lead an silver.

DOLOMITE

$CaMg(CO_3)_2$	Hexagonal
Hardness $3\frac{1}{2}$–4	Sp. Gr. 2.85±

Summary

Commonly in small, pale pink, curved rhombohedral crystals with a pearly luster ("pearl spar," Plate 26(*E*), facing p. 162), or in saddle-shaped aggregates of curved crystals. *Color:* usually pale pink; also white, colorless, gray, greenish, yellowish, brown, black. *Luster:* vitreous to pearly. *Streak:* white. *Cleavage:* perfect rhombohedral; transparent to translucent; brittle. Powdered dolomite will effervesce in cold HCl, but large fragments dissolve slowly. RESEMBLES CALCITE, but calcite readily dissolves with effervescence in cold acid. SMALL, PALE PINK, CURVED CRYSTALS ARE CHARACTERISTIC.

Composition. A carbonate of calcium and magnesium, $CaMg(CO_3)_2$. Ferrous iron commonly substitutes for magnesium. When the amount of iron is greater then magnesium the mineral is called *ankerite,* $Ca(Fe, Mg)(CO_3)_2$, and is brownish in color.

Tests. Infusible. Dissolves with effervescence in hot HCl. Powder dissolves with effervescence in cold acid, but large fragments are only slowly attacked.

Crystallization. Crystals are usually rhombohedral with curved faces; also saddle-shaped aggregates of curved crystals. Commonly in coarse, granular to fine-grained masses.

Alteration. From calcite and aragonite by the action of magnesium-rich solutions. Pseudomorphs of dolomite after calcite or aragonite are common. Metamorphosed dolomite frequently dissociates into calcite ($CaCO_3$)

and periclase (MgO); may also combine with silica to form calcium magnesium silicates such as diopside, $CaMg(Si_2O_6)$.

Occurrence. 1. Occurs as massive sedimentary deposits, often intimately mixed with calcite (dolomitic limestones). The sedimentary dolomite deposits are generally believed to be formed by the alteration of the original calcite rocks. 2. As dolomitic marbles formed by the metamorphism of the dolomitic limestones. 3. In hydrothermal veins commonly associated with galena, sphalerite, calcite, fluorite, barite, siderite, and quartz. 4. In cavities of limestones associated with calcite, celestite, gypsum, and quartz. 5. As embedded crystals in serpentine and talcose rocks. 6. As crystals in quartz geodes.

Good crystals have come from Traversella, Piedmont, Italy; from Binnenthal, Switzerland; near Djelfa in Algeria; and from the Guanajuato silver veins in Mexico. Crystals of optical quality occur at Brumada, Bahia, Brazil. In the United States, pale pink, curved crystal aggregates are common in the Joplin (Missouri) lead-zinc deposits. Large rhombohedral crystals occur in talc at Roxbury, Vermont. Found in cavities of dolomite at Rochester and Niagara Falls, New York. Thick beds of sedimentary dolomite occur in many parts of the world.

Use. As a building stone; in cements; in the manufacture of magnesia for refractories; as source of magnesium.

MALACHITE

$Cu_2CO_3(OH)_2$ Monoclinic
        Hardness $3\frac{1}{2}$–4   Sp. Gr. $4.0\pm$

Summary

Usually in bright green botryoidal masses with a radiating silky fibrous inner structure (Plate 27(*A*), facing p. 163). Light and dark green color bands are common (Plate 27(*C*)). Also in dull earthy masses. Crystals are rare. *Color:* bright green to blackish

green. *Streak:* pale green. *Cleavage:* basal, but rarely observed because of fibrous habit. *Luster:* silky (fibrous varieties), vitreous (crystals), dull (earthy varieties); brittle; translucent to opaque; effervesces in cold HCl. A secondary copper mineral found in the weathered zone of copper ore deposits, usually associated with blue azurite. BRIGHT GREEN COLOR, BOTRYOIDAL HABIT, AND EFFERVESCENCE IN ACID ARE DISTINCTIVE. Resembles green GARNIERITE, ANTLERITE, and BROCHANTITE but these minerals are not carbonates and do not effervesce in acid.

Composition. A basic carbonate of copper, $Cu_2CO_3(OH)_2$ (71.9% CuO, 19.9% CO_2, 8.2% H_2O).

Tests. Dissolves with effervescence in cold dilute HCl. Fuses on charcoal, colors flame green, and finally yields a globule of metallic copper. Gives much water in closed tube.

Crystallization. Monoclinic. Usually in botryoidal aggregates with a radiating fibrous inner structure, frequently banded. Crusts sometimes have a velvet-like surface. Commonly in dull earthy or granular masses. Acicular crystals are rare and are usually twinned. Sometimes pseudomorphous after azurite crystals.

Alteration. Pseudomorphs of malachite after azurite and cuprite are common.

Associated Minerals. Commonly associated with azurite, cuprite, limonite, copper, and chrysocolla.

Occurrence. A common secondary ore of copper found in the upper zone of copper deposits, where it alters from other copper minerals by the action of carbonated waters.

Among the many important localities is Tsumeb, South-West Africa, where malachite is commonly found as banded masses and as pseudomorphs after azurite crystals. Large compact banded masses from Siberia, U.S.S.R., have long been used for decorative purposes. It is also common in Rhodesia; in Katanga, Congo; and in New South Wales, Australia. In the United States, it is common in Arizona and New Mexico.

Malachite is frequently seen as a green coating on the copper roofs of buildings, which have been partially altered to malachite by the action of the carbon dioxide and moisture in the atmosphere.

Use. An ore of copper, and sometimes as a decorative material.

AZURITE

$Cu_3(CO_3)_2(OH)_2$ Monoclinic
     Hardness $3\frac{1}{2}$–4   Sp. Gr. 3.77

Summary

A blue mineral, commonly occurring in glassy highly modified tabular crystals. Crystals are dark blue in color but botryoidal aggregates are often bright azure blue (Plate 27(*B, D*), facing p. 163). Frequently, it is partially altered to green malachite. *Streak:* blue. *Luster:* vitreous to dull; fracture conchoidal; brittle; transparent to opaque; effervesces in cold dilute HCl.

Azurite, like malachite, is a secondary copper mineral found in the upper zone of copper ore deposits. AZURE BLUE COLOR, ASSOCIATION WITH GREEN MALACHITE, AND EFFERVESCENCE IN ACID ARE CHARACTERISTIC. Resembles azure blue LINARITE, $PbCu(SO_4)(OH)_2$, which does not effervesce in acid.

Composition. A basic carbonate of copper, $Cu_3(CO_3)_2$-$(OH)_2$ (69.2% CuO, 25.6% CO_2, 5.2% H_2O).

Tests. Dissolves with effervescence in cold dilute HCl. Fuses on charcoal, colors flame green, and finally yields a globule of metallic copper. Yields much water in closed tube.

Crystallization. Monoclinic; prismatic class, $\frac{2}{m}$. Complex tabular crystals are common. Crystals are varied in habit and are frequently distorted. Frequently forms botryoidal aggregates. Also as crusts and earthy masses.

Alteration. To malachite.

Occurrence. Azurite is less common than malachite but has the same origin and mineral associations. Both are formed by the oxidation and carbonation of other copper minerals, and are commonly associated with each other.

Beautiful crystals of unusual size, some over 9 in. long, occur in Tsumeb, South-West Africa. Some have been completely altered to malachite. Good crystals also occur at Chessy, France; in Siberia, U.S.S.R.; and in South Australia. In the United States, fine crystals of azurite were formerly found in Clifton and Bisbee, Arizona.

Use. A minor ore of copper; sometimes used as a decorative material.

AURICHALCITE

$(Zn,Cu)_5(OH)_6(CO_3)_2$ Orthorhombic
     Hardness 1–2   Sp. Gr. 3.6

Summary

Usually in crusts of soft, pale greenish-blue pearly scales (Plate 27(*F*), facing p. 163). *Color:* pale greenish-blue. *Luster:* pearly. *Cleavage:* micaceous; flexible; translucent; effervesces in HCl. A secondary mineral found in the weathered zone of copper and zinc deposits associated with limonite, malachite, azurite, cuprite, smithsonite, and hydrozincite. PALE GREENISH-BLUE PEARLY SCALES AND EFFERVESCENCE IN ACID ARE CHARACTERISTIC.

Composition. A carbonate-hydroxide of zinc and copper, $(Zn,Cu)_5(OH)_6(CO_3)_2$.

Tests. Infusible. Colors flame green. Dissolves with effervescence in HCl.

Crystallization. Orthorhombic. Usually in crusts of small fragile pearly scales, often coating limonite.

Occurrence. A secondary mineral found in the oxidized zone of zinc and copper deposits. Good specimens have come from Laurium, Greece; Chessy, France; Roumania; Mexico; Derbyshire, England; and Leadhills, Scotland. Found in the United States in Arizona; in New Mexico; and in Utah.

HYDROZINCITE

$Zn_5(OH)_6(CO_3)_2$
  Hardness 2–2½

Monoclinic
Sp. Gr. 3.5–3.8

Summary

Usually in *dull white* to *gray* earthy compact crusts (Plate 27(E), facing p. 163) that luminesce blue under ultraviolet light. Occurs as a secondary mineral in the weathered zone of zinc deposits, associated with sphalerite, smithsonite, hemimorphite, aurichalcite, cerussite, and limonite. BLUE LUMINESCENCE, EFFERVES- CENCE IN HCl, AND DULL WHITE CRUSTS ON OTHER ZINC MINERALS OR ON LIMONITE ARE CHARACTERISTIC.

Composition. A carbonate-hydroxide of zinc, $Zn_5(OH)_6$-$(CO_3)_2$.

Crystallization. Monoclinic. Occurs in earthy to compact masses; sometimes colloform crusts with a concentrically banded fibrous inner structure; also stalactites; and chalk-like masses.

Alteration. Commonly alters from sphalerite (ZnS), smithsonite ($ZnCO_3$), and hemimorphite ($Zn_4Si_2O_7$-$(OH)_2 \cdot H_2O$).

Associated Minerals. Sphalerite, smithsonite, hemimorphite, cerussite, aurichalcite, limonite, and calcite.

Occurrence. Hydrozincite is a secondary mineral found in the oxidized zone of zinc deposits. Found abundantly near Santander, Spain; in Carinthia (Austria); and Algeria. In the United States, good specimens occur at Goodsprings, Nevada, and near Socorro, New Mexico.

Use. A minor ore of zinc.

NITRATES

The nitrates are largely soluble in water and are thus found only in arid climates such as in the deserts of northern Chile, and in Death Valley, California. With the exception of soda-niter ($NaNO_3$), which oc- curs in large beds in Chile, the nitrates are rare and are therefore not an important mineral group.

Nitrate Minerals

Soda Niter, $NaNO_3$
Niter, KNO_3

SODA NITER (Chile Saltpeter)

$NaNO_3$
  Hardness 1–2

Hexagonal
Sp. Gr. 2.2

Summary

A soft, white, water-soluble mineral with a cooling taste. Usually in granular masses or crusts; seldom in rhombohedral crystals. Occurs as surface deposits in extremely arid regions; commonly associated with niter, gypsum, and other salts. *Color:* white to color- less; sometimes tinted reddish-brown or yellowish by impurities. *Streak:* white. *Luster:* vitreous. *Cleavage:* rhombohedral; transparent to translucent; somewhat sectile. WHITE GRANULAR MASSES WITH A COOLING TASTE ARE CHARACTERISTIC. May resemble NITER or HALITE, but niter gives a violet potassium flame, and halite has a salty taste that is not cooling.

Composition. Sodium nitrate, $NaNO_3$ (36.5% Na_2O, 63.5% N_2O_5).

Tests. Easily soluble in water. On charcoal, fuses read- ily and gives a yellow sodium flame. Tastes cooling.

Crystallization. Hexagonal; hexagonal-scalenohedral class, $\bar{3}\frac{2}{m}$. Usually as granular masses; sometimes as rhombohedral crystals. Also as oriented overgrowths on calcite. Calcite and soda-niter have similar struc- tures.

Occurrence. The most important deposits are in the deserts of northern Chile, where soda-niter occurs in a huge area (some 450 miles long and 10 to 50 miles wide); associated with halite, gypsum, polyhalite, and other salts. The mixture of soda-niter and its associated minerals is called *caliche.* In the United States, small deposits of soda-niter are found in California and in Nevada.

Use. As a fertilizer; in explosives; and in the manufacture of nitric acid. In recent years, nitrates have been artificially made by fixation of atmospheric nitrogen.

NITER (Saltpeter)

KNO_3	Orthorhombic
Hardness 2	Sp. Gr. 2.1

Summary

A soft, white, water-soluble mineral with a cooling taste; usually in thin crusts or silky needle-like crystals. Occurs on cave walls, as thin encrustations in arid regions, and as a constituent of soils containing excrement. Commonly associated with soda-niter, gypsum, and epsomite. *Color:* white to colorless; sometimes tinted by impurities. *Luster:* vitreous to silky. *Streak:* white. *Cleavage:* one good, two imperfect; transparent to translucent; brittle. SALTY COOLING TASTE AND BEHAVIOR ON CHARCOAL ARE DISTINCTIVE. Resembles SODA-NITER but explodes more violently on charcoal and colors flame violet.

Composition. Potassium nitrate, KNO_3 (46.5% K_2O, 53.5% N_2O_5).

Tests. Soluble in water. Tastes salty and cooling. Explodes violently (like gunpowder on glowing charcoal); colors flame violet.

Crystallization. Orthorhombic; dipyramidal class, $\frac{2}{m}\frac{2}{m}\frac{2}{m}$. Usually as thin crusts, silky fibers, and as aggregates of needle-like crystals. Niter is isostructural with aragonite ($CaCO_3$). At 129°C niter inverts to a hexagonal form, which is isostructural with soda-niter.

Occurrence. Occurs as an efflorescence on soils in Spain, India, Algeria, Egypt, Persia, and Italy. Associated with soda-niter in the deserts of northern Chile. In the

United States, occurs in caves in Kentucky, Tennessee, Alabama, and New Mexico.

Use. A minor source of nitrogen compounds.

HYDROUS BORATES

The hydrous borates occur in arid regions and are commonly formed by the evaporation of boron-rich surface waters. Huge bedded deposits occur in the Mojave Desert at Searles Lake and Kramer, California; and in the salt lakes and playas in the desert regions of Tibet, Argentina, Chile, Peru, Turkey, and Bolivia. A unique borax deposit occurs near Pisa, Italy, where volcanic steam vents carrying boric acid yield borax. The most abundant borates are borax, colemanite, kernite, and ulexite.

Hydrous Borates

Borax, $Na_2B_4O_7 \cdot 10H_2O$
Kernite, $Na_2B_4O_7 \cdot 4H_2O$
Ulexite, $NaCaB_5O_9 \cdot 8H_2O$
Colemanite, $Ca_2B_6O_{11} \cdot 5H_2O$

BORAX

$Na_2B_4O_7 \cdot 10H_2O$	Monoclinic
Hardness $2-2\frac{1}{2}$	Sp. Gr. 1.7

Summary

A soft, *colorless to white,* water-soluble mineral with a sweetish-alkaline taste. Commonly in large, colorless crystals that lose water and become chalky white on exposure to air. Occurs as crusts as well as in porous masses. *Streak:* white. *Cleavage:* two imperfect, and one perfect parallel to front pinacoid. *Luster:* vitreous to earthy. *Fracture:* conchoidal; brittle. Commonly associated with ulexite, halite, and gypsum. SLIGHTLY SWEETISH TASTE, CRYSTALS, SOLUBILITY IN WATER, AND VERY LOW SPECIFIC GRAVITY ARE CHARACTERISTIC.

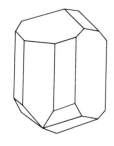

Fig. 8-32. Borax crystal.

Composition. Hydrous sodium borate, $Na_2B_4O_7 \cdot 10H_2O$ (16.2% Na_2O, 36.6% B_2O_3, 47.2% H_2O).

Tests. Readily soluble in water. Sweetish alkaline taste. Swells and fuses easily on charcoal to a clear borax glass bead, colors flame yellow (sodium). Gives green (boron) flame when moistened with sulfuric acid, covered with alcohol, and set on fire.

Crystallization. Monoclinic; prismatic class, $\frac{2}{m}$. Frequently in large short prismatic crystals (Figure 8-32). Also as encrustations and porous masses.

Alteration. Colorless crystals lose water on exposure to the atmosphere and alter to powdery white *tincalconite* ($Na_2B_4O_7 \cdot 5H_2O$).

Associated Minerals. Halite, ulexite, colemanite, kernite, gypsum, calcite, thenardite, soda-niter.

Occurrence. 1. Borax occurs in evaporite deposits of salt lakes and playas in desert regions; as, for example, Kramer and Searles Lake in the Mojave Desert (California). In ancient times, borax was brought to Europe from the salt lakes of Tibet. 2. From the evaporation of water from hot springs (as in Pisa, Italy). 3. As an efflorescence on arid soils.

Use. Borax is used as a cleaning agent; in medicine; as a flux in soldering and welding; as a preservative; and in the manufacture of such high-temperature-resistant glass as Pyrex. Boron carbide is used as an abrasive. Boron has the ability to absorb neutrons and is used for control rods in nuclear reactors. Boron hydride is used as a constituent of high-energy fuels.

ULEXITE ("Cotton Balls")

$NaCaB_5O_9 \cdot 8H_2O$ Triclinic
Hardness 1   Sp. Gr. 1.9

Summary

Usually as short, silky, white fibers that form small, rounded, loose textured masses or "cotton balls" (Figure 8-33). *Color:* white. *Luster:* silky; brittle; tasteless; slightly soluble in hot water. Occurs in dry lake beds in very arid regions and is commonly associated with borax and halite. SOFT, LOOSE-TEXTURED, WHITE, SILKY COTTON-BALL AGGREGATES ARE CHARACTERISTIC.

Composition. Hydrous sodium calcium borate, $NaCaB_5O_9 \cdot 8H_2O$ (7.7% Na_2O, 13.8% CaO, 43.0% B_2O_5, 35.5% H_2O).

Tests. Swells and fuses easily to a clear glass, coloring flame yellow (sodium). Gives a green (boron) flame when moistened with sulfuric acid, covered with alcohol, and set on fire. Slightly soluble in hot water. Tasteless.

Fig. 8-33. Ulexite "cottonball" aggregates: Peru (x 0.8).

Crystallization. Triclinic; pinacoidal class, $\bar{1}$. Usually in loose-textured rounded masses of hair-like crystals (cotton balls).

Alteration. To gypsum.

Associated Minerals. Borax, halite, gypsum, colemanite, glauberite ($Na_2Ca(SO_4)_2$) and soda-niter.

Occurrence. Found in arid regions in dry salt lake beds. Commonly occurs with borax in the Mojave Desert. With soda-niter in the desert region of northern Chile. Also in Argentina and Peru.

Use. As a source of borax.

KERNITE

$Na_2B_4O_7 \cdot 4H_2O$ Monoclinic
Hardness $2\frac{1}{2}$ Sp. Gr. 1.9

Summary

A soft, colorless to white mineral commonly in elongated cleavage fragments; resembles platy gypsum, but usually has a white opaque surface alteration, and is soluble in hot water (Figure 8-34). *Color:* colorless but alters white. *Streak:* white. *Luster:* vitreous on a fresh surface. *Cleavage:* perfect basal and front pinacoid.

Kernite is found in great quantities at Kramer, Kern County, in the Mojave Desert, California, where it occurs with borax in a huge buried lake deposit; it is believed to have been derived from borax under elevated temperature and pressure. ELONGATED CLEAVAGE FRAGMENTS WITH A PARTIAL WHITE OPAQUE SURFACE ALTERATION, LOW SPECIFIC GRAVITY, AND SOLUBILITY IN WATER ARE CHARACTERISTIC. Resembles COARSE FIBROUS GYPSUM.

Composition. Hydrous sodium borate, $Na_2B_4O_7 \cdot 4H_2O$ (22.7% Na_2O, 51.0% B_2O_3, 26.3% H_2O).

Tests. Swells and fuses to a clear borax glass, coloring flame yellow (sodium). Slowly soluble in cold water; readily soluble in hot.

Crystallization. Monoclinic; prismatic class, $\frac{2}{m}$. Usually in coarse elongated cleavage fragments. Crystals are usually large and heavily striated.

Occurrence. Found with borax in a large lake deposit buried several hundred feet below the surface at Kramer, Kern County, California. Large crystals of kernite, some several feet long, are abundant in the clay shales. Kernite is believed to have been derived from borax that was heated under igneous activity.

Kernite was discovered by drilling in Kern County in 1926. Thousands of tons of kernite are mined, and today it is a major mineral of borax in the United States. The discovery of kernite was unique, since most new minerals today are found in small amounts.

Use. A major source of borax.

Fig. 8-34. Kernite: Kramer, Kern County, California (x 0.7).

COLEMANITE

$Ca_2B_6O_{11} \cdot 5H_2O$ Monoclinic
Hardness $4\frac{1}{2}$ Sp. Gr. 2.42

Summary

Frequently in glassy white complex equidimensional crystals that resemble datolite (Figure 8-35).

A   x 0.80

LEUCITE
K(AlSi$_2$O$_6$)

Trapezohedral leucite crystals with a dull surface alteration: Roccamonfina, Rome, Italy.

B   x 0.53

LAZURITE (LAPIS LAZULI)
Na$_8$Al$_6$Si$_6$O$_{24}$S$_2$

Bright blue lazurite with white calcite and minute pyrite inclusions (lapis lazuli): Iran.

C   x 0.50

SCAPOLITE
(Na, Ca)$_4$Al$_3$(Al, Si)$_3$Si$_6$O$_{24}$(Cl, CO$_3$, SO$_4$)

Large square prismatic scapolite crystal with typical dull "wood-like" surface alteration.

D   x 0.50

SODALITE
Na$_8$(Al$_6$Si$_6$O$_{24}$)Cl$_2$

Massive blue sodalite.

E   x 0.66

SODALITE AND CANCRINITE

Massive blue sodalite and yellow cancrinite in a nepheline-syenite rock: Litchfield, Maine.

F   x 0.70

NEPHELINE
Na$_3$K(Al$_4$Si$_4$O$_{16}$)

Massive reddish nepheline showing greasy luster on fresh surface: Magnet Cove, Arkansas.

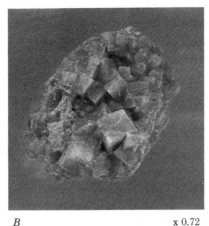

A   x 0.90   *B*   x 0.72   *C*   x 0.50

NATROLITE

$Na_2(Al_2Si_3O_{10}) \cdot 2H_2O$

White needle-like crystals with green prehnite: Paterson, New Jersey.

CHABAZITE

$Ca(Al_2Si_4O_{12}) \cdot 6H_2O$

Rhombohedral crystals of chabazite which resemble slightly distorted cubes: Swan Creek, Nova Scotia.

ANALCITE

$Na(AlSi_2O_6) \cdot H_2O$

Trapezohedral crystals of analcite: Cape Blomidon, Nova Scotia.

D   x 1.0   *E*   x 0.81   *F*   x 0.70

STILBITE

$(Ca, Na_2, K_2)(Al_2Si_7O_{18}) \cdot 7H_2O$

Characteristic sheaf-like aggregates: Wasson's Bluff, Bay of Fundy, Nova Scotia.

STILBITE

Reddish sheaf-like aggregates: Nova Scotia.

HEULANDITE

$(Ca, Na_2)(Al_2Si_7O_{18}) \cdot 6H_2O$

White tabular crystal with tapering ends and pearly luster on cleavage face: Iceland.

Fig. 8-35. Colemanite: Death Valley, California (x 0.7).

Color: colorless to white; sometimes gray or yellowish. *Streak:* white. *Luster:* vitreous. *Cleavage:* perfect side pinacoid; brittle; transparent to translucent. Colemanite is a secondary mineral, found in buried lake beds in California and Nevada, and is believed to have been formed by the action of meteoric waters on borax and ulexite. COMPLEX GLASSY WHITE CRYSTALS ASSOCIATED WITH BORAX AND ULEXITE, AND BEHAVIOR ON CHARCOAL ARE CHARACTERISTIC.

Composition. Hydrous calcium borate, $Ca_2B_6O_{11} \cdot 5H_2O$ (27.2% CaO, 50.9% B_2O_3, 21.9% H_2O).

Tests. Decrepitates and crumbles on charcoal, giving a green (boron) flame. Soluble in hot HCl; on cooling, precipitates boric acid crystals.

Crystallization. Monoclinic; prismatic class, $\frac{2}{m}$. Crystals are commonly equidimensional; usually complex and varied; frequently in geodes. Cleavable to fine-grained masses are common.

Occurrence. Found in buried ancient lake beds at Kramer and in Death Valley, California and Nevada. Associated minerals include ulexite, gypsum, borax, calcite, and celestite.

Use. A source of borax.

SULFATES

Anhydrous Sulfates

Thenardite	Orth.	Na_2SO_4
Barite Group	Orth.	
Barite		$BaSO_4$
Celestite		$SrSO_4$
Anglesite		$PbSO_4$
Anhydrite	Orth.	$CaSO_4$

Hydrated Sulfates

Gypsum	Mon.	$CaSO_4 \cdot 2H_2O$
Polyhalite	Tri.	$K_2Ca_2Mg(SO_4)_4 \cdot 2H_2O$
Chalcanthite	Tri.	$CuSO_4 \cdot 5H_2O$
Epsomite	Orth.	$MgSO_4 \cdot 7H_2O$
Copiapite	Tri.	$(Fe,Mg)Fe_4(SO_4)_6(OH)_2 \cdot 20H_2O$

Anhydrous Sulfates Containing Hydroxyl

Linarite	Mon.	$PbCu(SO_4)(OH)_2$
Brochantite	Mon.	$Cu_4(SO_4)(OH)_6$
Antlerite	Orth.	$Cu_3(SO_4)(OH)_4$
Alunite Group	Hex.	
Alunite		$KAl_3(SO_4)_2(OH)_6$
Jarosite		$KFe_3(SO_4)_2(OH)_6$

The sulfates comprise a large family of minerals, but only the more common species are included. Some, such as BARITE, CELESTITE, and ANGLESITE, are comparatively heavy nonmetallic minerals. Others, such as THENARDITE, POLYHALITE, CHALCANTHITE, EPSOMITE, and COPIAPITE, are soluble in water. The sulfates do not effervesce in warm acid, a feature that distinguishes them from the carbonates.

THENARDITE

Na_2SO_4	Orthorhombic
Hardness $2\frac{1}{2}$–3	Sp. Gr. 2.67

Summary

A soft, water-soluble mineral with a salty taste; commonly in tabular cross-twinned crystals (Plate

Table 8-1

Some Properties of the Sulfates

MINERAL	SYSTEM	COLOR	HARDNESS	SP. GR.	REMARKS
Thenardite (Na_2SO_4)	Orthorhombic	White; yellowish	$2\frac{1}{2}$–3	2.67	Soluble in water; tastes salty; tabular cross-twin crystals
Barite ($BaSO_4$)	Orthorhombic	Colorless; pale blue, yellow, or red tints	3–$3\frac{1}{2}$	4.5	Heavy; tabular crystals; yellow-green (Ba) flame
Celestite ($SrSO_4$)	Orthorhombic	Colorless; pale blue or red tints	3–$3\frac{1}{2}$	4.0	Red (Sr) flame; heavy; tabular crystals
Anglesite ($PbSO_4$)	Orthorhombic	Colorless; white; pale tints	$2\frac{1}{2}$–3	6.38	Alters from galena (PbS)
Anhydrite ($CaSO_4$)	Orthorhombic	Colorless; white; pale tints	$3\frac{1}{2}$	3.0	Cubic cleavage; commonly associated with gypsum
Gypsum ($CaSO_4 \cdot 2H_2O$)	Monoclinic	Colorless; white; yellowish	2	2.3	Easily scratched by fingernail
Polyhalite ($K_2Ca_2Mg(SO_4)_4 \cdot 2H_2O$)	Triclinic	Colorless; white; red	$3\frac{1}{2}$	2.78	Usually massive; bitter taste; commonly associated with halite and anhydrite
Chalcanthite ($CuSO_4 \cdot 5H_2O$)	Triclinic	Deep blue	$2\frac{1}{2}$	2.28	Blue color; powdery pale green alteration; soluble in water
Epsomite ($MgSO_4 \cdot 7H_2O$)	Orthorhombic	Colorless; white	2–$2\frac{1}{2}$	1.67	Hair-like crusts; bitter taste
Copiapite ($(Fe,Mg)Fe_4(SO_4)_6(OH)_2 \cdot 20H_2O$)	Triclinic	Yellow	$2\frac{1}{2}$–3	2.1±	Yellow color; pearly luster; soluble in water; scaly aggregates
Linarite ($PbCu(SO_4)(OH)_2$)	Monoclinic	Azure blue	$2\frac{1}{2}$	5.3	Azure blue color
Brochantite ($Cu_4SO_4(OH)_6$)	Monoclinic	Emerald green; pale to blackish green	$3\frac{1}{2}$–4	4.0±	Green color; vitreous luster

Table 8-1   *(continued)*

Some Properties of the Sulfates

MINERAL	SYSTEM	COLOR	HARDNESS	SP. GR.	REMARKS
Antlerite ($Cu_3SO_4(OH)_4$)	Orthorhombic	Emerald green; pale to blackish green	$3\frac{1}{2}$	$3.9\pm$	Green color; vitreous luster
Alunite ($KAl_3(SO_4)_2(OH)_6$)	Hexagonal	White; grayish; reddish	$3\frac{1}{2}$–4	2.6–2.9	Usually massive; may resemble limestone
Jarosite ($KFe_3(SO_4)_2(OH)_6$)	Hexagonal	Yellow ocher to brown	$2\frac{1}{2}$–$3\frac{1}{2}$	3.0	Resembles limonite

29(*A*), facing p. 179). *Color:* colorless, white, yellowish, brownish. *Luster:* vitreous. *Cleavage:* good side pinacoid; transparent to translucent. Commonly formed by the evaporation of salt lakes in arid regions. CROSS-TWINNED TABULAR CRYSTALS, AND SOLUBILITY IN WATER ARE CHARACTERISTIC.

Composition. Sodium sulfate, Na_2SO_4 (43.7% Na_2O, 56.3% SO_3).

Tests. Fuses easily giving yellow (sodium) flame. Readily soluble in water. Tastes slightly salty. Gives tests for sulfate and sulfur.

Crystallization. Orthorhombic; dipyramidal class, $\frac{2}{m}\frac{2}{m}\frac{2}{m}$. Crystals are usually tabular. Cross-twins are common. Also as earthy crusts and efflorescences on arid soils. Prismatic crystals are rare.

Associated Minerals. Gypsum, halite, epsomite, glauberite, $Na_2Ca(SO_4)_2$, and mirabilite, $Na_2SO_4 \cdot 10H_2O$.

Occurrence. Found in arid regions in dry salt lake beds. Also as efflorescences on arid soils. Sometimes around fumaroles, and as crusts on recent lavas (Vesuvius and Etna, Italy). Large bedded deposits occur in Spain, Siberia, and Africa. Occurs in the desert regions of northern Chile. In the United States, occurs in beds at Searles Lake, and at other locations in California; also in Arizona and Nevada.

BARITE

$BaSO_4$ Orthorhombic
  Hardness 3–$3\frac{1}{2}$ Sp. Gr. 4.5

Summary

The word barite is derived from the Greek word meaning heavy. Barite is a soft, heavy nonmetallic mineral, commonly in large tabular crystals; also in divergent plates forming "crested barite" (Plate 28(*A, B, C*), facing p. 178) or in rosette-like aggregates of tabular crystals ("desert roses"). See Plate 2(*D*), facing p. 3. Also in granular to compact masses. *Color:* colorless, white, bluish, reddish, brown, and yellow tints. *Luster:* vitreous. *Streak:* white. *Cleavage:* good basal and prismatic at right angles, with less distinct side pinacoid; transparent to translucent; brittle; sometimes luminescent. TABULAR CRYSTALS, CLEAVAGE, AND HIGH SPECIFIC GRAVITY FOR A NONMETALLIC MINERAL ARE CHARACTERISTIC. Resembles CELESTITE and CALCITE.

Composition. Barium sulfate, $BaSO_4$ (65.7% BaO, 34.3% SO_3). Strontium substitutes for barium and a com-

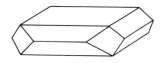

Fig. 8-36. Barite.

plete substitutional solid solution probably extends to celestite ($SrSO_4$). Small amounts of lead or calcium may substitute for barium.

Tests. Insoluble in acids. Decrepitates and fuses with difficulty, coloring flame yellowish-green (Ba). When fused with a flux of sodium carbonate and then moistened and placed on a silver coin, a dark brown stain of silver sulfide is produced on the coin (S-test).

Crystallization. Orthorhombic; dipyramidal class, $\frac{2}{m}\frac{2}{m}\frac{2}{m}$. Commonly in thin to thick tabular crystals (Figure 8-36). In divergent plates ("crested barite"); in rosette-like aggregates, in granular to compact masses, and in concretionary forms with a fibrous inner structure. Transforms at 1149°C to a monoclinic(?) form. Melting point 1580°C.

Minerals of Similar Appearance

Celestite. Orthorhombic crystals closely resemble barite. Distinguished by crimson (strontium) flame. Celestite is often associated with yellow sulfur.

Calcite. Effervesces in cold HCl, and has a specific gravity of only 2.7. Barite is heavy (Sp. Gr. 4.5) and insoluble in acid.

Alteration. To witherite ($BaCO_3$).

Associated Minerals. Calcite, dolomite, fluorite, siderite, quartz, stibnite, galena, and silver.

Occurrence. Barite is the most common barium mineral. It occurs in various types of deposits. (1) As a gangue mineral in low- to moderate-temperature hydrothermal sulfide veins. (2) In veins, cavities, and replacement deposits of limestones and other sediments. (3) As a hot spring deposit. (4) In cavities of basic igneous rocks. (5) In sedimentary iron and manganese deposits. (6) As the petrifying mineral in some fossils.

A few of the important occurrences and locations are: In large crystals, some several feet in length, at Cumberland, Cornwall, Westmoreland, and elsewhere in England. With stibnite at Felsöbánya, Roumania. As concretions in marl at Mount Paterno near Bologna, Italy. These concretions are often thermoluminescent. In the United States, deposits of massive barite are mined in California, Georgia, Tennessee, Missouri, and Arkansas. "Desert roses" containing large amounts of sand are found at Norman, Cleveland County, Oklahoma (Plate 2, facing p. 3). In the Bad Lands of South Dakota, yellowish-brown barite crystals occur in cracks and cavities of large concretions; also in fossils.

Use. Much barite is ground and made into a heavy sludge for deep oil- and gas-well drilling. Barite is used in the manufacture of lithophone, a white pigment that is formed by the reaction of barium sulfide and zinc sulfate; this reaction forms a mixture of zinc sulfide and barium sulfate. Barite is also used as a filler in glazed paper, in cosmetics, and in medicine. A source of barium.

CELESTITE

$SrSO_4$ Orthorhombic
        Hardness 3–3½   Sp. Gr. 3.97

Summary

A soft, heavy, transparent to translucent mineral, commonly in thick to thin tabular crystals that closely resemble barite. Crystals are sometimes elongated parallel to the *a* axis. Also fibrous and massive. *Color:* colorless, white, commonly bluish; sometimes with blue tints irregularly distributed. The first specimens described were celestial blue—hence the name celestite. Also reddish, brownish, yellowish. *Streak:* white. *Luster:* vitreous to pearly. *Cleavage:* good basal and

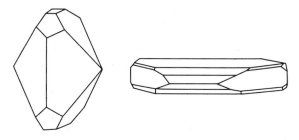

Fig. 8-37. Celestite crystals.

prismatic, poor pinacoidal; brittle; occasionally luminescent. Often associated with yellow sulfur (Plate 28(*D*), facing p. 178, and Plate 11(*E*), facing p. 82). CRYSTALS, HIGH SPECIFIC GRAVITY, ASSOCIATION WITH SULFUR, AND TESTS FOR STRONTIUM AND SULFUR ARE CHARACTERISTIC. Flame test for strontium usually distinguishes celestite from BARITE (BaSO₄).

Composition. Strontium sulfate, $SrSO_4$ (56.4% SrO, 43.6% SO_3). Barium substitutes for strontium, and a complete solid solution probably extends to barite ($BaSO_4$). Small amounts of calcium may substitute for strontium.

Tests. Decrepitates and fuses with difficulty, coloring flame crimson (Sr-test). When fused with sodium carbonate and then moistened and placed on a silver coin, a dark brown stain of silver sulfide is produced on the coin (S-test). Only slightly soluble in hot concentrated acids.

Crystallization. Orthorhombic; dipyramidal class, $\frac{2}{m}\frac{2}{m}\frac{2}{m}$. Commonly in tabular crystals; sometimes elongated parallel to the *a* axis. Also fibrous and radiating; sometimes in granular masses. Transforms at about 1152°C to hexagonal α-strontianite. Melting point 1605°C.

Alteration. To strontianite, $SrCO_3$.

Associated Minerals. Sulfur, gypsum, anhydrite, halite, strontianite, fluorite, calcite, dolomite.

Occurrence. Celestite is usually found in sedimentary limestones, and only rarely is it found as a gangue mineral in hydrothermal veins, or in cavities in basalt. In sedimentary deposits, celestite occurs in (**1**) bedded deposits of gypsum, halite, and aragonite, commonly associated with sulfur; (**2**) in veins, cavities, and disseminations in limestones and dolomites; (3) found in small amounts in evaporite deposits of potassium salts and borates; and (**4**) disseminated in marl, shale, and sandstone.

A few of the notable occurrences and localities are celestite crystals associated with sulfur at Girgenti, Sicily; large flat crystals in England near Bristol; numerous celestite crystals, some over a foot long, in a dolomite cave at Put-in-Bay, Lake Erie; abundant crystals with fluorite in cavities of dolomite at Clay Center, Ohio; large blue crystals at Lampasas, Texas; with gypsum and halite in lake bed deposits in San Bernardino County, California.

Use. Source of strontium. Strontium is used in beet sugar refining. Strontium nitrate is used in fireworks and in the manufacture of signal flares because it produces a bright crimson flame.

ANGLESITE

$PbSO_4$	
Hardness 2½–3	Orthorhombic
	Sp. Gr. 6.38

Summary

A soft, heavy, secondary, nonmetallic mineral, usually formed by the weathering of galena. *Color:* colorless, white, gray, pale tints. *Luster:* adamantine to resinous (crystals), dull (fine-grained, massive varieties). Crystals are common and frequently varied in habit, usually tabular, prismatic, or equant. Also commonly in light gray masses or concentric bands that replace dark galena "woody anglesite" (Plate 28(*E*), facing p. 178). Commonly associated with cerussite $PbCO_3$, which also alters from galena. *Cleavage:* good basal. *Fracture:* conchoidal; transparent to opaque; brittle; often luminesces yellow. HIGH SPECIFIC GRAVITY, ADAMANTINE LUSTER IN CRYSTALS, AND ASSOCIATION WITH GALENA ARE CHARACTERISTIC. Resembles CERUSSITE $PbCO_3$, but is easily distinguished by lack of effervescence in warm acid.

Composition. Lead sulfate, $PbSO_4$ (73.6% PbO, 26.4% SO_3).

Tests. Fuses very easily. Produces a lead globule when heated on charcoal with sodium carbonate. Gives sulfur test on silver coin.

Crystallization. Orthorhombic; dipyramidal class, $\frac{2}{m}\frac{2}{m}\frac{2}{m}$. Crystals are common, and often varied in habit, usually tabular, prismatic, or equant. Also commonly granular to compact masses. Often in concentric massive bands around a core of galena. Transforms at 864°C to a monoclinic polymorph.

Alteration. From galena (PbS). To cerussite ($PbCO_3$).

Associated Minerals. Galena, cerussite, gypsum, native sulfur, wulfenite, pyromorphite, linarite, cerargyrite, and smithsonite.

Occurrence. Anglesite is a common secondary lead mineral that is usually formed by the oxidation of galena. Some notable localities are Derbyshire, England; Broken Hill, New South Wales; Dundas, Tasmania; Sardinia; Tunis; and Tsumeb, South-West Africa. In the United States, it is found at Coeur d'Alene District, Idaho; in the Tintic District, Utah; in Inyo County, California; and in Eureka, Nevada. Good specimens were once found at the Wheatley Mine, Chester County, Pennsylvania.

Use. A minor ore of lead.

ANHYDRITE

CaSO₄   Orthorhombic
    Hardness 3½   Sp. Gr. 2.98–3.0

Summary

Usually in cleavable masses that break up into rectangular cleavage fragments. *Color:* white, gray, bluish, reddish. *Luster:* vitreous to pearly on cleavage. *Streak:* white. *Cleavage:* good parallel to three pinacoids; transparent to translucent; brittle. Commonly associated with gypsum in bedded limestone or salt deposits. RECTANGULAR OR CUBIC CLEAVAGE FRAGMENTS, LOW SPECIFIC GRAVITY, AND TEST FOR SULFUR ARE CHARACTERISTIC. May resemble massive CALCITE, CELESTITE, or GYPSUM.

Composition. Calcium sulfate, $CaSO_4$ (41.2% CaO, 58.8% SO_3).

Tests. Fuses with difficulty. Soluble in HCl. When fused with sodium carbonate flux, the residue produces a dark brown stain on a silver coin (test for S). Gives orange-red flame test (Ca).

Crystallization. Orthorhombic; dipyramidal class, $\frac{2}{m}\frac{2}{m}\frac{2}{m}$. Usually in cleavable masses that break up into pseudocubic cleavage fragments. Also fibrous or granular masses. Rarely in thick tabular or prismatic crystals.

Minerals of Similar Appearance

Calcite, $CaCO_3$. Distinguishable from anhydrite by rhombohedral cleavage and effervescence in HCl.

Celestite, $SrSO_4$. Distinguishable from massive anhydrite by higher specific gravity (Sp. Gr. celestite 3.9–4.0), blocky cleavage, and crimson strontium flame test. Flame test for anhydrite is orange-red (Ca). Both anhydrite and celestite give positive test for sulfur on a silver coin.

Gypsum, $CaSO_4 \cdot 2H_2O$. Distinguishable from massive anhydrite by hardness. Gypsum has a hardness of only 2, and is thus easily scratched by a finger nail.

Alteration. Anhydrite readily hydrates to gypsum. Under some conditions, gypsum may alter to anhydrite.

Associated Minerals. Gypsum, halite, dolomite, calcite, celestite, and quartz.

Occurrence. Anhydrite is commonly deposited by the evaporation of sea water. It frequently occurs in large sedimentary beds associated with gypsum, limestone, dolomite, or halite. Anhydrite is common in the caprock of many salt domes in Texas and Louisiana; also found in cavities of basalt with zeolite minerals and prehnite. Hollow rectangular molds after bladed anhydrite crystals are common in the zeolite-filled cavities in the traprock of Paterson, New Jersey. The origin of these hollow rectangles in quartz was a mystery until some residual anhydrite was found. Anhydrite sometimes occurs as a gangue mineral in hydrothermal veins.

Anhydrite is common in the salt beds of Krakow, Poland; in Punjab, India; Salzburg, Austria; Stassfurt, Germany, and in the Pyrénées. Occurs in violet crystals and cleavable masses in Bex, Switzerland; also in large bedded deposits in Nova Scotia.

GYPSUM

CaSO$_4$ · 2H$_2$O   Monoclinic
  Hardness 2   Sp. Gr. 2.3

Summary

A soft, transparent to translucent mineral that can be scratched by a finger nail; commonly in elongated tabular crystals (Plate 7(*A*), facing p. 50). Crystals several feet long have been found. Also commonly in "fish-tail" contact twins (Figure 8-38); in a fibrous variety known as *satin spar;* in a transparent variety known as *selenite;* and in a fine-grained massive variety known as *alabaster. Color:* colorless, white, gray, yellowish, brownish. *Streak:* white. *Cleavage:* three directions, one very good {010}, yielding flattened rhombic fragments that can be bent but are not elastic. *Luster:* vitreous to silky, often pearly on cleavage faces; sometimes luminescent. Gypsum aggregates are frequently bent (Figure 8-39).

Commonly formed by the evaporation of salt water, or by the hydration of anhydrite (CaSO$_4$). Associated with halite, anhydrite, and limestone. THE

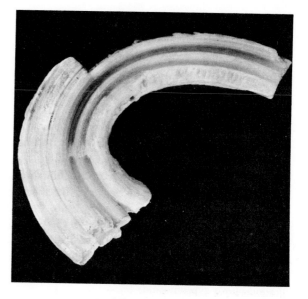

Fig. 8-39. Curved gypsum.

HARDNESS OF 2, WHICH ENABLES GYPSUM TO BE SCRATCHED BY A FINGER NAIL, DISTINGUISHES IT FROM SUCH MINERALS AS CALCITE, ANHYDRITE, AND DOLOMITE.

Composition. Hydrous calcium sulfate, CaSO$_4$ · 2H$_2$O (32.6% CaO, 46.5% SO$_3$, 20.9% H$_2$O).

Tests. Fusible. Soluble in hot dilute HCl, the addition of barium chloride solution gives a white precipitate of barium sulfate. Gives silver coin test for sulfur. Gives water in closed tube and turns white.

Crystallization. Monoclinic; prismatic class, $\frac{2}{m}$. Commonly thick to thin tabular crystals; also prismatic. Crystals often have curved surfaces. Twinned crystals are common, often twinned on the front pinacoid, yielding fish-tailed contact twins. Also fibrous, granular, compact, or foliated masses; rosette-like aggregates; sand crystals (Figure 2-27, p. 21), and curved twisted aggregates.

Alteration. Readily changes to and from anhydrite (CaSO$_4$).

Associated Minerals. Anhydrite, halite, calcite, dolomite, sulfur, aragonite, and quartz.

Fig. 8-38. "Fish-tail" contact twin.

Occurrence. Gypsum is the most common sulfate mineral. (1) It commonly occurs in large, bedded sedimentary deposits associated with limestone, sandstones, red shale, clay, and halite. When deposited by the evaporation of sea water, gypsum is usually deposited first; and as the salt concentration increases may be followed by anhydrite, halite, and polyhalite. (2) Gypsum may be formed by the hydration of anhydrite beds. (3) Found in the caprock of salt domes; (4) with sulfur around fumaroles; and (5) in gossans.

Very large gypsum crystals occur in Chihuahua, Mexico; in the sulfur mines of Sicily; in the Braden Mine, Chile; and in Wayne County, Utah. Large, bedded sedimentary deposits are mined in the United States—in New York, Michigan, California, Texas, Iowa, New Mexico, and Colorado. The sand dunes at White Sands near Alamogordo, New Mexico, consist of wind-blown gypsum sand.

Much of the gypsum mined is used for the building trade and in the manufacture of plaster of Paris. Large massive deposits of gypsum are mined in the Paris basin in France, the origin of the name plaster of Paris. Extensive beds occur in Nova Scotia; these have altered from anhydrite.

Use. In the manufacture of plaster of Paris; in plaster board; as a retarder in Portland cement; as a fertilizer; as a flux in glass manufacturing; and as an ornamental material (albaster and satin spar).

POLYHALITE

$K_2Ca_2Mg(SO_4)_4 \cdot 2H_2O$ Triclinic
Hardness $3\frac{1}{2}$ Sp. Gr. 2.78

Summary

A soft, colorless to brick red mineral with a bitter taste; usually massive; commonly associated with halite (Plate 29(*E*), facing p. 179). *Luster:* resinous. *Streak:* white. *Cleavage:* one direction distinct; translucent. BRICK-RED COLOR, BITTER TASTE, AND TEST FOR SULFUR ARE CHARACTERISTIC. Resembles SYLVITE and other associated bedded salts.

Composition. A hydrous sulfate of potassium, calcium, and magnesium, $K_2Ca_2Mg(SO_4)_4 \cdot 2H_2O$ (15.6% K_2O, 18.6% CaO, 6.7% MgO, 53.1% SO_3, 6.0% H_2O). Iron oxide inclusions that cause red color are common.

Tests. Fuses easily, giving violet (potassium) flame. Dissolves in water, with the separation of an insoluble calcium sulfate residue. Tastes bitter.

Crystallization. Triclinic; pinacoidal class, $\bar{1}$. Usually in fine-grained masses; also fibrous or foliated. Small, tabular, or elongated crystals are rare.

Occurrence and Associated Minerals. Usually in bedded deposits formed by the evaporation of sea water. Commonly associated with halite, anhydrite, sylvite, and carnallite ($KMgCl_3 \cdot 6H_2O$). Large, bedded deposits occur in Stassfurt, Germany; and in the Saratov region of the U.S.S.R. In the United States, polyhalite is common in the salt beds of New Mexico and western Texas.

Use. A source of potassium.

CHALCANTHITE

$CuSO_4 \cdot 5H_2O$ Triclinic
Hardness $2\frac{1}{2}$ Sp. Gr. 2.28

Summary

A soft, water-soluble, *deep azure blue* mineral that alters readily to a pale greenish powder. Crystals are commonly thick tabular (Plate 28(*F*), facing p. 178). *Luster:* vitreous. *Streak:* white. *Cleavage:* three directions poor; transparent to translucent; alteration opaque; taste metallic (poisonous). DEEP BLUE COLOR, GREENISH ALTERATION, AND SOLUBILITY IN WATER ARE CHARACTERISTIC. Not likely to be confused with other minerals.

Composition. Hydrous copper sulfate, $CuSO_4 \cdot 5H_2O$ (31.8% CuO, 32.1% SO_3, 36.1% H_2O). Small amounts of Fe^{2+} may be present.

Tests. Soluble in water. Tastes metallic but is poisonous. Loses water in closed tube and turns white.

Crystallization. Triclinic; pinacoidal class, $\bar{1}$. Crystals are commonly short prismatic to thick tabular. May also be stalactitic, botryoidal, fibrous, or massive.

Alteration. Dehydrates to an opaque greenish aggregate of hydrous copper sulfate.

Occurrence. Chalcanthite is not a common mineral. It is a secondary mineral that occurs chiefly in arid regions in the oxidized zone of copper sulfide deposits. The major chalcanthite deposits occur in the desert region of northern Chile (Chuquicamata, Copaquire, and Quetena). Antlerite and brochantite are associated with chalcanthite in Chile. Ions of copper and sulfate are frequently present in mine waters, and thus stalactites of chalcanthite often crystallize on the mine timbers and walls. Copper is sometimes recovered from mine waters that are rich in copper sulfate by the addition of metallic iron. The copper from the solution replaces the iron and forms a copper-plated surface. The copper-coated iron is then smelted.

Use. A minor ore of copper.

EPSOMITE

$MgSO_4 \cdot 7H_2O$ Orthorhombic
    Hardness 2–2$\frac{1}{2}$ Sp. Gr. 1.67

Summary

A white, water-soluble mineral with a bitter taste, commonly in hair-like crusts or cottony efflorescences; found on the walls of limestone caves and mine workings. *Color:* colorless to white. *Luster:* vitreous, silky, or earthy. *Cleavage:* one perfect and one distinct; transparent to translucent; brittle. **BITTER TASTE, SOLUBILITY IN WATER, AND HABIT ARE CHARACTERISTIC.**

Composition. Hydrous magnesium sulfate, $MgSO_4 \cdot 7H_2O$ (16.3% MgO, 32.5% SO_3, 51.2% H_2O). Nickel or zinc may substitute for magnesium, forming complete solid-solution series in artificial crystals extending to morenosite ($NiSO_4 \cdot 7H_2O$, pale green) and goslarite ($ZnSO_4 \cdot 7H_2O$). Limited amounts of Fe, Co, and Mn may substitute for Mg.

Tests. Soluble in water. Tastes bitter. Melts in its own water of crystallization. Gives silver-coin test for sulfur.

Crystallization. Orthorhombic; disphenoidal class 2 2 2. Usually as hair-like aggregates. Also as wooly efflorescences and botryoidal masses. Crystals are rare.

Alteration. Alters to and from hexahydrite ($MgSO_4 \cdot 6H_2O$). Alters from kieserite ($MgSO_4 \cdot H_2O$).

Associated Minerals. Melanterite ($FeSO_4 \cdot 7H_2O$) and other iron sulfates, also gypsum, anhydrite, and mirabilite ($Na_2SO_4 \cdot 10H_2O$).

Occurrence. (1) As fibrous crusts and cottony efflorescences in limestone caves and on the walls of coal or metal mines; also in sheltered places on outcrops of magnesium-rich rocks, gypsum, and limestone. (2) In the oxidized zone of pyrite-rich sulfide deposits in arid regions. (3) In solution in mineral spring waters. (4) In salt lake and oceanic deposits. (5) In small amounts around fumaroles.

Epsomite was named after the deposit found in mineral waters at Epsom, Surrey, England. Large crystals of epsomite, some several feet long, occur in the salt lakes of Kruger Mountain near Oroville, Washington. Also occur in the salt deposits of Albany County, Wyoming; Carlsbad, New Mexico; Nevada; and Stassfurt, Germany. Found as efflorescences in limestone caves in Kentucky, Tennessee, and Indiana; also in pyritic sulfide mines in California. In small amounts around the fumaroles on Vesuvius.

Use. Epsom salt is used for medical purposes; in dyeing; in tanning; and as a filler in cotton goods. It was first processed from the waters of the mineral springs at Epsom, England, but is now manufactured for commercial use from other magnesium-bearing minerals.

COPIAPITE

$(Fe,Mg)Fe_4^{3+}(SO_4)_6(OH)_2 \cdot 20H_2O$ Triclinic
    Hardness 2$\frac{1}{2}$–3   Sp. Gr. 2.08–2.17

Summary

A soft, *yellow* to *greenish-yellow*, water-soluble mineral with a metallic taste; usually in scaly aggregates with a *pearly luster* (Plate 29(*D*), facing p. 179). *Cleavage:* micaceous; translucent. Copiapite is a secondary mineral usually formed by the oxidation of pyrite. YELLOWISH COLOR, PEARLY LUSTER, SCALY HABIT, METALLIC TASTE, AND SOLUBILITY IN WATER ARE CHARACTERISTIC, but cannot be distinguished from similar water-soluble iron sulfates without special tests, such as x-ray examination.

Composition. A hydrated basic ferric iron sulfate with Fe^{2+}, Mg, Cu, or Zn. Small amounts of Mn^{2+}, Ca, Ni, and Al may also be present.

Tests. Soluble in water, with solution becoming cloudy on heating. Tastes metallic.

Crystallization. Triclinic; pinacoidal class, $\bar{1}$. Usually in loose masses of pearly scales. Also in crusts of granular masses. Tabular crystals are rare.

Occurrence. Copiapite is formed by the oxidation of pyrite, and other iron sulfides. It occurs near Copiapó, Chile; in coal mines in France; and in the Harz Mountains, Germany. An interesting occurrence of copiapite was at the United Verde copper mine in Jerome, Arizona. A portion of the mine burned, causing the rapid oxidation of the sulfide minerals and the subsequent formation of copiapite.

LINARITE

PbCu(SO₄)(OH)₂   Monoclinic
Hardness 2½   Sp. Gr. 5.35

Summary

A soft, *deep azure blue* mineral that is often mistaken for azurite. Usually in crusts (Plate 29(*B*), facing p. 179); small, elongated prismatic, or tabular crystals; or indistinct crystal aggregates. *Luster:* vitreous to subadamantine. *Streak:* pale blue. *Cleavage:* one perfect, and one imperfect. *Fracture:* conchoidal; brittle; translucent. Linarite is a secondary mineral found in the weathered zone of copper and lead deposits, and is commonly associated with cerussite, anglesite, malachite, aurichalcite, chrysocolla, and hemimorphite. DISTINGUISHABLE FROM AZURITE $Cu_3(OH)_2(CO_3)_2$ BY LACK OF EFFERVESCENCE IN ACID.

Composition. A basic sulfate of lead and copper, PbCu(SO₄)(OH)₂ (19.8% CuO, 55.7% PbO, 20.0% SO₃, 4.5% H₂O).

Tests. Fuses easily on charcoal. Gives silver-coin test for sulfur. Soluble in dilute nitric acid with the separation of an insoluble residue of lead sulfate.

Crystallization. Monoclinic; prismatic class, $\frac{2}{m}$. Crystals are small and are commonly elongated or tabular; frequently in aggregates of small indistinct prismatic crystals; also commonly in crusts.

Alteration. To antlerite $(Cu_3(SO_4)(OH)_4)$ and cerussite $(PbCO_3)$ with malachite $(Cu_2CO_3(OH)_2)$.

Associated Minerals. Cerussite, anglesite, malachite, aurichalcite, chrysocolla, antlerite, brochantite, hemimorphite, and cerargyrite.

Occurrence. Linarite is a widespread secondary mineral found in small amounts in the weathered zone of copper, lead, and silver ore deposits. Named after Linares, Spain, where it was discovered. Also found in Tsumeb, South-West Africa; Cumberland, England; Leadhills, Scotland; Broken Hill, New South Wales; Peru; Chile; Argentina; Germany; and Siberia. In the United States, elongated crystals of linarite, some several inches long, are found in the Mammoth Mine, Tiger, Arizona. Found also in Inyo County, California; in Tintic, Utah; in Butte, Montana; in Nevada; and in Idaho.

BROCHANTITE

Cu₄SO₄(OH)₆   Monoclinic
Hardness 3½–4   Sp. Gr. 4.0±

Summary

An *emerald green* to *blackish-green* mineral that resembles malachite and antlerite. Occurs in crusts of loosely coherent needle-like crystals, small glassy prismatic to tabular crystals, masses, and fibrous veinlets (Plate 29(*F*), facing p. 179). *Luster:* vitreous to slightly pearly on cleavage faces. *Streak:* pale green. *Cleavage:* perfect front pinacoid; transparent to translucent. Brochantite is a secondary mineral found in arid regions in the oxidized zone of copper deposits, and is commonly associated with malachite, antlerite, cuprite, chrysocolla, and limonite. DISTINGUISHABLE FROM MALACHITE BY LACK OF EFFERVESCENCE IN ACID AND FREQUENTLY A MORE GLASSY APPEARANCE. Indistinguishable from ANTLERITE, $Cu_3SO_4(OH)_4$, without special tests. Also resembles ATACAMITE, $Cu_2Cl(OH)_3$.

Composition. A basic sulfate of copper, $Cu_4SO_4(OH)_6$ (70.4% CuO, 17.7% SO_3, 11.9% H_2O).

Tests. Fusible. Dissolves in HCl and gives white precipitate of barium sulfate with the addition of barium chloride. May also give a white precipitate of calcium sulfate with the addition of calcium carbonate (calcite, aragonite).

Crystallization. Monoclinic; prismatic class, $\frac{2}{m}$. Crystals are small prismatic, needle-like, or tabular. Commonly in twinned pseudo-orthorhombic crystals. Also as crusts of loosely coherent needle-like crystals, fibrous veinlets, and masses.

Alteration. Alters to chrysocolla, and found as pseudomorphs after azurite and malachite.

Associated Minerals. Azurite, malachite, chrysocolla, limonite, linarite, cerussite, cuprite, atacamite.

Occurrence. Brochantite is a secondary mineral found in arid regions in the oxidized zone of copper deposits. Occurs abundantly in Chuquicamata in the desert region of northern Chile. Also found in England, the U.S.S.R., Roumania, Germany, France, Italy, Africa, and Australia. In the United States, found in Arizona, Nevada, Utah, California, Colorado, New Mexico, and Idaho.

Use. An ore of copper.

crystals; also in velvety crusts of needle-like crystals, and in fibrous veinlets that resemble malachite. *Luster:* vitreous. *Streak:* pale green. *Cleavage:* perfect side pinacoid, poor front pinacoid; translucent; brittle. Antlerite is a secondary mineral found in arid regions in the oxidized zone of copper deposits, and is commonly associated with malachite, brochantite, atacamite ($Cu_2Cl(OH)_3$), chalcanthite, linarite, and gypsum. DISTINGUISHED FROM MALACHITE ($Cu_2CO_3(OH)_2$) BY LACK OF EFFERVESCENCE IN ACID. Indistinguishable from BROCHANTITE ($Cu_4SO_4(OH)_6$) without special tests. Also resembles ATACAMITE ($Cu_2Cl(OH)_3$).

Composition. A basic sulfate of copper, $Cu_3SO_4(OH)_4$ (67.3% CuO, 22.5% SO_3, 10.2% H_2O).

Tests. Fusible. Dissolves in HCl and gives white precipitate of barium sulfate when barium chloride is added; or gives a white precipitate of calcium sulfate when calcium carbonate (calcite, aragonite) is added.

Crystallization. Orthorhombic; dipyramidal class, $\frac{2}{m}\frac{2}{m}\frac{2}{m}$. Commonly in fibrous masses; aggregates of needle-like crystals with a velvety surface; granular masses; and small short prismatic to tabular crystals.

Associated Minerals. Malachite, brochantite, atacamite, gypsum, linarite, chalcanthite, and limonite.

Occurrence. Antlerite is a secondary mineral found in arid regions in the weathered zone of copper deposits. It is one of the chief ore minerals of copper at Chuquicamata, in the desert region of northern Chile. Named from the Antler Mine in Mohave County, Arizona, where it was first observed.

Use. An ore of copper.

ANTLERITE

$Cu_3SO_4(OH)_4$	Orthorhombic
Hardness $3\frac{1}{2}$–4	Sp. Gr. $3.9\pm$

Summary

An *emerald green* to *very dark green* mineral, commonly in small short prismatic to tabular glassy

ALUNITE

$KAl_3(SO_4)_2(OH)_6$	Hexagonal
Hardness $3\frac{1}{2}$–4	Sp. Gr. 2.6–2.9

Summary

Usually in granular to dense masses that resemble limestone, dolomite, magnesite, or feldspar. Often intermixed with silica. Rarely in small pseudocubic crystals. *Color:* white, gray, yellowish, flesh-red. *Streak:* white. *Luster:* vitreous. *Cleavage:* imperfect basal, poor rhombohedral. *Fracture:* concoidal to uneven; translucent or transparent; brittle. Alunite is a common mineral, generally formed by the action of water (containing sulfuric acid) on rocks that are rich in orthoclase feldspar. The process is called "alunitization." FINE-GRAINED, WHITE, OR FLESH-COLORED MASSES (Plate 29(*C*), facing p. 179) ARE TYPICAL.

Composition. A basic sulfate of aluminum and potassium in which sodium often substitutes in part for potassium, $(K,Na)Al_3(SO_4)_2(OH)_6$. Where Na exceeds K, the mineral is called *natroalunite*. Natroalunite is much less common than alunite.

Tests. Infusible, decrepitates and colors flame violet (K). In closed tube, gives acid water (blue litmus paper turns pink). Soluble in sulfuric acid. After heating, soluble in nitric acid. Moistened with cobalt nitrate, solution turns blue (aluminum) on heating.

Crystallization. Hexagonal; ditrigonal-pyramidal class, 3 *m*. Usually massive granular, dense, or fibrous. Commonly mixed with quartz or kaolin. Small pseudocubic rhombohedral crystals are rare. Crystals sometimes found in fissures and cavities in massive alunite.

Occurrence. Alunite occurs widely, usually formed by the action of water (containing sulfuric acid) on rocks that are rich in orthoclase feldspar. Sometimes accompanied by the action of sulfuric acid, which is produced by the oxidation of pyrite. Alunite is common in the altered volcanic rocks of western United States. Large deposits are found in Marysvale, Utah; Goldfield District, Nevada; and in Mineral County and Hinsdale County, Colorado.

Use. For the production of alum. As a filler in fertilizer.

JAROSITE

$KFe_3(SO_4)_2(OH)_6$ Hexagonal
Hardness $2\frac{1}{2}$–$3\frac{1}{2}$ Sp. Gr. 2.91–3.26

Summary

A soft, *yellow ocher* to *brown* mineral, commonly massive or in crusts of minute hexagonal scales or pseudocubes. Earthy masses of jarosite strongly resemble limonite. *Luster:* vitreous, resinous to earthy. *Streak:* pale yellow. *Cleavage:* good basal; translucent. A secondary mineral, commonly associated with limonite, hematite, and pyrite. MINUTE HEXAGONAL OR RHOMBOHEDRAL PLATES, PSEUDOCUBES, AND SILVER-COIN TEST FOR SULFUR ARE CHARACTERISTIC. Frequently mixed with limonite in gossans.

Composition. Basic hydrous sulfate of potassium and iron, $KFe_3(SO_4)_2(OH)_6$. Sodium commonly substitutes for potassium and a series extends towards *natrojarosite*, $NaFe_3(SO_4)_2(OH)_6$.

Tests. Fuses on charcoal to a black magnetic mass. Soluble in HCl. Insoluble in water. Partly fused jarosite gives silver coin test for sulfur.

Crystallization. Hexagonal; ditrigonal-pyramidal class, 3 *m*. Commonly as crusts or coatings of minute or microscopic crystals. Also as granular, fibrous, or earthy masses, and nodules. Crystals are minute and indistinct; usually pseudocubic, or as hexagonal or rhombohedral plates.

Alteration. Alters from iron sulfides. To iron oxides.

Associated Minerals. Limonite, hematite, pyrite, and quartz.

Occurrence. A widespread secondary mineral, found as crusts and coatings on iron ores. Often a common constituent of limonitic gossans or "iron hats." Also found in veins and cavities in rocks surrounding iron minerals. Jarosite is common in the desert region of northern Chile, at Chuquicamata. In the United States, minute hexagonal scales of jarosite occur on hematite at Bisbee, Arizona. Thick, tabular crystals have been found in Custer County, Idaho. Good crystals occur in Chaffee County, Colorado. Common in limestone veins in Lincoln, Nevada, and at numerous other locations in the United States.

TUNGSTATES, MOLYBDATES, CHROMATES

Wolframite Series
    Wolframite $(Fe,Mn)WO_4$
    Huebnerite $MnWO_4$
    Ferberite $FeWO_4$
  Scheelite $CaWO_4$
  Wulfenite $PbMoO_4$
  Crocoite $PbCrO_4$

WOLFRAMITE

$(Fe,Mn)WO_4$ Monoclinic
Huebnerite, $MnWO_4$
Ferberite, $FeWO_4$
    Hardness $4-4\frac{1}{2}$ Sp. Gr. 7.12–7.51

Summary

A *heavy* mineral, commonly in bladed crystals associated with quartz. *Color:* wolframite is brownish-black to black; huebnerite is reddish-brown (Plate 30(*D, E*), facing p. 186); and ferberite is black. *Streak:* black to reddish-brown. *Luster:* submetallic to resinous. *Cleavage:* perfect side pinacoid; brittle. HIGH SPECIFIC GRAVITY, BLADED CRYSTALS, DARK COLOR, AND TEST FOR TUNGSTEN ARE CHARACTERISTIC. Difficult to distinguish from COLUMBITE-TANTALITE, which may contain small amounts of tungsten.

Composition. *Wolframite* is a tungstate of manganese and iron, $(Fe,Mn)WO_4$. Iron and manganese substitute for each other and a complete solid solution series exists between *huebnerite* ($MnWO_4$) and *ferberite* ($FeWO_4$). May contain small amounts of calcium, columbium, or tantalum.

Tests. Fuses to a magnetic globule. Gives good tungsten test: fuse the mineral with sodium carbonate, dissolve the mixture in hydrochloric acid, add a small piece of tin, then boil. The solution turns blue (tungsten test).

Crystallization. Monoclinic; prismatic class, $\frac{2}{m}$. Crystals are commonly tabular to bladed. Also in granular masses.

Associated Minerals. Quartz, cassiterite, molybdenite, beryl, topaz, apatite, tourmaline, pyrrhotite, pyrite, arsenopyrite, native bismuth, chalcopyrite, galena, scheelite, and fluorite.

Occurrence. Wolframite frequently occurs in high- to moderate-temperature hydrothermal quartz veins.

Table 8-2

Some Properties of the Tungstates, Molybdates, and Chromates

MINERAL	SYSTEM	COLOR	HARDNESS	SP. GR.	REMARKS
Wolframite	Monoclinic	Brownish-black	$4-4\frac{1}{2}$	7.1–7.5	Heavy bladed crystals
Huebnerite	Monoclinic	Reddish-brown	$4-4\frac{1}{2}$	7.1	Bladed crystals, heavy
Ferberite	Monoclinic	Black	$4-4\frac{1}{2}$	7.5	Bladed crystals, heavy
Scheelite	Tetragonal	White, tan, yellowish	$4\frac{1}{2}-5$	6.1	Luminesces pale blue
Wulfenite	Tetragonal	Yellow, orange	2.75–3	6.5–7.0	Yellow to orange tabular crystals
Crocoite	Monoclinic	Deep orange	$2\frac{1}{2}-3$	6.0–6.1	Deep orange; brilliant elongated crystals

Wolframite, cassiterite, and molybdenite are among the earliest metallic minerals to crystallize. Black-bladed crystals of wolframite are often attached to the sides of the quartz veins and extend inward in a "comb-like" structure. The richest deposits of wolframite occur in the Nanling range in southern China. Other notable deposits occur in the Malayan peninsula and in Burma; in Bolivia; Cornwall, England; Portugal; New Brunswick, Canada; and Zinnwald, Bohemia. In the United States, found in Boulder County, Colorado; in the Black Hills of South Dakota; in Nevada and other western states.

Use. Major ore of tungsten.

SCHEELITE

CaWO₄   Tetragonal
    Hardness 4½–5   Sp. Gr. 6.1

Summary

A *heavy,* nonmetallic mineral that usually *luminesces pale blue* under short-wave ultraviolet light (Plate 9, facing p. 66). Commonly in tetragonal bipyramidal crystals (Figure 8-40; Plate 30(*F*), facing p. 186). *Color:* white, tan, gray, yellowish, greenish. *Luster:* vitreous to adamantine. *Streak:* white. *Cleavage:* good pyramidal; transparent to translucent; brittle. Scheelite is a high-temperature mineral frequently found in contact metamorphic deposits associated with granet, epidote, quartz, tremolite, molybdenite, and calcite. Also in high-temperature quartz veins often associated with cassiterite, wolframite, topaz, fluorite, apatite, and molybdenite. HIGH SPECIFIC GRAVITY, PALE BLUE LUMINESCENCE, AND PSEUDO-OCTAHEDRAL CRYSTALS ARE CHARACTERISTIC. Resembles QUARTZ.

Composition. Calcium tungstate, CaWO₄ (19.4% CaO, 80.6% WO₃). Molybdenum often substitutes for part of the tungsten, and a partial solid solution series extends toward *powellite* (CaMoO₄), which luminesces yellow.

Tests. Usually luminesces pale blue under short-wave ultraviolet light. The luminescence becomes white or yellowish with increasing molybdenum substitution. Decomposes in HCl, and gives tungsten test when solution is boiled with metallic tin. Solution turns blue, and later, brown. Sometimes thermoluminescent.

Crystallization. Tetragonal; dipyramidal class, $\frac{4}{m}$. Usually in bipyramidal crystals. Also in granular masses.

Minerals of Similar Appearance

Quartz, feldspar. Distinguishable from quartz and feldspar by high specific gravity, vulnerability to scratch of a pen knife, and usual pale blue luminescence.

Apatite. Grains of apatite that sometimes luminesce yellow may resemble scheelite or powellite. Distinguishable by test for tungsten or molybdenum.

Alteration. To yellow *tungstite* (WO₃) or green *hydrotungstite* (WO₃·nH₂O). Also to greenish *cuprotungstite* (CuWO₄).

Occurrence. Scheelite is an important ore of tungsten and a typically high-temperature mineral. (1) It is found in contact-metamorphic limestones where granite rocks have intruded into the limestones. Commonly associated with garnet, epidote, diopside, tremolite, idocrase, epidote, fluorite, and molybdenite. (2) In hydrothermal quartz veins associated with cassiterite, wolframite, topaz, tourmaline, apatite, beryl, molybdenite, and arsenopyrite. (3) In pegmatites.

Scheelite occurs widely. A few of the numerous important localities are Mill City, Nevada; Pine Creek near Bishop, and Atolia in California. Crystals are usually small, but one of over a foot long has been reported from Japan, and 5½-in. crystals occur in

Fig. 8-40. Scheelite crystal.

Taehwa, Korea. Scheelite also occurs in Brazil; Zinnwald, Bohemia; Burma; Nova Scotia; England; Malaysia; and the U.S.S.R.

Use. An ore of tungsten.

WULFENITE

$PbMoO_4$ Tetragonal
    Hardness 2.75–3   Sp. Gr. 6.5–7.0

Summary

Usually in yellow to orange tabular crystals that are sometimes very thin (Plate 30(*A, B*), facing p. 186; and Plate 1(*C*), facing p. 2). *Color:* yellow, orange, red, brown, gray, grayish-white. *Streak:* white. *Luster:* adamantine to resinous. *Cleavage:* distinct prismatic; transparent to translucent. Wulfenite is a secondary mineral found in the oxidized zone of lead veins; commonly associated with pyromorphite, vanadinite, cerussite, galena, and limonite. Distinguishable from CROCOITE by tabular habit. YELLOW TO ORANGE COLOR, AND TABULAR HABIT ARE DISTINCTIVE.

Composition. Lead molybdate, $PbMoO_4$ (60.8% PbO, 39.2% MoO_3). Some calcium may substitute for lead, and tungsten or vanadium for molybdenum. May also contain small amounts of chromium or arsenic.

Tests. Fuses easily. On fusing with sodium carbonate on charcoal, gives lead globule.

Crystallization. Tetragonal; pyramidal class, 4. Crystals are usually tabular and often very thin (Figure 8-41). Less frequently in pyramidal or prismatic crystals. Also in granular masses.

Occurrence. Wulfenite is a secondary mineral found in the oxidized zone of lead deposits and is frequently

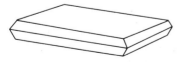

Fig. 8-41. Thin tabular wulfenite crystal.

associated with pyromorphite, vanadinite, cerussite, limonite, calcite, and galena. Found in brilliant orange-red tabular crystals—some 2 in. across—at the Red Cloud Mine in Yuma County, Arizona; in beautiful crystals at the Mammoth Mine, Pinal County, Arizona; and at the Glove Mine, Santa Cruz County, Arizona. Also found in New Mexico, Nevada, and Utah. Other notable localities are Příbram, Bohemia; Broken Hill district, New South Wales, Australia; and Durango, Mapimi, Sonora, Chihuahua, and other localities in Mexico.

Use. A minor ore of molybdenum.

CROCOITE

$PbCrO_4$ Monoclinic
    Hardness $2\frac{1}{2}$–3   Sp. Gr. 6.0–6.1

Summary

A bright, *reddish-orange* mineral with a *brilliant luster;* commonly in slender elongated crystals which are deeply striated, parallel to their length (Plate 30, facing p. 186). *Streak:* orange-yellow. *Cleavage:* poor prismatic; translucent. Crocoite is a rare secondary mineral found in oxidized lead deposits, and is commonly associated with pyromorphite and cerussite. Resembles *wulfenite.* COLOR, BRILLIANT LUSTER, HIGH SPECIFIC GRAVITY, AND ELONGATED PRISMATIC CRYSTALS (FREQUENTLY HOLLOW) ARE CHARACTERISTIC.

Composition. Lead chromate, $PbCrO_4$ (69.1% PbO, 30.9% CrO_3).

Tests. Fuses very easily. Gives green (chromium) borax bead.

Crystallization. Monoclinic; prismatic class, $\frac{2}{m}$. Usually in elongated prismatic crystals; commonly vertically striated; sometimes hollow.

Minerals of Similar Appearance

Wulfenite, $PbMoO_4$. Crocoite is distinguishable from orange-red wulfenite by its elongated prismatic habit, high specific gravity, and borax bead test for chromium. Wulfenite crystals are usually thin and tabular (Plate 30(*A, B*), facing p. 186).

Occurrence. A rare but very attractive secondary mineral, found in oxidized lead veins altered by chromium-bearing solutions. Associated with pyromorphite, cerussite, wulfenite, and vanadinite. Notable localities are Dundas, Tasmania; Ekaterinburg, Ural Mountains, the U.S.S.R.; and Goyaberia, Minas Gerais, Brazil. Found in the United States at the Darwin Mines, Inyo County, California; and at the El Dorado Mine, Riverside County, California. Also at the Mammoth Mine in Arizona.

Use. Too rare to be of commercial use, but forms very attractive mineral specimens.

PHOSPHATES, ARSENATES, VANADATES

The phosphates, arsenates, and vanadates comprise a large group of minerals, most of which are rare. The only really common phosphate is APATITE and its cryptocrystalline variety COLLOPHANE. The minerals AMBLYGONITE, LAZULITE, TURQUOISE, and MONAZITE have hardnesses of $5\frac{1}{2}$–6, but many phosphates, arsenates, and vanadates are soft, and can be scratched with a knife blade. Arsenic, phosphorus, and vanadium frequently substitute for each other; as, for example, in pyromorphite, mimetite, and vanadinite.

Phosphates, Arsenates, Vanadates

Monazite	Mon.	$(Ce,La,Y,Th)PO_4$
Apatite Group	Hex.	
Apatite		$Ca_5(PO_4)_3(F,Cl,OH)$
Pyromorphite		$Pb_5(PO_4)_3Cl$
Mimetite		$Pb_5(AsO_4)_3Cl$
Vanadinite		$Pb_5(VO_4)_3Cl$
Amblygonite	Tri.	$(Li,Na)Al(PO_4)(F,OH)$
Lazulite	Mon.	$(Mg,Fe)Al_2(PO_4)_2(OH)_2$
Wavellite	Orth.	$Al_3(OH)_3(PO_4)_2 \cdot 5H_2O$
Vivianite	Mon.	$Fe_3(PO_4)_2 \cdot 8H_2O$
Erythrite	Mon.	$(Co,Ni)_3(AsO_4)_2 \cdot 8H_2O$
Annabergite	Mon.	$(Ni,Co)_3(AsO_4)_2 \cdot 8H_2O$
Descloizite	Orth.	$(Zn,Cu)Pb(VO_4)(OH)$
Mottramite	Orth.	$(Cu,Zn)Pb(VO_4)(OH)$
Variscite	Orth.	$Al(PO_4) \cdot 2H_2O$
Strengite	Orth.	$Fe(PO_4) \cdot 2H_2O$
Turquoise	Tri.	$CuAl_6(PO_4)_4(OH)_8 \cdot 4H_2O$
Autunite	Tet.	$Ca(UO_2)_2(PO_4)_2 \cdot 10\text{–}12H_2O$
Torbernite	Tet.	$Cu(UO_2)_2(PO_4)_2 \cdot 8\text{–}12H_2O$
Carnotite		$K_2(UO_2)_2(VO_4)_2 \cdot 3H_2O$
Brazilianite	Mon.	$NaAl_3(PO_4)_2(OH)_4$

MONAZITE

$(Ce,La,Y,Th)PO_4$ Monoclinic
     Hardness 5–$5\frac{1}{2}$ Sp. Gr. 4.6–5.4

Summary

Commonly in yellowish to reddish-brown grains disseminated in granites and gneisses; also as a constituent of heavy beach sands. Crystals are often small and flattened. Large monazite crystals occur in pegmatites. *Color:* yellowish or reddish-brown to brown. *Luter:* resinous, waxy, or vitreous. *Cleavage:* good in one direction. *Parting:* good basal; brittle; translucent. RADIOACTIVITY, BROWN COLOR, CRYSTAL FORM, AND TEST FOR PHOSPHATE ARE DISTINCTIVE. Resembles ZIRCON ($ZrSiO_4$) in color and occurrence, but differs in crystal shape (Figure 8-42) and hardness.

Composition. A phosphate of rare-earth metals (cerium and lanthanum). Thorium and yttrium usually substitute for (Ce,La), $(Ce,La,Y,Th)PO_4$. The radioactivity of thorium causes structural damage; thus, many specimens are found in the METAMICT state.

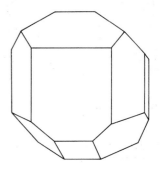

Fig. 8-42. Monazite crystal.

A   x 0.53

SERPENTINE
$Mg_3(Si_2O_5)(OH)_4$

Massive green serpentine showing typical waxy luster: Montville, New Jersey.

B   x 0.50

SERPENTINE
(CHRYSOTILE ASBESTOS)

Veinlets of chrysotile asbestos in massive serpentine: Morris County, New Jersey.

C   x 0.50

MUSCOVITE
$KAl_2(AlSi_3O_{10})(OH)_2$

Muscovite showing micaceous cleavage with blue tourmaline.

D   x 0.57

PYROPHYLLITE
$Al_2(Si_4O_{10})(OH)_2$

Radiating aggregates of pyrophyllite: Chesterfield County, South Carolina.

E   x 0.6

TALC
$Mg_3(Si_4O_{10})(OH)_2$

Light green talc showing foliated structure and pearly luster.

F   x 0.31

PREHNITE
$Ca_2Al(AlSi_3O_{10})(OH)_2$

Yellow-green rounded aggregates with ridged surfaces: West Paterson, New Jersey.

A x 0.4

BIOTITE

K(Mg, Fe)$_3$(AlSi$_3$O$_{10}$)(OH)$_2$

Black biotite in graphic granite: Bedford, New York.

B x 0.9

PHLOGOPITE

KMg$_3$(AlSi$_3$O$_{10}$)(OH)$_2$

Brown pseudohexagonal crystals showing micaceous cleavage: St. Lawrence County, New York.

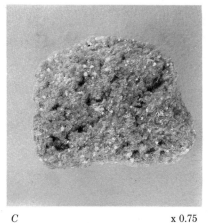

C x 0.75

LEPIDOLITE

K$_2$Li$_2$Al$_3$(AlSi$_7$O$_{20}$)(OH, F)$_4$

Scaly aggregate of lilac lepidolite: Pala, California.

D x 0.33

RHODONITE

MnSiO$_3$

Rough tabular rose red rhodonite crystal with typical black manganese oxide surface alteration, associated with white calcite: Franklin, New Jersey.

E x 0.6

PECTOLITE

Ca$_2$NaH(SiO$_3$)$_3$

Aggregate of very fine radiating needle-like pectolite crystals with a silky luster.

F x 0.5

WOLLASTONITE

CaSiO$_3$

Coarse fibrous aggregate: Nevada.

Tests. Infusible, but turns gray. Radioactive. When fused with sodium carbonate and dissolved in nitric acid, and a few drops of dissolved mineral are added to a solution of ammonium molybdate, a yellow precipitate is produced (test for phosphate).

Crystallization. Monoclinic; prismatic class, $\frac{2}{m}$. Usually granular, or as rounded grains in heavy beach sands. Crystals are usually small and often flattened. Also equant, elongate, prismatic,. and wedge-shaped forms. Contact twins are common. Large crystals occur in pegmatites.

Associated Minerals. Zircon, magnetite, apatite, ilmenite, columbite-tantalite, smarskite, and xenotime (YPO_4).

Occurrence. Monazite is widely disseminated as an accessory mineral in granites, syenites, and gneisses. Also as large crystals in pegmatites, as in Colorado. Because of its resistance to weathering, monazite is often concentrated in heavy beach sands or stream placers. Large deposits of monazite occur in Brazil, India, and Ceylon. Found in heavy sands in the United States in North Carolina, Florida, and Idaho.

Use. Major source of thorium oxide. Up to 12% ThO_2 may be present.

APATITE

Ca₅(PO₄)₃(F,Cl,OH)   Hexagonal
    Hardness 5   Sp. Gr. 3.1–3.2

Summary

Commonly in six-sided prismatic crystals (Figure 8-43) that often contain many small cracks, giving the crystals a shattered appearance (Plate 6(*B*, *C*), facing p. 35; Plate 31(*A*), facing p. 187). The edges of some crystals appear to have been partially dissolved. *Color:* usually shades of green, blue-green, or brown; also colorless, yellow, violet, red, deep blue. *Luster:* vitreous. *Streak:* white. *Cleavage:* imperfect basal; brittle; transparent to opaque. Apatite is frequently luminescent. CRYSTAL HABIT, AND HARDNESS OF 5 (CAN

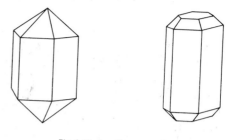

Fig. 8-43. Apatite crystals.

JUST BE SCRATCHED BY A KNIFE BLADE) ARE CHARACTERISTIC. Distinguishable from hexagonal crystals of BERYL or TOURMALINE by hardness. The name apatite comes from the Greek word meaning to *deceive*, because apatite has often been mistaken for other minerals, notably beryl, tourmaline, and diopside.

Composition. Calcium fluorine-chlorine-hydrozyl phosphates. Calcium fluorophosphate (fluor-apatite), $Ca_5(PO_4)_3F$; calcium chlorophosphate (chlor-apatite), $Ca_5(PO_4)_3Cl$; (hydroxyl-apatite) $Ca_5(PO_4)_3OH$. F, Cl, and OH can substitute for each other to form a complete series, with fluor-apatite, chlor-apatite, and hydroxyl-apatite as end members. Fluor-apatite is the most common variety. Calcium may be replaced in part by Mn, Sr, Mg, or Fe.

Tests. Fuses with difficulty. Violet varieties lose color on heating. Many specimens luminesce deep yellow under ultraviolet light. Others luminesce after heating (thermoluminescent). Soluble in acids. A few drops of dissolved mineral in dilute nitric acid added to a solution of ammonium molybdate give a yellow precipitate of ammonium phosphomolybdate (phosphorus test).

Crystallization. Hexagonal; dipyramidal class, $\frac{6}{m}$. Commonly in well-formed prismatic to tabular crystals, often terminated by pyramids. Also in granular to dense masses, and in colloform or botryoidal crusts. The name *collophane* is given to massive impure cryptocrystalline apatite that may make up phosphate rock (Plate 31(*B*), facing p. 187).

Occurrence. Apatite is a common and widespread mineral. It is found (**1**) as an accessory mineral in igneous rocks; (2) as large, well-formed crystals in pegmatites; (3) in hydrothermal ore veins; (4) in magnetite deposits; (5) in metamorphic rocks; (6) in bedded marine deposits; (7) as a component of fossil bone, shell, and pellets; (8) as replacements of limestone derived from the leaching of guano deposits; and (9) as late magmatic segregations from alkalic igneous rocks.

Deep violet tabular crystals are found in the pegmatites at Mount Apatite, Auburn, Maine. Fine crystals occur at Knappenwand, Austria. Very large greenish or brownish crystals, some weighing a few hundred pounds, occur in Ontario, Canada. Transparent yellow crystals occur in the magnetite deposits at Cerro Mercado, Durango, Mexico. The largest granular apatite deposit occurs in the Kola peninsula, the U.S.S.R. In the United States, large deposits of phosphate rock occur in Florida, North Carolina, Tennessee, Idaho, Wyoming, and Utah. Other large deposits occur in Morocco, Spanish Morocco, Algeria, Tunisia, Togo, and Egypt.

Use. Fertilizers. Transparent apatite is sometimes used as a gem stone.

■   **PYROMORPHITE SERIES:**

Pyromorphite, $Pb_5(PO_4)_3Cl$;
Mimetite, $Pb_5(AsO_4)_3Cl$

PYROMORPHITE

$Pb_5(PO_4,AsO_4)_3Cl$ Hexagonal
Hardness $3\frac{1}{2}$–4   Sp. Gr. 7.04

Summary

Commonly in aggregates of short, six-sided prisms, which are often hollow (Plate 31(*D, E*), facing p. 187; Figure 8-44); also in rounded or barrel-shaped crystals. *Color:* usually green, also light brown; sometimes gray or yellowish. *Streak:* white. *Luster:* resinous to adamantine; brittle; translucent. Pyromorphite is a secondary mineral found in the oxidized zone of lead deposits frequently associated with cerussite and

Fig. 8-44. Hollow pyromorphite crystals.

other lead minerals. GREEN COLOR, SMALL SIX-SIDED HOLLOW CRYSTALS, BRIGHT LUSTER, AND HIGH SPECIFIC GRAVITY ARE CHARACTERISTIC. Crystals of VANADINITE strongly resemble pyromorphite, but vanadinite is commonly ruby red.

Composition. A lead chlorophosphate, $Pb_5(PO_4)_3Cl$; but arsenic substitutes for phosphorus, and a complete solid solution series extends to mimetite, $Pb_5(AsO_4)_3Cl$. Calcium may substitute in part for lead, and vanadium for limited amounts of phosphorus or arsenic.

Tests. Fuses easily on charcoal to a globule, which, on cooling, assumes an angular crystal-like shape. Dissolves in acids. A few drops of dissolved mineral in nitric acid, added to a solution of ammonium molybdate, gives a yellow precipitate of ammonium phosphomolybdate (test for phosphorus).

Crystallization. Hexagonal; dipyramidal class, $\frac{6}{m}$. Often in short, hollow hexagonal prisms; or in rounded, barrel-shaped forms. Also globular, botryoidal crusts, and granular masses.

Alteration. Alters from galena and cerussite.

Occurrence. Pyromorphite is a secondary mineral found in the oxidized portions of lead deposits; associated with cerussite, limonite, hemimorphite, smithsonite, wulfenite, anglesite, and vanadinite. Crystals are commonly small, usually less than $\frac{1}{2}$ in. in length. In the United States, occurs in good specimens at the Wheatley Mine, Phoenixville, Pennsylvania; Coeur d'Alene district, Idaho; and Davidson County, North Carolina. Also in fine specimens at Friedrichsegen, Germany;

Poullaouen and Huelgoat, France; and Mapimi, Mexico.

The name pyromorphite is derived from two Greek words meaning "fire" and "form," because a molten globule assumes an angular crystal-like shape when it cools.

MIMETITE

$Pb_5(AsO_4,PO_4)_3Cl$ Hexagonal
Hardness $3\frac{1}{2}$–4 Sp. Gr. 7.24

Summary

Commonly in rounded, brownish-orange melon-shaped aggregates (campylite variety, Plate 31(*C*), facing p. 187); also in small, six-sided prismatic crystals and botryoidal crusts. *Color:* colorless, yellow, orange, brown. *Streak:* white. *Luster:* resinous to adamantine; brittle; translucent. Mimetite is less common than pyromorphite and is formed by secondary processes in the oxidized portions of lead deposits. THE NAME MIMETITE COMES FROM THE GREEK WORD MEANING "IMITATOR," BECAUSE ITS CRYSTALS RESEMBLE PYROMORPHITE. Also resembles some varieties of VANADINITE.

Composition. A lead chloroarsenate, but phosphorus substitutes for arsenic, and a complete solid solution series extends to pyromorphite, $Pb_5(PO_4)_3Cl$. As $>$ P $=$ mimetite; P $>$ As $=$ pyromorphite.

Tests. Fuses easily on charcoal and gives off garlic fumes (arsenic).

Crystallization. Hexagonal; dipyramidal class, $\frac{6}{m}$. Commonly in rounded, barrel- or melon-shaped crystals. Also in short prismatic and needle-like crystals; and in botryoidal crusts.

Occurrence. Mimetite is a rare secondary mineral found in the oxidized portions of lead deposits. Crystals are usually small. Found in good crystals at Johanngeorgenstadt, Saxony. Campylite crystals occur in Cornwall and Cumberland, England.

VANADINITE

$Pb_5(VO_4)_3Cl$ Hexagonal
Hardness 2.75–3 Sp. Gr. 6.88

Summary

Commonly in small, red, six-sided prisms with a brilliant luster (Plate 31(*F*), facing p. 187). Crystals frequently have curved sides and are hollow in the center. *Color:* bright red, orange-red, brown, yellow. *Luster:* adamantine to resinous. *Streak:* white to yellowish; transparent to translucent; brittle. Vanadinite is a secondary mineral formed by alteration in the oxidized portion of lead deposits, and is commonly associated with pyromorphite, wulfenite, descloizite, and cerussite. CRYSTAL HABIT, BRILLIANT LUSTER, RED COLOR, AND HIGH SPECIFIC GRAVITY ARE CHARACTERISTIC. Crystals of VANADINITE closely resemble pyromorphite-mimetite crystals but vanadinite is distinguishable when bright red in color.

Composition. Lead chlorovanadate, $Pb_5(VO_4)_3Cl$. Limited amounts of phosphorus and arsenic substitute for vanadium. The light yellow variety (*endlichite*) contains As in substitution for V in a As:V ratio of 1:1. Small amounts of copper, zinc, and calcium may substitute for lead.

Tests. Fuses easily, and gives a lead globule when fused on charcoal with sodium carbonate. Gives a yellow-green borax bead (vanadium). The addition of silver nitrate to a solution of vanadinite in dilute nitric acid gives a white silver chloride precipitate.

Crystallization. Hexagonal; dipyramidal class, $\frac{6}{m}$. Commonly in small prismatic crystals; sometimes hollow prisms; also needle-like crystals and globular masses.

Alteration. Alters to descloizite.

Occurrence. Vanadinite, like pyromorphite and mimetite, is a secondary mineral formed by alteration in the oxidized portion of lead deposits. Vanadinite is commonly associated with descloizite, pyromorphite-mimetite, wulfenite, cerussite, and limonite. Found in large crystals in Grootfontein, South-West Africa; and in large crystals over an inch long at Djebel Mahseur, Morocco. In the United States, brilliant orange-red crystals occur in the Old Yuma Mine, Pima County, Arizona; and also at the Red Cloud Mine in Yuma County, Arizona. Bright yellow crystals (rich in arsenic) occur in the Hillboro and Lake Valley districts of New Mexico.

Use. As a source of vanadium.

AMBLYGONITE

$(Li,Na)Al(PO_4)(F,OH)$ Triclinic
     Hardness $5\frac{1}{2}$–6   Sp. Gr. 3.0–3.1

Summary

Usually in coarse cleavable masses found in lithium-bearing pegmatites. *Color:* white with tints of yellow, blue, violet, green, gray, pink. *Luster:* vitreous to greasy, pearly on base. *Cleavage:* perfect basal, imperfect parallel to front pinacoid; brittle; transparent to translucent; may luminesce orange in long wave ultraviolet light. Commonly associated with spodumene, lepidolite, apatite, tourmaline, and quartz. RESEMBLES MASSIVE FELDSPAR, BUT SLIGHTLY GREASY TO PEARLY LUSTER IS DISTINCTIVE.

Composition. A basic phosphate of lithium, sodium, and aluminum with fluorine and hydroxyl $(Li,Na)Al(PO_4)$-(F,OH). Sodium substitutes for lithium, and hydroxyl for fluorine. Lithium is usually present in greater amounts than is sodium. When OH is greater than F the mineral is called *montebrasite*.

Tests. Fuses easily to a white globule. Gives red (Li) flame. Dissolves in acids and gives phosphate test with ammonium molybdate.

Crystallization. Triclinic; pinacoidal class, $\bar{1}$. Commonly in coarse cleavable masses. Crystals are usually large, rough, and equant; sometimes short prismatic or lath-like.

Alteration. Alters to kaolin and mica. Also to turquoise, wardite, and wavellite.

Occurrence. Found (1) in granite pegmatites rich in lithium and phosphorus. Usually associated with spodumene, apatite, lepidolite, tourmaline, and quartz. (2) Less commonly, in high-temperature tin veins associated with cassiterite, topaz, and mica. In the United States, large crystals occur in pegmatites in the Black Hills of South Dakota. Also in Pala, California; and in Maine. Unusual colorless transparent crystals are found at Newry, Oxford County, Maine. Yellowish 6-in. crystals occur in Minas Gerais, Brazil. Montebrasite occurs in a pegmatite in São Paulo, Brazil. Both amblygonite and montebrasite are found at Montebras, France; Sweden; and Australia.

Use. A source of lithium.

LAZULITE

$(Mg,Fe)Al_2(PO_4)_2(OH)_2$ Monoclinic
     Hardness $5\frac{1}{2}$–6   Sp. Gr. 3.1–3.39

Summary

Usually in azure blue masses. Steep pyramidal crystals, often imbedded in quartz masses, are distinctive but uncommon (Plate 32(*A*), facing p. 194). *Color:* azure blue to light blue, dark blue (scorzalite). *Streak:* white. *Luster:* vitreous. *Cleavage:* indistinct prismatic; translucent; brittle. Lazulite is an uncommon mineral, usually found in pegmatites, quartz veins, or quartzite; frequently associated with corundum, kyanite, rutile, andalusite, and quartz. BLUE COLOR AND STEEP PYRAMIDAL CRYSTALS ARE DISTINCTIVE. Massive lazulite resembles blue SODALITE and LAZURITE.

Composition. A basic phosphate of aluminum, magnesium, and iron, $(Mg,Fe)Al_2(PO_4)_2(OH)_2$. A complete isomorphous series exists between end members containing magnesium or iron. When the iron content exceeds the magnesium content the mineral is called SCORZALITE $(Fe,Mg)Al_2(PO_4)_2(OH)_2$.

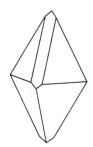

Fig. 8-45. Lazulite crystal.

Tests. Infusible, but loses color, swells, cracks, and falls apart. Only slightly soluble in hot acids. After fusion with sodium carbonate, a few drops of a nitric acid solution of the fused lazulite added to a solution of ammonium molybdate give a yellow precipitate (phosphorus test).

Crystallization. Monoclinic; prismatic class, $\frac{2}{m}$. Usually in granular to compact masses. Steep pyramidal (Figure 8-45) or tabular crystals are rare.

Minerals of Similar Appearance

Sodalite and Lazurite. Massive lazulite is distinguished from the blue silicate minerals sodalite and lazurite by a chemical test for phosphorus. Both sodalite and lazurite are fusible and soluble in hydrochloric acid. Lazurite (lapis lazuli) is almost always associated with pyrite.

Occurrence. Lazulite is not a common mineral; is usually found in granite pegmatites, quartz-rich metamorphic rocks, and quartz veins in metamorphic rocks. Associated minerals include quartz, kyanite, andalusite, dumortierite, corundum, rutile, muscovite, and garnet. Good crystals have been found near Zermatt, Valais, Switzerland, and at various localities in Austria. Dark blue scorzalite occurs with brazilianite in the Corrego Frio pegmatite, Minas Gerais, Brazil. In the United States, steep pyramidal crystals, up to $1\frac{1}{2}$ in. long, occur with rutile and kyanite imbedded in quartzite at Graves Mountain, Georgia.

Use. Minor gem stone.

WAVELLITE

$Al_3(OH)_3(PO_4)_2 \cdot 5H_2O$	Orthorhombic
Hardness $3\frac{1}{2}$–4	Sp. Gr. 2.36

Summary

Frequently in rounded aggregates with a fibrous radiating structure (Plate 32(*B*), facing p. 194). *Color:* white, green, yellow, brown, black. *Luster:* vitreous to silky. *Cleavage:* three directions; brittle; translucent. RADIATING SPHERULITIC AGGREGATES ARE CHARACTERISTIC.

Composition. A hydrous basic aluminum phosphate, $Al_3(OH)_3(PO_4)_2 \cdot 5H_2O$ (37.1% Al_2O_3, 34.5% P_2O_5, 28.4% H_2O). Small amounts of fluorine often substitute for OH.

Tests. Infusible, but whitens and exfoliates. Fused with sodium carbonate and dissolved in nitric acid, gives yellow precipitate when a few drops of nitric acid solution are added to an ammonium molybdate solution (test for phosphorus). When moistened with cobalt nitrate and heated, gives blue color (aluminum).

Crystallization. Orthorhombic. Commonly in radiating spherulitic aggregates. Also in crusts. Rarely in small striated prismatic crystals.

Occurrence. Wavellite is a secondary mineral found in aluminous and phosphate rocks. Commonly associated with apatite and turquoise. Abundant in the tin veins of Bolivia. Interesting radiating aggregates are found near Hot Springs and in Montgomery County, Arkansas.

VIVIANITE

$Fe_3(PO_4)_2 \cdot 8H_2O$	Monoclinic
Hardness $1\frac{1}{2}$–2	Sp. Gr. 2.68

Summary

Commonly in small, soft, dark blue tabular crystals; also radiating aggregates, and bright blue earthy masses. *Color:* colorless and transparent when unaltered, but usually altered to dark blue, green, blue-green, or almost black. *Streak:* white, changing to bright blue or brown on exposure to light. *Luster:* vitreous to earthy, pearly on cleavage faces. *Cleavage:* perfect micaceous; parallel to side pinacoid with flexible cleavage lamellae. Vivianite is an uncommon secondary mineral and is not likely to be confused with other minerals. THE SOFT, DARK BLUE, OR GREENISH TABULAR CRYSTALS WITH PERFECT CLEAVAGE IN ONE DIRECTION ARE DISTINCTIVE (Plate 32(C), facing p. 194).

Composition. Hydrous iron phosphate, $Fe_3(PO_4)_2 \cdot 8H_2O$.

Tests. Fuses easily to a magnetic globule. A solution of ammonium molybdate plus a few drops of vivianite dissolved in nitric acid give a bright yellow precipitate (test for phosphate).

Crystallization. Monoclinic; prismatic class, $\frac{2}{m}$. Crystals are commonly prismatic to tabular; also in globular masses, fibrous crusts, radiating aggregates, or earthy masses. Commonly known as blue-iron earth.

Alteration. The change in color from colorless to dark blue is caused by the partial oxidation of the ferrous iron (Fe^{2+}) to ferric iron (Fe^{3+}) on exposure to light.

Associated Minerals and Occurrence. Vivianite occurs as a secondary mineral with pyrrhotite, pyrite, and limonite in the oxidized zone of copper and tin deposits; also as a weathering product of iron-manganese phosphates in pegmatites; found in bones, shells, fossils, and clays. Important localities are Leadville, Colorado; Bingham, Utah; the tin deposits of Cornwall (England) and Bolivia; and the pegmatites of Hagendorf, Bavaria. Blue to green tabular crystals, up to three inches in length, sheaf-like and stellate aggregates, and nodules of vivianite are found in Richmond, Virginia.

ERYTHRITE (Cobalt Bloom)

$(Co,Ni)_3(AsO_4)_2 \cdot 8H_2O$	Monoclinic
Hardness $1\frac{1}{2}$–$2\frac{1}{2}$	Sp. Gr. 3.1

Summary

Commonly in pink, earthy crusts formed as a surface alteration on such cobalt minerals as cobaltite and skutterudite. Crystals are not common but sometimes occur in radiating groups (Plate 32(E), facing p. 194). *Color:* crimson red, bluish-pink, pale pink. *Luster:* vitreous to dull, pearly on cleavage. *Streak:* pink. *Cleavage:* perfect parallel to side pinacoid; flexible laminae; sectile; translucent. PINK COLOR AND ASSOCIATION WITH COBALT ARSENIDES ARE DISTINCTIVE.

Composition. Hydrous cobalt arsenate. Nickel substitutes for cobalt to form a complete isomorphous series to apple green *annabergite,* $(Ni,Co)_3(AsO_4)_2 \cdot 8H_2O$.

Tests. Fuses easily to a gray globule. Gives off arsenic fumes with a garlic odor. Gives deep blue color to borax bead (cobalt).

Crystallization. Monoclinic; prismatic class, $\frac{2}{m}$. Commonly in pink botryoidal crusts. Slender striated prismatic crystals are less common. Sometimes in radiating groups.

Occurrence. Erythrite is a secondary mineral formed as an alteration product on cobalt arsenides. It usually occurs in small quantities as crusts on cobalt minerals. Its pink color is a good guide to cobalt ores. Hence erythrite is often referred to as "cobalt bloom." Specimens of radiating groups of crystals have been found in Schneeberg, Saxony (Plate 32). Pinkish crusts are common in the cobalt-silver veins of Cobalt, Ontario, Canada. Associated with some uranium ores on the Colorado Plateau in the United States.

ANNABERGITE (Nickel Bloom)

$(Ni,Co)_3(AsO_4)_2 \cdot 8H_2O$	Monoclinic
Hardness $1\frac{1}{2}$–$2\frac{1}{2}$	Sp. Gr. 3.0

Summary

Usually in pale, apple green crusts formed as an alteration product of nickel-cobalt minerals (Plate 32(D), facing p. 194). *Color:* pale apple green. *Luster:*

vitreous to dull. *Streak:* light green; translucent to earthy. PALE GREEN CRUSTS ON NICKEL-COBALT ARSENIDES ARE DISTINCTIVE.

Composition. Hydrous nickel arsenate. Cobalt substitutes for nickel and a complete isomorphous series extends to *erythrite*, $(Co,Ni)_3(AsO_4)_2 \cdot 8H_2O$.

Tests. Fuses to a magnetic globule in the reducing flame. Gives nickel test: dissolve mineral in nitric acid and neutralize with ammonia; then add a few drops of dimethylglyoxime solution to obtain a scarlet precipitate; this indicates nickel.

Crystallization. Monoclinic; prismatic class, $\frac{2}{m}$. Usually in pale green crusts. Small needle-like crystals in radiating groups are rare.

Occurrence. Annabergite is a secondary mineral, formed as an alteration product on arsenides of nickel and cobalt such as nickel-skutterudite, $(Ni,Co)As_3$. Usually found as pale green crusts, and is often referred to as "nickel bloom." Occurs at Cobalt, Ontario; in Annaberg and Schneeberg, Saxony; and in Laurium, Greece.

DESCLOIZITE, $(Zn,Cu)Pb(VO_4)(OH)$;
MOTTRAMITE, $(Cu,Zn)Pb(VO_4)(OH)$

Orthorhombic
Hardness 3–3½   Sp. Gr. 5.9 (mottramite);
6.2 (descloizite)

Summary

Usually in drusy crusts of minute crystals (Plate 32(*F*), facing p. 194), commonly associated with vanadinite and pyromorphite. *Color:* reddish to blackish-brown; also red, yellowish-brown, black, green. *Luster:* greasy. *Streak:* yellowish orange to brownish red; brittle; transparent to translucent. BROWNISH CRUSTS OF MINUTE CRYSTALS WITH ASSOCIATED MINERALS ARE CHARACTERISTIC.

Composition. A basic vanadate of lead, copper, and zinc, $(Zn,Cu)Pb(VO_4)(OH)$. Zinc and copper substitute for

each other, forming a complete isomorphous series between descloizite and mottramite. When zinc is more abundant than copper the mineral is called descloizite; and when copper is greater than zinc the mineral is mottramite. Mottramite is often green in color.

Tests. Fuses easily to a lead globule, surrounded by a black mass. Dissolves in acids. Gives V, Pb, Zn, Cu, and H_2O tests.

Crystallization. Orthorhombic; dipyramidal class, $\frac{2}{m}\frac{2}{m}\frac{2}{m}$. Usually in crusts of minute intergrown crystals. Also in botryoidal or fibrous masses.

Alteration. Occurs as pseudomorphs after vanadinite.

Occurrence. The descloizite-mottramite series are secondary minerals found in the weathered zone of ore deposits where they are commonly associated with vanadinite, pyromorphite, cerussite, and wulfenite. Found in large deposits in the Octavi region of South-West Africa. In the United States, descloizite and mottramite are common in New Mexico and in Arizona.

VARISCITE, $Al(PO_4) \cdot 2H_2O$;
STRENGITE, $Fe(PO_4) \cdot 2H_2O$

Orthorhombic
Hardness 3½–4½   Sp. Gr. 2.2–2.5 (variscite);
2.87 (strengite)

Summary

Variscite commonly occurs in dense green masses (Plate 33(*A*), facing p. 195). Strengite is usually in deep pink to violet botryoidal crusts with a radial fibrous structure. *Color:* light green to emerald green (variscite); pink, peach, bluish-red, violet (strengite). *Luster:* vitreous (crystals) to waxy (masses). *Fracture:* smooth to conchoidal. *Cleavage:* prismatic, but not usually seen; brittle; transparent to translucent. VARISCITE RESEMBLES TURQUOISE, BUT IS USUALLY GREENER IN COLOR AND IS SOFTER THAN TURQUOISE.

Composition. Hydrated phosphates of aluminum and iron, $(Al,Fe)(PO_4) \cdot 2H_2O$. Aluminum and iron substitute for each other forming a complete isomorphous series between variscite and strengite.

Tests. Variscite is infusible. Usually turns lavender when heated at a low temperature. Becomes soluble in HCl after heating, and gives phosphorus test with ammonium molybdate. Strengite fuses easily to a black bead. Soluble in HCl.

Crystallization. Orthorhombic; dipyramidal class, $\frac{2}{m}\frac{2}{m}\frac{2}{m}$. Variscite usually occurs as fine-grained masses. Commonly as nodules, crusts, or veinlets. Small pseudo-octahedral crystals are rare. Strengite usually occurs in botryoidal aggregates with a radial fibrous inner structure. Small crystals occur in varied shapes.

Alteration. Variscite commonly alters to wardite and yellowish crandallite.

Associated Minerals. Variscite is associated with wavellite, apatite, wardite, millisite, gordonite, crandallite, limonite and chalcedony. Strengite is commonly associated with vivianite, dufrenite, and fine-grained apatite.

Occurrence. Variscite is a secondary mineral formed near the surface by the alteration of aluminum-bearing rocks that are rich in phosphates. Strengite, the iron-rich member of the series, is derived from the alteration of iron-rich phosphates such as triphylite ($LiFePO_4$) or dufrenite ($Fe_5(PO_4)_3(OH)_5 \cdot 2H_2O$). In the United States, variscite occurs in large, rounded nodules, some over a foot in diameter, near Fairfield, Utah. The outer portions of the nodules are commonly altered to other phosphates, notably green to colorless WARDITE, pale gray MILLISITE, colorless GORDONITE, and yellowish CRANDALLITE (Plate 33, facing p. 195). Purple strengite is found in iron mines in Germany; as an alteration if iron phosphates in pegmatite at Pleystein, Bavaria; and in altered tripylite at Pala, San Diego county, California.

Use. Variscite is sometimes used in jewelry.

TURQUOISE

$CuAl_6(PO_4)_4(OH)_8 \cdot 4H_2O$	Triclinic
Hardness 5–6	Sp. Gr. 2.6–2.9

Summary

Commonly in bluish-green fine-grained masses with a dull porcelain-like or waxy luster (Plate 33(*B*), facing p. 195). *Color:* sky blue, blue-green, light green. *Luster:* waxy, vitreous (crystals). *Streak:* white to pale green. *Fracture:* conchoidal to smooth; brittle; thin edges are translucent. Turquoise is a secondary mineral usually found in narrow veins formed by the alteration of rocks that are rich in aluminum in arid regions. BLUE TO DULL GREENISH PORCELAIN-LIKE MASSES ARE CHARACTERISTIC. Resembles CHRYSOCOLLA (Plate 42, facing p. 258), VARISCITE (Plate 33, facing p. 195), and CHRYSOPRASE (apple green chalcedony, Plate 35, facing p. 211).

Composition. A basic hydrous phosphate of aluminum with copper, $CuAl_6(PO_4)_4(OH)_8 \cdot 4H_2O$. Iron substitutes for aluminum to form a complete isomorphous series to chalcosiderite, $CuFe_6(PO_4)_4(OH)_8 \cdot 4H_2O$.

Tests. Infusible. After heating, soluble in hot HCl. A few drops of solution added to ammonium molybdate solution give a yellow precipitate (phosphate test). Also give tests for Cu and H_2O.

Crystallization. Triclinic; pinacoidal class, $\bar{1}$. Usually in cryptocrystalline masses. In nodules and thin veins. Minute, short prismatic crystals are extremely rare.

Minerals of Similar Appearance

Chrysocolla, $CuSiO_3 \cdot 2H_2O$. Distinguishable from turquoise by inferior hardness (H. = 2–4, chrysocolla). Chrysocolla, impregnated with quartz, is distinguishable from turquoise by phosphate test.

Chrysoprase, SiO_2. Apple green chalcedony—phosphate test distinguishes turquoise from chrysoprase.

Variscite, $Al(PO_4) \cdot 2H_2O$. Softer than turquoise (H. = $3\frac{1}{2}$–$4\frac{1}{2}$) and usually greener in color.

Occurrence. Turquoise is a secondary mineral formed in arid regions from the alteration of rocks that are rich in aluminum containing apatite and chalcopyrite, accessory minerals. In ancient times, turquoise was mined on the Sinai Peninsula in Egypt, and near Nishâpûr, Khorasan, in Persia. The Persian (Iranian) deposits are still productive, and yield the most highly prized turquoise, which is deep sky-blue. The Iranian turquoise occurs as patches and thin seams in brecciated trachite and clay slate. The greenish varieties, or those containing thin dendritic veinings of matrix rock in the turquoise, are less valuable. In the United States, found in Nevada, California, New Mexico, Arizona, and Colorado. Drusy crusts of pale blue microscopic crystals occur on quartz in Campbell County, Virginia.

Use. As a gem stone.

AUTUNITE

$Ca(UO_2)_2(PO_4)_2 \cdot 10\text{--}12H_2O$ Tetragonal
     Hardness $2\text{--}2\frac{1}{2}$ Sp. Gr. 3.1–3.2

Summary

Commonly in bright yellow micaceous flakes that luminesce brilliant yellowish-green in ultraviolet light. Also in aggregates of crystals standing on edge (meta-autunite, Plate 33(*E, F*), facing p. 195). *Color:* yellow to greenish yellow. *Luster:* vitreous to pearly. *Cleavage:* perfect basal and prismatic; translucent. Autunite is a secondary mineral derived from the weathering of uraninite and other uranium-bearing minerals. AGGREGATES OF THIN SQUARE YELLOW PLATES, AND BRILLIANT YELLOWISH-GREEN LUMINESCENCE IN ULTRAVIOLET LIGHT AND RADIOACTIVITY ARE CHARACTERISTIC.

Composition. Hydrous calcium uranium phosphate, $Ca(UO_2)_2(PO_4)_2 \cdot 10\text{--}12H_2O$. On drying, autunite passes reversibly to meta-autunite(I) with $6\frac{1}{2}\text{--}2\frac{1}{2}H_2O$; when heated to about 80°C, autunite changes irreversibly to orthorhombic meta-autunite(II) with $0\text{--}6H_2O$.

Tests. Luminesces bright yellowish-green in ultraviolet light. Radioactive. Tests for Ca, U, P_2O_5, and H_2O.

Crystallization. Tetragonal; ditetragonal-dipyramidal class, $\frac{4}{m}\frac{2}{m}\frac{2}{m}$. Commonly in thin, square, yellow to greenish-yellow plates. Also as scaly aggregates, and as aggregates of crystals standing on edge and forming a serrated surface.

Occurrence. A secondary mineral usually found in the weathered zone of uranium-bearing pegmatites and hydrothermal veins, and is derived from the weathering of uranium minerals. Notable localities are Autun, France; Johanngeorgenstadt district, Germany; Katanga district, Congo; at Mount Painter, South Australia; and Sabugal, Portugal. In the United States, autunite is found at Spruce Pine, North Carolina; near Spokane, Washington; and in scattered localities on the Colorado Plateau.

TORBERNITE

$Cu(UO_2)_2(PO_4)_2 \cdot 8\text{--}12H_2O$ Tetragonal
     Hardness $2\text{--}2\frac{1}{2}$ Sp. Gr. 3.2–3.7

Summary

Commonly in emerald green micaceous flakes that are nonluminescent; also in scaly aggregates (Plate 33(*D*), facing p. 195). *Color:* emerald green to apple green. *Luster:* vitreous, pearly on base. *Cleavage:* perfect basal; cleavage laminae fairly brittle; transparent to translucent. Frequently associated with yellow autunite, gummite, and other secondary alteration products of uraninite. GREEN MICACEOUS FLAKES WITH SQUARE OUTLINES AND ASSOCIATED URANIUM MINERALS AND RADIOACTIVITY ARE CHARACTERISTIC. Resembles green ZEUNERITE, $Cu(UO_2)_2(AsO_4)_2 \cdot 8\text{--}12H_2O$.

Composition. Hydrous copper uranium phosphate, $Cu(UO_2)_2(PO_4)_2 \cdot 8\text{--}12H_2O$. Small amounts of As substitute for P. Part of the H_2O evaporates at room temperature. Meta-torbernite has $8H_2O$.

Tests. Fuses easily to a black globule. Sodium fluoride bead luminesces bright yellow under ultraviolet light (U test). Radioactive. Tests for Cu, P_2O_5, and H_2O.

Crystallization. Tetragonal; ditetragonal dipyramidal class, $\frac{4}{m}\frac{2}{m}\frac{2}{m}$. Commonly in micaceous flakes; also scaly or foliated aggregates. Crystals are usually tabular with a square outline.

Occurrence. Torbernite is a secondary uranium mineral formed by the alteration of uraninite. Good crystals are found in Cornwall, England; Shinkolobwe in the Congo; Schneeberg, Saxony; Mount Painter, South Australia; and Joachimsthal and Zinnwald, Bohemia.

CARNOTITE

$K_2(UO_2)_2(VO_4)_2 \cdot 3H_2O$ Orthorhombic(?)
                                        Monoclinic(?)
Hardness 1–2   Sp. Gr. 4–5

Summary

Usually as bright yellow, powdery masses (Plate 33(*C*), facing p. 195), common in the sandstones of the Colorado Plateau area. The carnotite is disseminated in the sandstones or localized around petrified tree trunks or other vegetable matter. *Color:* bright lemon yellow to greenish-yellow. *Luster:* earthy. *Cleavage:* perfect basal; strongly radioactive; nonluminescent. YELLOW COLOR, EARTHY HABIT, AND STRONG RADIOACTIVITY ARE CHARACTERISTIC. Occurs with, and resembles TYU-YAMUNITE, $Ca(UO_2)_2(VO_4)_2 \cdot nH_2O$, which luminesces yellow-green and fuses easily. Difficult to distinguish from many yellow secondary uranium minerals without optical, chemical, or x-ray diffraction tests.

Composition. A hydrous potassium uranium vanadate, $K_2(UO_2)_2(VO_4)_2 \cdot 3H_2O$. The water content varies with humidity at room temperature.

Tests. Infusible. Strongly radioactive. Borax or sodium fluoride bead with dissolved carnotite luminesces green in ultraviolet light (U test). Gives tests for V, K, and H_2O.

Crystallization. Usually as a powdery or loosely coherent aggregate. Sometimes compact. Rarely as small, platy, six-sided crystals.

Occurrence. Carnotite is usually found disseminated in sandstones, or concentrated around petrified tree trunks or carbonized vegetable matter. Carnotite, tyuyamunite, and other secondary uranium minerals are widespread in the sandstones of the Colorado Plateau area in Colorado, and in adjoining parts of Utah, New Mexico, and Arizona.

Use. Ore of uranium and vanadium.

BRAZILIANITE

$NaAl_3(PO_4)_2(OH)_4$ Monoclinic
Hardness $5\frac{1}{2}$ Sp. Gr. 2.98

Summary

Brazilianite is an uncommon but attractive mineral found in cavities of pegmatites in Minas Gerais, Brazil, and in New Hampshire. It usually occurs in complex yellowish-green equant to elongated crystals (Figure 8-46). *Color:* yellowish-green to pale yellow. *Luster:* vitreous. *Streak:* colorless. *Cleavage:* good side pinacoid; brittle; transparent. COMPLEX YELLOW-

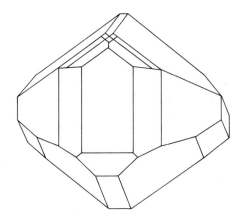

Fig. 8-46. Brazilianite crystal, Brazil.

ISH-GREEN CRYSTALS WITH GOOD CLEAVAGE IN ONE DIREC-
TION ARE DISTINCTIVE.

Composition. A hydrous sodium aluminum phosphate,
$NaAl_3(PO_4)_2(OH)_4$.

Tests. Difficult to fuse. Loses yellowish color when
heated to a low temperature. After fusion with sodium
carbonate, gives phosphate test with ammonium
molybdate.

Crystallization. Monoclinic; prismatic class, $\frac{2}{m}$. Usually
in complex equant to elongated crystals. Also in botryoidal masses with a radial fibrous inner structure.

Occurrence. An uncommon mineral found in complex
equant crystals, some over 3 in. wide, in cavities of a
decomposed pegmatite at Pamarol, Minas Gerais,
Brazil. Most of the crystals are imbedded in clay. Associated minerals include feldspar, mica, deep blue
apatite, and blue scorzalite (the end member of the
lazulite series). Also found in elongated crystals in
cavities of pegmatites at the Palermo Mine and Smith
Mine, New Hampshire.

SILICATES

The silicates form the largest and most diverse
groups of minerals. Most are transparent to translucent in thin slivers, but range widely in other physical
characteristics such as mode, habit, color, hardness,
and cleavage. The most common silicate minerals are
quartz and feldspar.

The silicates may be classified according to internal structure into six divisions. The basic structural
unit is a silicon atom, surrounded by four oxygens in
a tetrahedral arrangement (Figure 8-47). These units
may share one or more oxygens with adjacent SiO_4
tetrahedra to form the "skeleton" structure of a silicate mineral. The silica skeletal units are frequently
held together by positive ions lying between and
within them. However, in minerals such as quartz, net-
works of SiO_4 tetrahedra form the complete structure.
Table 8-3 diagrammatically illustrates the arrangements of the SiO_4 tetrahedra in each of the major

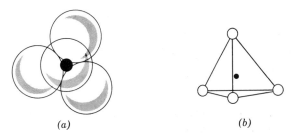

| (a) | (b) |

Fig. 8-47. (*a*) SiO_4 tetrahedron showing a silicon atom surrounded by four oxygens. (*b*) Size ratio of atoms is reduced to
show tetrahedral arrangement more clearly.

silicate divisions: (1) FRAMEWORKS; (2) SHEETS; (3)
CHAINS; (4) RINGS; (5) UNITS; and (6) PAIRS.

FRAMEWORK STRUCTURES

SILICA GROUP
   Quartz, SiO_2

FELDSPAR GROUP
   Orthoclase, $KAlSi_3O_8$
   Microcline, $KAlSi_3O_8$

   PLAGIOCLASE SERIES
      Albite, $Na(AlSi_3O_8)$
      Oligoclase
      Andesine   to
      Labradorite
      Anorthite, $Ca(Al_2Si_2O_8)$

FELDSPATHOID GROUP
   Leucite, $K(AlSi_2O_6)$
   Nepheline, $Na_3K(Al_4Si_4O_{16})$
   Sodalite, $Na_8(Al_6Si_6O_{24})Cl_2$
   Lazurite, $Na_8(Al_6Si_6O_{24})S_2$
   Cancrinite, $Na_3Ca(Al_3Si_3O_{12})CO_3(OH)_2$

SCAPOLITE SERIES
   $(Na,Ca)_4Al_3(Al,Si)_3Si_6O_{24}(Cl,CO_3,SO_4)$

ZEOLITE GROUP
   Stilbite, $(Ca,Na_2,K_2)(Al_2Si_7O_{18}) \cdot 7H_2O$
   Natrolite, $Na_2(Al_2Si_3O_{10}) \cdot 2H_2O$
   Chabazite, $Ca(Al_2Si_4O_{12}) \cdot 6H_2O$
   Heulandite, $(Ca,Na_2)(Al_2Si_7O_{18}) \cdot 6H_2O$
   Analcite, $Na(AlSi_2O_6) \cdot H_2O$

Table 8-3

Structural Classification of the Silicates

STRUCTURAL ARRANGEMENTS OF SiO$_4$ TETRAHEDRA (• = silicon; ∘ = oxygen)	RATIO Si:O	EXAMPLE	
1. **FRAMEWORK STRUCTURE:** Three dimensional framework of tetrahedra. Each tetrahedron shares all *four* oxygens.	1:2	Quartz, SiO$_2$	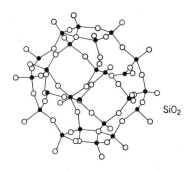

(After Harry Berman, Constitution and Classification of the Natural Silicates, Am. Min., 1937)

2. **SHEET STRUCTURE:** Continuous sheets of tetrahedra. Each tetrahedron shares *three* oxygens.	1:2½	Talc, Mg$_3$(Si$_4$O$_{10}$)(OH)$_2$	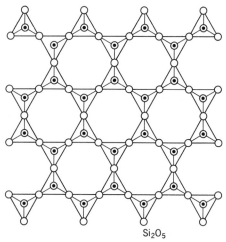

3. **CHAIN STRUCTURE:** *Single Chains:* Continuous single chains of tetrahedra. Each tetrahedron shares *two* oxygens.	1:3	Pyroxenes	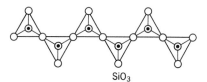

Table 8-3   *(continued)*

STRUCTURAL ARRANGEMENTS OF SiO₄ TETRAHEDRA	RATIO Si:O	EXAMPLE	
Double Chains: Continuous chains of tetrahedra. Tetrahedra share alternately *two* and *three* oxygens.	$1:2\frac{3}{4}$	Amphiboles	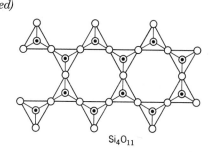 Si_4O_{11}

4. **RING STRUCTURE:**

Rings of tetrahedra. Each tetrahedron shares *two* oxygens.   1:3

Rings of 6 tetrahedra	1:3	Beryl, $Be_3Al_2Si_6O_{18}$	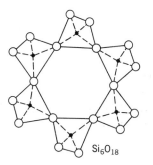 Si_6O_{18}
Rings of 3 tetrahedra	1:3	Benitoite, $BaTiSi_3O_9$	Si_3O_9
Rings of 4 tetrahedra	1:3	Axinite, $(Ca,Mn,Fe)_3Al_2BO_3\text{-}(Si_4O_{12})(OH)$	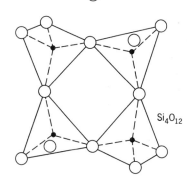 Si_4O_{12}

Table 8-3   *(continued)*

STRUCTURAL ARRANGEMENTS OF SiO_4 TETRAHEDRA	RATIO Si:O	EXAMPLE	
5. UNIT STRUCTURE: Independent tetrahedral units.	1:4	Olivine, $(Mg,Fe)_2SiO_4$	SiO₄
SUBSATURATES: Independent SiO_4 tetrahedra plus additional oxygen ions not in the silica tetrahedra.	$1:(4 + x)$	Andalusite, $Al_2(SiO_4)O$	
6. PAIR STRUCTURE: Two tetrahedra sharing *one* oxygen.	$1:3\frac{1}{2}$	Hemimorphite, $Zn_4Si_2O_7(OH)_2 \cdot H_2O$	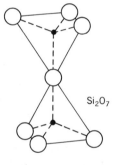 Si₂O₇

QUARTZ (Low Quartz)

SiO_2 Hexagonal

Hardness 7   Sp. Gr. 2.65
(Chalcedony 2.57–2.64.
The lower specific gravity of chalcedony is
caused primarily by
submicroscopic pores.)

Summary

Quartz is common, and of widespread occurrence. Two distinct types are found: (1) *Coarse-crystallized varieties,* including well-formed crystals and irregular masses (Plate 34, facing p. 210). (2) *Compact microcrystalline varieties* (*chalcedony;* Plates 35, facing p. 211, and 36(*E, F*), facing p. 218). Quartz crystals commonly occur as six-sided prisms terminated by two rhombohedrons, which, when equally developed, resemble a hexagonal bipyramid. Prism faces are frequently horizontally striated (Figure 2-12, p. 15). The *color* of quartz ranges widely from colorless or white to purple, violet, blue, pink, yellow, red, brown, green, or black. Many varieties of quartz are based on color. *Luster:* vitreous; waxy for chalcedony. *Streak:* white. *Fracture:* conchoidal; rhombohedral. *Cleavage* or *parting:* rarely observed; transparent to opaque; brittle. Twinning is common. Strong piezoelectric properties. Ordinary quartz (low or alpha quartz) transforms to a high-temperature form (high quartz or beta quartz) at about 573°C. CRYSTAL HABIT, HARDNESS, AND USUAL LACK OF CLEAVAGE ARE DISTINCTIVE. DULL, WAXY LUSTER IS CHARACTERISTIC OF CHALCEDONY. Quartz is easily distinguished from CALCITE by hardness, and from FELDSPAR by lack of cleavage.

Composition. Silicon dioxide (SiO_2). Normally close to 100% SiO_2. Trace amounts of Fe, Mg, Al, Ca, Li, Na, K, and Ti may occur in clear colorless quartz.

Tests. Infusible. Insoluble in acids except hydrofluoric.

Crystallization. Hexagonal; trigonal trapezohedral class, 3 2. Crystals and masses are common. Crystals are commonly prismatic, terminated by pyramidal faces (two rhombohedrons), usually unequally developed. Crystals can be right or left handed (Figure 8-48). The right-hand crystals have an *x* face (trigonal trapezohedron) to the *right* above the prism face *m*, and the left-hand crystals have a similar *x* face to the *left* above the *m* face. Many quartz crystals are twinned, but the twins are interpenetrated, and appear to be single crystals.

Polymorphs. HIGH QUARTZ; high, middle, low TRIDYMITE; high and low CRISTOBALITE; COESITE; STISHOVITE; MELA-NOPHLOGITE; SILICA GLASS (lechatelierite).

Hydrothermal Synthesis of Quartz. Quartz dissolves in water under high pressures and temperatures. Synthetic quartz is grown under pressures of about 21,000 lb. per square in., and may grow as fast as $\frac{1}{4}$ in. per day on the basal plane (0001). Nutrient quartz (crushed quartz or quartz-glass) is placed in the *bottom* part of a steel vessel, which is the *hot dissolving zone* (about 400°C). A sodium hydroxide or sodium carbonate solution is used as a solvent. Quartz seed crystals, which are cut in the proper orientation for the desired electrical properties, are attached to a wire frame and placed in the *upper, cooler crystallizing zone* of the vessel (about 360°C). A perforated metal disk or baffle

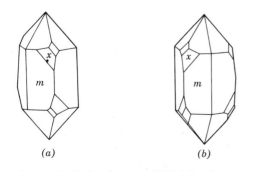

Fig. 8-48. (*a*) Right-hand quartz; (*b*) Left-hand quartz.

separates the two temperature zones, and convection currents transport newly saturated quartz solution to the upper crystallizing zone and depleted solution away. The seed crystals (thin plates) begin to grow as the supersaturated solution deposits solid quartz on the seeds in the upper cooler portion of the vessel. Crystals weighing over $1\frac{1}{2}$ lb. have been grown by this method.

Coarse-Crystallized Varieties of Quartz

■ MILKY QUARTZ

Milky white quartz is the most common type, frequently occurring as crystals and masses in pegmatites and hydrothermal veins. The milky white color is usually caused by microscopic liquid inclusions or flaws that scatter light. Large crystals are common. A giant milky white crystal weighing about 13 tons and measuring close to $11\frac{1}{2}$ ft. in length by $5\frac{1}{2}$ ft. in diameter was found in Siberia.[7]

■ ROCK CRYSTAL

Colorless transparent quartz is often referred to as rock crystal. Considerable amounts of colorless quartz crystals, from the very minute ones to the larger ones that weigh over 5 tons, occur in vugs or pockets in milky white quartz veins in Minas Gerais, Brazil. Rock crystal is also common in the Alps and in Madagascar. Small doubly terminated colorless quartz crystals are abundant in Herkimer County, New York, and are known as "Herkimer diamonds."

■ SMOKY QUARTZ

Smoky quartz ranges from pale smoky brown to almost black. It is generally believed to be secondary, produced from colorless quartz by exposure to radioactive material. Most smoky quartz crystals have an uneven color distribution.

Quartz can be artificially colored smoky brown by exposure to x-rays. However, a wide variation occurs in the degree of darkening in different crystals and in different parts of a single crystal. Impurity ions tend to be selectively distributed in solid solution in different zones of a crystal. Some selectively distributed impurities impart no color to a nonirradiated crystal, but produce color centers (structural defects which absorb light) when the colorless crystal is irradiated with x-rays. The resulting smoky color is unevenly distributed in zones corresponding to the distribution of the impurities.

The color centers in a good many smoky quartz crystals are believed to be related to the presence of aluminum ions (Al^{3+}) plus interstitial lithium ions (Li^+) in substitution for silicon ions (Si^{4+}), which cause a disturbance in the electronic configuration.[1,4] The intensity of the color has been found to rise with the amount of the aluminum substitution. Synthetic colorless quartz crystals containing Li^+ and Al^{3+} ions have been darkened by exposure to x-rays. The darkening occurs in zones corresponding to the concentration of the impurity ions.

Natural smoky quartz generally becomes colorless when heated above 225°C. Some smoky quartz assumes a yellowish to yellowish-brown color before it becomes colorless. Most of the gem material sold as *citrine* is really heat-treated smoky or amethyst quartz.

Crystals of smoky quartz are common. Some giant crystals weigh over 1000 lb. Smoky quartz generally occurs in HIGH-TEMPERATURE deposits such as pegmatites, hydrothermal cavities in pegmatites, and hydrothermal veins. Fine specimens are found in Brazil, Switzerland, the U.S.S.R., Madagascar, Cairngorm, Scotland; and in New England and Texas.

■ AMETHYST QUARTZ

The color of amethyst quartz ranges from violet, to reddish-violet, to purple; and is commonly distributed in patches or bands (Plate 34, facing p. 210). Phantoms of amethyst sometimes occur in colorless quartz. The ferric iron content of amethyst is large in comparison to that of smoky and colorless quartz; and the amethyst color is believed to be produced by a highly dispersed ferric iron compound. The small amounts of Mn and Ti present were found to bear no relationship to the depth of color.[5]

Amethyst can be decolorized by heating to about 290°C. The rate of bleaching increases at higher temperatures. Further heating at about 500–600°C often produces a yellow-brown (citrine) color. However, all specimens of amethyst do not respond alike to heat treatment. Some never show a colorless stage, and others may become colorless but not citrine.

Decolorized amethyst can be made purple again by irradiation with radium or x-rays. The purple is often accompanied by a smoky tint. Amethyst color has also been produced by irradiation of synthetic quartz containing ferric iron.[12] Some amethyst from Montezuma, Minas Gerais, Brazil, turns green when heated to 510°C. The purple color is bleached and the green residual color is believed to be caused by a separate green phase.[1] The color of some amethyst can be deepened by exposure to x-rays (reference 12, Chapter 2).

Amethyst crystals are moderate in size, usually not over 5 in. long. Most amethyst crystals show fine polysynthetic twinning striations on the rhombohedral faces. Amethyst commonly forms at RELATIVELY LOW TEMPERATURES in cavities in volcanic flows where it is frequently associated with stilbite, apophyllite, natrolite, heulandite, and agate. Some amethyst crystals contain needles of goethite. Notable localities include Brazil, Uruguay, Siberia, Ceylon, Nova Scotia, and the Tunder Bay region of Ontario.

■ BLUE QUARTZ

Blue quartz is commonly milky or grayish-blue, and usually occurs as grains in igneous and metamorphic rocks. Blue quartz commonly contains numerous oriented microscopic inclusions of rutile that precipitated from solid solution. The blue color may be caused by the rutile inclusions selectively scattering blue wavelengths of light. Blue quartz, which owes its color to tiny needles of tourmaline, has been reported by Parker (1961). Cr^{2+} ions are present in some synthetic blue quartz.

■ CITRINE QUARTZ

Citrine is a yellow to yellowish-brown variety of quartz; it is much less common than white, amethyst, or smoky quartz. The color of citrine quartz may be caused by a submicroscopic distribution of colloidal ferric hydroxide.[6] On heating, citrine becomes bleached, and turns smoky brown when irradiated with x-rays. Most of the gem material sold as citrine

A   x 0.85

DIOPSIDE
$CaMg(Si_2O_6)$

Light green monoclinic diopside crystal.

B   x 0.5

PYROXENE

Green pyroxene crystal with characteristic square cross section in white microcline: Pitcairn, New York.

C   x 0.6

GARNIERITE
$(Ni, Mg)(SiO_3) \cdot nH_2O$

Earthy mass of bright green garnierite:
New Caledonia.

D   x 0.6

SPODUMENE
$LiAl(Si_2O_6)$

Flat elongated deeply striated crystal
with steep terminations and "wood-like"
surface alteration: Norwich, Massachusetts.

E   x 0.75

SPODUMENE (GEM VARIETIES)

Pink transparent spodumene (kunzite):
Pala, San Diego County, California; and
yellow gem spodumene from Minas
Gerais, Brazil.

F   x 0.5

CHRYSOCOLLA
$Cu(SiO_3) \cdot H_2O$

Massive blue-green chrysocolla: Globe,
Arizona.

A   x 0.5

TREMOLITE (HEXAGONITE)

Fibrous violet colored manganiferous variety of tremolite: Edwards, New York.

B   x 0.5

TREMOLITE (ACTINOLITE)

$Ca_2(Mg, Fe)_5Si_8O_{22}(OH)_2$

Fibrous aggregate of green actinolite: Zillerthal, Tyrol.

C   x 0.6

TREMOLITE

$Ca_2Mg_5Si_8O_{22}(OH)_2$

White prismatic tremolite crystals: Gouverneur, New York.

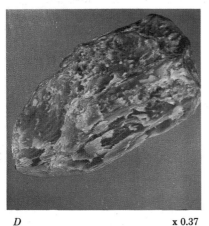

D   x 0.37

NEPHRITE (JADE)

Dense compact variety of actinolite: New Zealand.

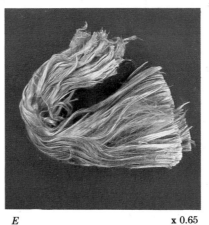

E   x 0.65

CROCIDOLITE

Long separable blue fibers (blue asbestos), an asbestiform variety of riebeckite: Cochabamba, Bolivia.

F   x 0.60

TIGERS EYE

Crocidolite replaced by quartz. Fibrous structure is preserved: South Africa.

is not natural citrine but heat-treated smoky or amethyst quartz. Yellow quartz is often sold as "topaz" but this usage is misleading and should not be confused with the mineral topaz.

Citrine quartz commonly occurs with amethyst in Brazil; Uruguay; Dauphiné, France; and near Mursinsk in the U.S.S.R.

■ TIGER-EYE

Quartz replacement of fibrous crocidolite retaining the original fibrous structure. Tiger-eye is yellow to yellow-brown in color and occurs in South Africa (Plate 43, facing p. 259).

■ ROSE QUARTZ

The color of rose quartz ranges from pale pink to deep rose, and may be caused by Mn^{3+} or Ti^{4+} ions in solid solution. Massive rose quartz sometimes occurs in considerable amounts in pegmatites. Crystals, however, are rare and small; usually less than $\frac{1}{2}$ in. in length. Aggregates of small pink quartz crystals grow in subparallel arrangement on large, slightly smoky quartz crystals in pegmatites in Capeunha, Minas Gerais, Brazil (Plate 34, facing p. 210). Because of their rarity, rose quartz crystals are prized by collectors. Small crystals are found in pegmatites in Pala and Mesa Grande, California; in Minas Gerais and Goias, Brazil; and in Newry, New England.

Rose quartz commonly contains microscopic needle-like oriented inclusions of rutile, which have been precipitated from solid solution. The star effect of asterated rose quartz is caused by the scattering of light from these oriented rutile needles (see asterism, p. 85).

Rose quartz becomes dark smoky gray to black when subjected to x-rays, and bleaches when heated in air to about 575°C.

Some white quartz, usually nonpegmatitic, is pigmented pink by microscopic minerals such as hematite or dumortierite.

Inclusions in Quartz

Needle-like crystals of rutile, tourmaline, goethite, actinolite, or hornblende are commonly included in quartz crystals during their growth. Included bright flaky crystals of mica, hematite, and chlorite are also common.

Liquid Inclusions. These may contain a gas phase; sometimes, floating inclusions of other minerals like halite or sylvite occur in quartz crystals. These liquid inclusions are either *primary* (trapped during the initial growth of the crystals) or *secondary* (introduced at a later time into cracks that subsequently are partially healed over by the deposition of silica).

Rutilated Quartz. Yellowish to red-brown hair-like rutile inclusions are common in colorless and smoky quartz crystals (Plate 34, facing p. 210). These randomly oriented inclusions were entrapped by the quartz crystals during their growth, and did not form by precipitation from solid solution (exsolution) as did the oriented microscopic rutile needles in rose and blue quartz. Fine specimens of rutilated quartz occur in Minas Gerais and other places in Brazil; in Madagascar; in Switzerland; and in Vermont and North Carolina.

Aventurine Quartz. Contains bright glistening scales of mica, sometimes hematite or goethite. Most aventurine is green or reddish to yellowish-brown, and is found in Brazil, India, China, Madagascar, and Siberia.

Cat's Eye Quartz. Contains parallel asbestiform fibers that reflect a wavy band of light (see chatoyancy, p. 85). Cat's eye quartz is commonly green, yellowish, or brownish, and may resemble the more valuable chrysoberyl cat's eye. Important localities are Brazil, Ceylon, and India.

Quartz Crystals Pigmented by Admixed Foreign Material*

Dark Gray to Black Quartz Crystals. Pigmented by disseminated carbon, sulfides, tourmaline, or other dark minerals. Should not be confused with blackish, smoky quartz that is colored by exposure to radiation.

Green Quartz. Pigmented by minute inclusions of chlorite, actinolite, green mica, etc.

*See also microcrystalline varieties.

Red to Pinkish Quartz Crystals (Ferruginous Quartz). Usually pigmented by minute hematite inclusions. Some pinkish ferruginous quartz crystals resemble rose quartz.

Compact Microcrystalline Quartz (Chalcedony)

Compact varieties of silica, composed of microscopic crystals of quartz with submicroscopic pores, are herein grouped together as *chalcedony*, although frequently considered as a distinct mineral. The colored varieties of chalcedony have special names, e.g., sard, carnelian, jasper, prase. Some opaque varieties such as jasper and plasma have, in part, especially in metamorphic types, a microgranular structure, possibly produced by the recrystallization of the original microfibrous chalcedonic silica.[3,11] Both microfibrous quartz and microgranular quartz may occur together in the same specimen.

Chalcedony usually has a waxy luster, and is commonly white, gray, bluish to greenish gray, yellowish, or brown, or also deep red and green in subvarieties. It is rarely bright blue, yellow, or pink. The slightly porous nature of chalcedony enables it to absorb various dyes, and many specimens of brightly colored chalcedony, including banded agates sold for ornamental purposes, have been artificially colored. A deep red-brown color may be produced by impregnating chalcedony with ferric hydroxide, followed by heat treatment. Black imitation jet may be made by soaking: first in sugar solution, then in sulfuric acid.

Chalcedony commonly occurs as botryoidal masses; as a replacement of fossils, other minerals, or wood fibers; in geodes; and as a cementing material. It is deposited under moderate- to low-temperature conditions.

Agate: A banded variety of chalcedony, commonly occurs as a cavity filling. Concentric banding frequently follows the outline of the cavity. When the cavity contains spike-like projections, the banding may also be concentric to the projections forming an *"eye agate."*

Onyx: Alternating layers of light and dark chalcedony.

Sard: Brown to yellowish, or reddish brown translucent chalcedony pigmented by iron oxide. Gradations occur between sard and carnelian.

Carnelian: Translucent red to orange-red chalcedony pigmented by finely dispersed hematite.

Chrysoprase: Apple green translucent chalcedony, colored by extremely fine disseminated particles of a hydrated nickel silicate. Chrysoprase occurs as a secondary mineral in veinlets in serpentine. The intensity of the green color is directly related to the amount of nickel present. The best specimens of chrysoprase are found in the Marlborough Area of Central Queensland, Australia. The deep apple green Australian chrysoprase resembles green jadeite.

Jasper: Opaque chalcedony commonly dark red, red-brown, or yellowish-brown; pigmented by finely divided hematite or goethite particles.

Prase: Dull green chalcedony, commonly colored by minute particles of chlorite or fibrous amphiboles.

Plasma: Dark dull green; similar to prase but opaque.

Bloodstone or Heliotrope: Green chalcedony with red spots of jasper resembling blood spots.

Chert: Opaque, dull chalcedony; black, white, gray, or yellowish.

Flint: Dark gray to black, opaque, dull chalcedony; usually occurring as nodules in chalk.

Moss Agate: Whitish chalcedony with black, brown, or reddish-brown branching moss-like aggregates of manganese or iron oxide. Some dendrites may form[10] by precipitation of manganese or iron oxides diffused into the gelatinous silica before solidification.

Enhydros: Chalcedony geodes that contain water. Common in Brazil and in Uruguay.

Silicified Wood: Wood that is commonly replaced by red jasper, and gray or brownish chalcedony. Opal may also replace wood fibers.

OPAL

$SiO_2 \cdot nH_2O$　　　　　Amorphous or near amorphous
Hardness $5\frac{1}{2}$–$6\frac{1}{2}$　　　　Sp. Gr. 2.0–2.2

Summary

Opal occurs in fine-grained masses; never in crystals. Commonly occurs as cavity fillings, irregular masses, veinlets, globular crusts, and as pseudomorphs after wood or fossils (Plate 36, facing p. 218). *Color:* usually colorless to milky or bluish-white; also black, and in tints of red, orange, yellow, brown, blue, green. *Luster:* greasy to resinous. *Fracture:* conchoidal; transparent to translucent; often luminesces yellow-green under ultraviolet light. Resembles CHALCEDONY, but opal has a lower hardness and a greasy luster. Chalcedony has a dull waxy luster.

Varieties of Opal

Precious Opal. The brilliant internal play of colors of precious opal is caused by the diffraction of light from the spaces between minute spheres of silica, stacked in an orderly three-dimensional pattern. The holes and the spacing of the amorphous silica spheres are about equal. (See cause of color in opal, p. 79.)

Fire Opal. Type of precious opal with a bright orange-red play of colors.

Common Opal. Without internal reflections. Occurs in a wide range of colors.

Hydrophane. An almost opaque white or lightly colored opal that contains cracks filled with air and becomes virtually transparent when immersed in water. Some specimens become iridescent.

Geyserite or Siliceous Sinter. Hot spring porous deposit. Usually white or gray.

Diatomite (diatomaceous earth). White, fine-grained masses that resemble chalk. Formed by the accumulation on the sea floor of the microscopic shells of diatoms. Porous and light.

Wood Opal. Petrified wood in which the mineral matter is opal.

Hyalite. Colorless, glassy, globular opal that frequently luminesces yellow-green under ultraviolet light.

Composition. Silicon dioxide with a variable content of water, $SiO_2 \cdot nH_2O$, and impurity ions such as Al, Ca, Mg, Na, K.

Tests. Infusible and insoluble. Gives water in closed tube.

Crystallization. Composed of minute spherical particles of amorphous or near amorphous silica. Contains variable amounts of water and impurity ions.

Alteration. Frequently transforms to chalcedony, the more stable form of silicon dioxide. Commonly dehydrates.

Occurrence. Precious opal was discovered in Hungary, where it commonly occurs as thin bands in massive milky white common opal. Black opals (black or dark background against which brilliant colors flash) occur at Lightning Ridge, New South Wales, Australia. Fire opals and other types of precious opal are found in Querétaro, Mexico. In the United States, precious opal occurs in petrified wood in Virgin Valley, Nevada. Opalized wood and geyserite are found in Yellowstone Park, Wyoming. Huge beds of diatomite, a few thousand feet thick, occur near Lompoc, California. Precious opal also occurs in California, Idaho, and Oregon.

Feldspar Group

The feldspars constitute the most important group of rock-forming minerals. All are aluminum silicates, and range in content of potassium, sodium, calcium, and occasionally barium. Two subgroups exist: (1) POTASH FELDSPARS (orthoclase, microcline, sanidine, adularia); and (2) THE SODIUM-CALCIUM FELDSPARS or PLAGIOCLASE SERIES (albite, oligoclase, andesine, labradorite, bytownite, and anorthite). All feldspars have good cleavage in two directions at or close to 90° (Plate 37, facing p. 219). Crystals often resemble each other, although some are monoclinic and others triclinic. Thin parallel twinning striations are almost always present (see oligoclase: Plate 8, facing p. 51, and Plate 37, facing p. 219) on plagioclase feldspar and distinguish it from potash feldspar.

ORTHOCLASE

$K(AlSi_3O_8)$	Monoclinic
Hardness 6	Sp. Gr. 2.56–2.59

Summary

Orthoclase is a common rock-forming mineral often associated with quartz and muscovite. Frequently found as crystals (Plate 7(*C*), facing p. 50), twinned crystals (Plate 8(*C*), facing p. 51), masses, and grains. *Color:* usually colorless, white, or flesh pink; rarely yellow. *Luster:* vitreous. *Streak:* white. *Cleavage:* good in two directions at 90°; transparent to translucent; brittle. CRYSTAL HABIT, BLOCKY CLEAVAGE, AND HARDNESS ARE CHARACTERISTIC. Distinguishable from PLAGIOCLASE feldspars by lack of twinning striations; from AMBLYGONITE by difficulty in fusing; and from QUARTZ by cleavage.

Polymorphs: Microcline, Sanidine, and Adularia (see order-disorder transformations, p. 54).

Composition. Potassium aluminum silicate, $K(AlSi_3O_8)$. Some sodium may replace potassium. In the high-temperature polymorph SANIDINE, as much as 50 percent of the potassium may be replaced by sodium. In HYALOPHANE, potassium is replaced in part by barium.

Tests. Fusible with difficulty. Insoluble in acids.

Crystallization. Monoclinic; prismatic class, $\frac{2}{m}$. Crystals are usually prismatic. Carlsbad, Baveno, and Manebach twins are common (Figure 8-49); also massive

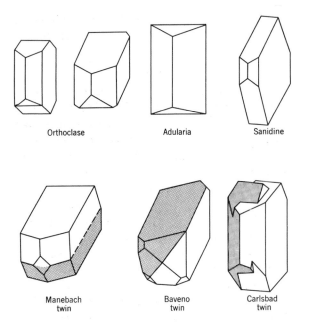

Orthoclase   Adularia   Sanidine

Manebach   Baveno   Carlsbad
twin   twin   twin

Fig. 8-49. Common crystal habits of orthoclase, adularia, and sanidine.

and in grains. Adularia commonly occurs in glassy colorless pseudo-orthorhombic crystals. Sanidine crystals are usually flat tabular and colorless. Varieties of orthoclase, adularia, sanidine, or albite with a bluish opalescent play of colors are called MOONSTONE. Microscopic to submicroscopic intergrowth of sodium-rich feldspar in a potassium-rich feldspar host (micro- or crypto-perthites) are believed to be responsible for the opalescence of some moonstone. Antiperthites are intergrowths of a potassium feldspar in a sodium-rich feldspar host.

Alteration. Commonly alters to sericite or kaolinite.

Occurrence. Orthoclase is a common rock-forming mineral, and is a major constituent of such igneous rocks as granites and syenites. Sanidine, the high-temperature polymorph, commonly occurs as glassy, flat, tabular crystals in such volcanic rocks as rhyolites and trachytes. Adularia forms colorless to white pseudo-orthorhombic crystals at comparatively low temperatures; these crystals are common in Alpine-type veins in Switzerland. Orthoclase also occurs in many types of metamorphic rocks; in veins; and in arkose sandstone. Sometimes occurs in pegmatites, but much less frequently than microcline.

Use. In the manufacture of porcelain, enamel, and glass.

MICROCLINE

$K(AlSi_3O_8)$ Triclinic
Hardness 6   Sp. Gr. 2.56–2.59

Summary

Microcline frequently occurs in pegmatites as large crystals and cleavable masses; often shows a coarse perthite structure (Plate 7(*E*), facing p. 50, and Plate 37(*A, B*), facing p. 219). The crystals are triclinic, but closely resemble monoclinic orthoclase, and are often twinned like orthoclase. *Color:* white, flesh, reddish-brown, green. MICROCLINE IS THE ONLY BRIGHT GREEN FELDSPAR. The plagioclase feldspar, oligoclase, occasionally has a pale greenish cast. Green microcline is called AMAZONITE or AMAZON STONE, and can be decolorized by heating above 270°C. Much white or flesh-colored feldspar called orthoclase is actually microcline. *Luster:* vitreous. *Streak:* white. *Cleavage:* good in two directions almost at right angles; translucent to transparent; brittle. *Microcline* is the potassium feldspar that crystallizes in the lowest temperature range. *Orthoclase* crystallizes at intermediate temperatures, and *sanidine* in the highest temperature range. BRIGHT GREEN COLOR, LARGE CRYSTALS, PERTHITIC TEXTURE, AND PEGMATITIC OCCURRENCE ARE CHARACTERISTIC OF MICROCLINE.

Composition. Potassium aluminum silicate, $K(AlSi_3O_8)$; like orthoclase. Sodium may replace some of the potassium (soda-microcline). If the amount of sodium is greater than potassium, then the mineral becomes ANORTHOCLASE. Anorthoclase generally forms at higher temperatures. Microcline PERTHITES (Plate 37, facing p. 219) are intergrowths of a sodium-rich feldspar like anorthoclase or albite in microcline (potassium-rich host). Orthoclase perthites also occur. Microperthites are visible with the aid of a microscope. The perthite structure may have such different origins as (1) the unmixing of an originally homogeneous potassium-sodium feldspar—generally accepted for most fine microperthites; (2) simultaneous crystallization of a sodium-rich and a potassium-rich feldspar; and (3) the replacement of potassium feldspar by sodium feldspar, especially in coarse pegmatite perthites.

Tests. Same as orthoclase.

Crystallization. Triclinic; pinacoidal class, $\bar{1}$. Large crystals are common. Carlsbad twins are common. Under the microscope in polarized light, most microcline crystals have a *"plaid" structure,* caused by twinning in which two sets of twin lamellae are approximately at right angles to each other. Commonly microcline is intimately intergrown with quartz in pegmatites. These intergrowths frequently resemble ancient writing, and are called GRAPHIC GRANITE. Perthites are also common.

Alteration. Commonly alters to sericite and kaolinite.

Occurrence. Commonly found as large crystals and cleavable masses in granite pegmatites associated with quartz and albite. Amazonstone is found at Amelia Court House, Virginia; and in large crystals with smoky quartz near Pike's Peak, Colorado. Also occurs in Brazil, the U.S.S.R., Norway, India, and Madagascar.

Use. Microcline, like orthoclase, is used for ceramics, ceramic glazes, and glass. Amazonstone is sometimes used in jewelry, and as an ornamental material.

PLAGIOCLASE SERIES

		Triclinic
		Percent Albite
Albite	$Na(AlSi_3O_8)$	100–90
Oligoclase		90–70
Andesine	to	70–50
Labradorite		50–30
Bytownite		30–10
Anorthite	$Ca(Al_2Si_2O_8)$	10–0

Hardness 6   Sp. Gr. 2.62 (albite)– 2.76 (anorthite)

Summary

The plagioclase feldspars are the most abundant rock-forming minerals. When crystallized at high temperatures, they form an almost complete solid solution series from ALBITE (sodium-rich end member) to ANORTHITE (calcium-rich end member). *Color:* white, gray; occasionally reddish; yellowish, greenish, black. *Luster:* vitreous. *Streak:* white. *Cleavage:* two good, at about 93°–94°; translucent to transparent; brittle. A brilliant, iridescent color play of blue and green is commonly seen in LABRADORITE (Plate 37(*E*), facing p. 219). A milky opalescent play of color occurs in some albite (ALBITE MOONSTONE). Some specimens of albite, oligoclase, and labradorite contain reflecting inclusions of hematite that give the feldspars a golden sparkle. Such feldspars are called AVENTURINE or SUNSTONE (Plate 37(*D*), facing p. 219). PARALLEL TWINNING STRIATIONS

(ALBITE TWINNING) ARE PRESENT ON MOST SPECIMENS OF PLAGIOCLASE (Figure 8-50).

Composition. Sodium and calcium aluminum silicates. The high-temperature series from albite $Na(AlSi_3O_8)$ to anorthite $Ca(Al_2Si_2O_8)$ is one of almost complete solid solution, but the low-temperature series is structurally complex (Deer, Howie, and Zussman).[2] Potassium is often present toward the albite end of the series.

Tests. Fuses with difficulty. Albite is insoluble in ordinary acids, but the calcium-rich members are attacked by hydrochloric acid and form a silica gel.

Crystallization. Triclinic; pinacoidal class, $\overline{1}$. Small crystals of albite are more common than other members of the series. Twinning is common, both contact and polysynthetic. Albite sometimes occurs in aggregates of flat, platy, white crystals known as CLEAVELANDITE. Usually in cleavable masses or irregular grains.

Albite. Common constituent of igneous rocks such as granite, syenite, rhyolite, and trachyte. Also found in pegmatites as platy cleavelandite, and as replacements of microcline sometimes forming a coarse perthite structure. Cleavelandite is common at Amelia Court House, Virginia (Plate 37, facing p. 219), and in pegmatites in Maine and Connecticut. Moonstone varieties are found in Ceylon and in Madagascar.

Oligoclase. Usually massive. Aventurine or sunstone oligoclase containing gleaming reddish hematite inclusions is common in Tvedestrand, Norway (Plate 37, facing p. 219) and in Canada. Greenish oligoclase contains ferric and ferrous iron.

Andesine. Chief feldspar of the fine-grained andesite lavas common in the Andes Mountains.

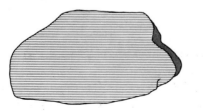

Fig. 8-50. Parallel twinning striations (albite twinning).

Labradorite. Common rock-forming mineral. Found in Labrador in large cleavable masses that show a brilliant blue and green iridescent play of color (see Plate 37(*E*), facing p. 219; and interference colors, p. 78).

Bytownite. A less common feldspar found as grains in calcium-rich igneous rocks.

Anorthite. A rare feldspar found in basic igneous rocks, in contact metamorphic limestones, and in meteorites.

Alteration. The plagioclase feldspars alter to sericite, kaolinite, and calcite.

Use. The same as for microcline and orthoclase, but not extensively.

Feldspathoid Group

The feldspathoids are a group of minerals with chemical compositions similar to the feldspars but showing a range in crystallization. They crystallize instead of feldspar minerals in silica-poor magmas. Nepheline is the most common feldspathoid.

LEUCITE

$K(AlSi_2O_6)$ Isometric (roughly above 620°C)
Hardness $5\frac{1}{2}$–6   Sp. Gr. 2.5

Summary

Leucite occurs in recent lavas in white trapezohedral crystals with a dull surface (Plate 38(*A*), facing p. 226). *Color:* white, gray. *Luster:* vitreous to dull. *Streak:* white. *Cleavage:* poor dodecahedral; translucent to opaque. Leucite is sometimes completely replaced by an intergrowth of nepheline and a potassium-rich feldspar; or by nepheline, analcite, and orthoclase. These pseudomorphs after leucite are called *pseudoleucite.* DULL WHITE TRAPEZOHEDRAL CRYSTALS (Figure 8-51) EMBEDDED IN A FINE-GRAINED ROCK MATRIX ARE CHARACTERISTIC. Resembles white ANALCITE, but analcite crystals are free-growing in cavities, generally have a glassy luster, and are fusible. Rare white GARNETS are harder, and have a higher specific gravity.

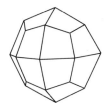

Fig. 8-51. Trapezohedral leucite crystal.

Composition. Potassium aluminum silicate, $K(AlSi_2O_6)$. Small amounts of potassium may be replaced by sodium. Traces of Cs, Rb, and Li may also be present.

Tests. Infusible. Decomposed by hydrochloric acid without gelatinization.

Crystallization. Isometric at somewhere above 620°C. Pseudo-isometric at lower temperatures. On cooling, the isometric structure probably changes to a tetragonal structure, but the external form remains. Usually in embedded trapezohedral crystals; also in disseminated grains.

Alteration. To kaolinite, sericite, analcite, nepheline.

Occurrence. Leucite is a fairly uncommon mineral, usually occurring in silica-poor igneous rocks, especially in recent lavas. Because leucite is easily altered, it is rarely found in lavas older than Tertiary. Commonly found in the lavas of Vesuvius. Large altered crystals occur in Brazil (pseudoleucite). In the United States, it is found in the Leucite Hills, Wyoming; in Montana; and in Magnet Cove, Arkansas.

NEPHELINE

$Na_3K(Al_4Si_4O_{16})$	Hexagonal
Hardness $5\frac{1}{2}$–6	Sp. Gr. 2.55–2.65

Summary

Usually massive, or as embedded grains in such silica-poor igneous rocks as nepheline syenites and phonolites. *Color:* colorless, white, gray, yellowish, reddish, greenish. *Luster:* greasy on cleavage surfaces;

crystals vitreous. *Streak:* white. *Cleavage:* poor prismatic; transparent to translucent. PECULIAR GREASY LUSTER (Plate 38(*F*), facing p. 226), AND ASSOCIATION WITH OTHER FELDSPATHOIDS ARE CHARACTERISTIC. Distinguished from QUARTZ by lower hardness; from FELDSPAR by poor cleavage and solubility in HCl; and from APATITE by greater hardness.

Composition. A sodium potassium aluminum silicate, $Na_3K(Al_4Si_4O_{16})$.

Tests. Fuses to a clear glass. Powdered mineral is soluble in hydrochloric acid. When evaporated, yields a silica gel.

Crystallization. Hexagonal; pyramidal class, 6. Usually as irregular grains and masses. Small flat prismatic crystals are rare. Sometimes occurs as large dull rough crystals in pegmatites.

Alteration. Alters to kaolinite, sodalite, cancrinite, sericite, and zeolites.

Occurrence. Usually as grains and masses in syenites, phonolites, and some basalts. Also in pegmatites associated with nepheline-syenites. Associated minerals include sodalite, cancrinite, zircon, corundum, biotite, and feldspar. Large amounts of nepheline are mined near Bancroft, Ontario; and in the Kola Peninsula, U.S.S.R. Large, rough crystals are found in the Bancroft region of Ontario. Small, glassy crystals occur in the lavas of Vesuvius.

Use. In the manufacture of glass and ceramics.

SODALITE

$Na_8(Al_6Si_6O_{24})Cl_2$	Isometric
Hardness $5\frac{1}{2}$–6	Sp. Gr. 2.27–2.33

Summary

Usually as masses in silica-poor igneous rocks. Commonly associated with nepheline, cancrinite, and leucite. *Color:* commonly blue; also white, colorless, gray, pink, yellow, and green. Some varieties of sodalite (HACKMANITE), when freshly fractured, have a pink hue that fades on exposure to light but which returns when the mineral is either exposed to ultraviolet light or is kept in the dark for a few weeks. *Luster:* vitreous. *Cleavage:* poor dodecahedral; transparent to translucent. Commonly luminesces bright yellow, orange, or orange-red under long-wave ultraviolet light. BLUE COLOR IS CHARACTERISTIC (Plate 38(*D*), facing p. 226). Decolorized when heated. Resembles the related mineral LAZURITE (lapis lazuli), but lazurite almost always contains associated pyrite.

Composition. Sodium aluminum silicate with chlorine. Small amounts of potassium are often present. Hackmanite contains small amounts of sulfur.

Tests. Turns white when heated and fuses to a white glass. Gives test for chlorine after decomposition in nitric acid.

Crystallization. Isometric; hexoctahedral class, $\frac{4}{m}\,\bar{3}\,\frac{2}{m}$. Commonly in large masses. Small dodecahedral crystals are rare.

Alteration. To zeolites, cancrinite, kaolinite, and sericite.

Occurrence. Occurs in silica-poor rocks associated with nepheline, cancrinite, leucite, and feldspars. Massive blue sodalite occurs near Bancroft, Ontario; Litchfeld, Maine; Minas Gerais, Brazil; and Ice River, British Columbia. Pink hackmanite that loses color on exposure to light is found near Bancroft, Ontario. Hackmanite commonly occurs in the Kola Peninsula, U.S.S.R. Turns green when heated at 700°–900°C, and blue when heated at 950°–1150°C. The color does not disappear on cooling or on exposure to light.

LAZURITE (Lapis Lazuli)

$Na_8(Al_6Si_6O_{24})S_2$ Isometric
        Hardness 5–5½   Sp. Gr. 2.4–2.45

Summary

Usually as compact masses in metamorphosed limestones. Almost always intimately associated with
pyrite (Plate 38(*B*), facing p. 226). *Color:* bright azure blue, violet-blue, bluish-green. *Luster:* vitreous. *Streak:* white. *Cleavage:* poor dodecahedral; translucent; brittle. Lapis lazuli is commonly a mixture of lazurite, pyrite, calcite, and pyroxene, and is used in jewelry and as an ornamental stone. Its powdered form was formerly used as the paint pigment *ultramarine,* which is now produced from synthetic lazurite. PYRITE ASSOCIATION DISTINGUISHES LAZURITE FROM SODALITE AND LAZULITE.

Composition. Sodium aluminum silicate with sulfur, $Na_8(Al_6Si_6O_{24})S_2$.

Tests. Fusible, but does not lose color upon heating like sodalite. Soluble in HCl. Test for S.

Crystallization. Isometric; hexoctahedral class, $\frac{4}{m}\,\bar{3}\,\frac{2}{m}$. Usually as masses or grains in limestone. Dodecahedral crystals are rare.

Occurrence. Lazurite is an uncommon mineral found in metamorphosed limestones. The best deep blue lapis lazuli comes from Afghanistan, where it occurs as masses and crystals in marble. Also found in Ovalle, Chile; at Lake Baikal, Siberia; near Mogok, Burma; and in San Bernardino County, California.

Use. Ornamental and gem stone.

CANCRINITE

$Na_3Ca(Al_3Si_3O_{12})CO_3(OH)_2$ Hexagonal
        Hardness 5–6   Sp. Gr. 2.42–2.51

Summary

Usually in lamellar masses or disseminated grains in such silica-poor rocks as nepheline syenites. Commonly associated with sodalite (Plate 38(*E*), facing p. 226), nepheline, feldspar, sphene, and apatite. *Color:* yellow, yellow ocher, gray, white, colorless, green, blue, reddish. *Luster:* vitreous to greasy. *Streak:* white. *Cleavage:* perfect prismatic; transparent to translucent; brittle. Cancrinite is commonly formed by the alteration of nepheline. YELLOW COLOR, MINERAL ASSOCIATION, LOSS OF COLOR ON HEATING, AND EFFERVESCENCE IN HOT HCl ARE CHARACTERISTIC.

Composition. The composition is usually variable. May contain some sulfur and chlorine, or potassium.

Tests. Loses color when heated. Effervesces in hot HCl, and forms a silica gel when heated.

Crystallization. Hexagonal. Usually massive. Rarely in prismatic crystals.

Alteration. Commonly alters from nepheline.

Occurrence. Cancrinite is sometimes a primary mineral in silica-poor igneous rocks, but is more frequently a secondary mineral formed by the alteration of nepheline under considerable carbon dioxide pressure. Associated with sodalite, nepheline, and other feldspathoids. Occurs with sodalite in Ontario, Canada; and in Litchfield, Maine.

SCAPOLITE SERIES

MARIALITE, $Na_4(AlSi_3O_8)_3Cl$ Tetragonal
MEIONITE, $Ca_4(Al_2Si_2O_8)_3CO_3$
     Hardness 5–6   Sp. Gr. 2.5–2.7

Summary

Commonly occurs as large, coarse prismatic crystals with a dull wood-like surface alteration (Plate 38(C), facing p. 226). Usually found in metamorphosed limestones, associated with apatite, sphene, zircon, diopside, amphiboles, garnet, and biotite. Pure MARIALITE and MEIONITE, the end members of the scapolite series, have not been found in nature. *Color:* usually white or gray; sometimes bluish, greenish, reddish, pink, yellow. *Luster:* vitreous; dull when altered. *Streak:* white. *Cleavage:* poor prismatic; transparent to translucent; brittle. May luminesce bright yellow or orange under long-wave ultraviolet light (Plate 9, facing p. 66). WOODY-LOOKING PRISMATIC CRYSTALS WITH A SQUARE CROSS-SECTION, YELLOW LUMINESCENCE, AND FUSION TO A WHITE BUBBLY GLASS ARE CHARACTERISTIC.

Composition. $(Na,Ca)_4Al_3(Al,Si)_3Si_6O_{24}(Cl,CO_3,SO_4)$. Complex sodium-calcium aluminum silicate with Cl, CO_3, and SO_4. Potassium, fluorine, and OH may also be present. MARIALITE is the sodium-rich end member of the series and MEIONITE is the calcium-rich end member. Scapolites containing more than 80 percent of the components of either marialite or meionite are rarely found in nature.

Tests. Fuses with intumescence to a white bubbly glass. Some specimens luminesce yellow, yellow-orange, or orange-red under long-wave ultraviolet light. Luminescence is usually stronger after heating. Imperfectly decomposed by HCl; does not yield a silica gel.

Crystallization. Tetragonal; dipyramidal class, $\frac{4}{m}$. Woody-looking prismatic crystals with a square cross section are common. Also massive.

Alteration. Readily alters to kaolinite, mica, zeolites, albite, talc, and epidote.

Occurrence. Commonly found in metamorphic rocks, especially in impure limestones that have been metamorphosed as the result of igneous intrusion. Frequently associated with diopside, amphiboles, sphene, zircon, apatite, garnet, and biotite. Scapolite also occurs in some unusual pegmatites. Large crystals, some over a foot long, have been found in metamorphosed limestones at Rossie and Pierrepont, St. Lawrence County, New York; and in Ontario, Canada, at Bedford and Renfrew; and near Grenville, Quebec. Transparent yellow crystals of gem quality occur in a pegmatite in Betroka, Madagascar. Bright yellow gem scapolite also occurs in the pegmatites of Minas Gerais, Brazil; and Tremorgio, Switzerland. Cat's eye varieties are found in Mogok, Burma.

Use. Minor gem stone.

Zeolite Group

The zeolites comprise a large group of related minerals. They are usually well crystallized, and commonly occur in cavities in such lavas as basalts. The following common zeolites are included in this section: NATROLITE, STILBITE, ANALCITE, CHABAZITE, and HEU-

LANDITE. All have characteristic crystal habits (Plate 39, facing p. 227), and are frequently associated with calcite, prehnite, pectolite, datolite, and apophyllite.

Most zeolite minerals intumesce (bubble up) when heated. The name zeolite (boiling stone), derives from the Greek words *zein* (to boil) and *lithos* (stone). When zeolites are heated, water is given off continuously as the temperature rises; not in separate stages at certain temperatures. Dehydrated zeolites can absorb other liquids in place of water. Most zeolites are relatively soft (can be scratched by a knife blade), and are light in weight (Sp. Gr. 2.0–2.3). Those containing barium may be heavier.

The zeolite structure is an open aluminosilicate framework composed of $(Si,Al)O_4$ tetrahedra. The wide channelways in the structure contain water molecules and cations, such as Ca, Na, and K, which balance the negative charge of the framework. The cations and water molecules do not fill all the cavities and can move about. They can also be replaced or removed without disrupting the strongly bonded framework structure. This property of easy ion substitution and the ability to strongly adsorb foreign ions on the surface of the zeolite grains is utilized in synthetic zeolite water softeners.

Water that contains calcium ions in solution is "hard," but when calcium is replaced by sodium, the water becomes soft and more suitable for washing. Hard water can be filtered through artificial zeolite grains, and in the process the calcium ions in the water are replaced by the zeolite's sodium ions. The reverse substitution can also take place, and the zeolites can be recharged by placing them in a concentrated NaCl solution that forces out the calcium ions and replaces them with sodium ions. Zeolites are also used in the manufacture of catalysts used in oil refining.

with calcite and other zeolites. Usually in radiating groups of white needle-like crystals (Plate 39(*A*), facing p. 227, and Plate 2(*A*), facing p. 3). The needles are frequently thick enough to show a square cross section. *Color:* usually colorless to white; also yellowish, reddish, greenish. *Luster:* vitreous to silky. *Streak:* white. *Cleavage:* perfect prismatic; transparent to translucent. Sometimes luminesces orange. RADIATING NEEDLE-LIKE HABIT AS WELL AS MINERAL ASSOCIATION ARE CHARACTERISTIC. DIFFICULT TO DISTINGUISH FROM OTHER FIBROUS ZEOLITES like mesolite, scolecite, thomsonite, and edingtonite. Distinguishable from associated PECTOLITE, $Ca_2NaSi_3O_8(OH)$, by test for aluminum.

Composition. A hydrous sodium aluminum silicate, $Na_2(Al_2Si_3O_{10}) \cdot 2H_2O$. Adsorbed surface ions are present. May contain small amounts of K or Ca.

Tests. Fuses easily to a bubbly colorless glass. Gives water in closed tube. Soluble in HCl and yields a silica gel on evaporation.

Crystallization. Orthorhombic; pyramidal class, *m m* 2. Commonly in thin prisms with square cross sections and in needle-like radiating groups. Frequently, it is vertically striated. Also massive with a radiating fibrous structure. Sometimes in cruciform twins.

Alteration. Alters from nepheline, sodalite, and plagioclase.

Occurrence. Commonly in cavities in basaltic rocks; associated with calcite, apophyllite, datolite, prehnite, and zeolite minerals. Also as an alteration product of nepheline, sodalite, and plagioclase. Common in cavities with other zeolites in Nova Scotia and at Paterson, New Jersey. Occurs with large, pale green apophyllite crystals and reddish stilbite aggregates in Rio Grande do Sul, Brazil.

NATROLITE

$Na_2(Al_2Si_3O_{10}) \cdot 2H_2O$	Orthorhombic
Hardness 5–5½	Sp. Gr. 2.2–2.26

Summary

Natrolite is one of the later zeolites to crystallize and is commonly found in cavities in basalt associated

STILBITE

$(Ca,Na_2,K_2)(Al_2Si_7O_{18}) \cdot 7H_2O$	Monoclinic
Hardness 3½–4	Sp. Gr. 2.1–2.2

Summary

Usually in sheaf-like aggregates found in cavities in basaltic rocks, commonly associated with heulandite and chabazite. *Color:* white, reddish, yellowish, reddish-brown. *Luster:* vitreous, pearly on cleavage face. *Streak:* white. *Cleavage:* one direction perfect; transparent to translucent. SHEAF-LIKE AGGREGATES ARE DISTINCTIVE (Plate 39(*D, E*), facing p. 227).

Composition. A hydrous calcium, sodium, potassium, aluminum silicate. Adsorbed ions are present.

Tests. Swells and fuses easily to a white enamel. Decomposed by HCl without the formation of a silica gel. Gives water in closed tube.

Crystallization. Monoclinic; prismatic class, $\frac{2}{m}$. Usually in sheaf-like aggregates; also radial and globular masses, and in cruciform twins.

Occurrence. Usually found in cavities in basalt, where it is associated with other zeolites, and with apophyllite, calcite, and prehnite. Notable locations are Paterson, New Jersey; Nova Scotia; Kilpatrick, Scotland; and Rio Grande do Sul, Brazil.

sionally red. *Luster:* vitreous, pearly on cleavage face. *Streak:* white. *Cleavage:* perfect side pinacoid; transparent to translucent. "COFFIN-SHAPED" CRYSTALS WITH PEARLY CLEAVAGE FACES, AND MINERAL ASSOCIATIONS ARE CHARACTERISTIC.

Composition. Hydrous calcium sodium aluminum silicate. Some potassium may be present, and adsorbed surface ions.

Tests. Fuses to a bubbly white glass. Gives water in the closed tube.

Crystallization. Monoclinic; prismatic class, $\frac{2}{m}$. Tabular elongated crystals with tapering ends.

Occurrence. Associated with other zeolites, calcite, prehnite, datolite, and quartz in cavities in igneous rocks. Good crystals are found in the trap rocks of Paterson, New Jersey. Also in Nova Scotia, in Iceland, and in Rio Grande do Sul, Brazil.

HEULANDITE

$(Ca,Na_2)(Al_2Si_7O_{18}) \cdot 6H_2O$ Monoclinic
Hardness $3\frac{1}{2}$–4   Sp. Gr. 2.1–2.2

Summary

Occurs in elongated tabular crystals, which narrow at the ends—"coffin-shaped" (Figure 8-52). Crystals have a pearly luster on cleavage surfaces (Plate 39(*F*), facing p. 227). Commonly found in cavities in basaltic-type rocks; associated with *calcite* and other *zeolites*. *Color:* colorless, white, pink, yellowish, occa-

Fig. 8-52. Heulandite crystal.

CHABAZITE

$Ca(Al_2Si_4O_{12}) \cdot 6H_2O$ Hexagonal
Hardness $4\frac{1}{2}$ Sp. Gr. 2.05–2.1

Summary

As rhombohedral crystals that resemble slightly distorted cubes (Plate 39(*B*), facing p. 227). Commonly found in cavities in basalt and andesite, associated with calcite, apophyllite, stilbite, and other zeolites. *Color:* colorless, white, pink, reddish, brownish. *Luster:* vitreous. *Cleavage:* distinct rhombohedral; transparent to translucent; brittle. PSEUDOCUBIC CRYSTALS, AND MINERAL ASSOCIATIONS ARE CHARACTERISTIC. Distinguished from associated CALCITE by lack of effervescence in hydrochloric acid; from APOPHYLLITE by lack of pearly cleavage faces.

Composition. Composition is variable. Essentially a hydrous calcium aluminum silicate. Some potassium and sodium often partially replace calcium. Also contains adsorbed surface ions.

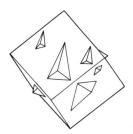

Fig. 8-53. Penetration twinned chabazite.

Tests. Swells and fuses to a bubbly glass. Gives water in closed tube. Decomposed by HCl without formation of a gel.

Crystallization. Hexagonal; scalenohedral class, $\bar{3}\frac{2}{m}$. Crystals are rhombohedral with angles close to cubes. Frequently in penetration twins (Figure 8-53).

Occurrence. Commonly found in cavities in basalts and andesites associated with other zeolites and calcite. Good crystals occur in West Paterson, New Jersey, and in Nova Scotia.

ANALCITE

$Na(AlSi_2O_6) \cdot H_2O$ Isometric
Hardness $5\frac{1}{2}$ Sp. Gr. 2.24–2.29

Summary

Usually in trapezohedral crystals (Plate 39(*C*), facing p. 227), that resemble leucite and garnet; sometimes in cubes. Commonly associated with zeolites, calcite, and prehnite. *Color:* usually colorless or white; sometimes gray, greenish-gray, yellowish, or reddish. *Luster:* vitreous. *Streak:* white. *Cleavage:* basal, very poor; transparent to translucent; brittle. WHITE, FREE-GROWING TRAPEZOHEDRAL CRYSTALS ASSOCIATED WITH OTHER ZEOLITES IN CAVITIES ARE DISTINCTIVE. Resembles LEUCITE but leucite crystals are embedded in a fine-grained rock matrix. Lightly colored GARNET crystals have a greater hardness and specific gravity.

Composition. Hydrous sodium aluminum silicate, $Na(AlSi_2O_6) \cdot H_2O$. Small amounts of potassium or calcium are often present; also adsorbed surface ions.

Tests. Swells and fuses easily to a colored glass. Gives water in the closed tube. Gelatinizes with hydrochloric acid.

Crystallization. Isometric; hexoctahedral class, $\frac{4}{m}\bar{3}\frac{2}{m}$. Usually in trapezohedral crystals. Sometimes in cubes with three trapezohedral faces at each corner. Also massive. The rare cesium-bearing mineral *pollucite* $(Cs(AlSi_2O_6) \cdot H_2O)$ has a structure similar to analcite.

Occurrence. Commonly occurs in cavities in basic lavas associated with calcite, prehnite, datolite, and zeolites. Also as a primary rock-forming mineral in the groundmass of basic igneous rocks. Found in lake sediments that are formed by the action of lake water containing sodium salts on volcanic ash and clay minerals.

Good crystals of analcite are found in Paterson, New Jersey; in Nova Scotia, and in the copper deposits of Michigan.

SHEET STRUCTURES

Prehnite (?)	Mica Group
Apophyllite	Muscovite
Serpentine	Biotite
Pyrophyllite	Phlogopite
Talc	Lepidolite
Clay Minerals	Margarite
	Chlorite Group

APOPHYLLITE

$KCa_4(Si_4O_{10})_2F \cdot 8H_2O$ Tetragonal
Hardness $4\frac{1}{2}$–5   Sp. Gr. 2.33–2.37

Summary

Usually in well-formed crystals that at times appear cubic but have a pearly luster on the base faces

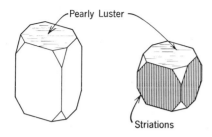

Fig. 8-54. Apophyllite crystals.

and a vitreous luster on the prisms faces (Figure 8-54). Commonly occurs as a secondary mineral in cavities in basalt associated with zeolites, prehnite, datolite, and calcite. *Color:* white, colorless, pale green, pale blue, yellowish, pink. *Luster:* pearly on base faces, others vitreous. *Cleavage:* perfect basal; transparent to translucent. PSEUDOCUBIC CRYSTALS WITH A PEARLY LUSTER ON THE TWO BASE FACES, PERFECT BASAL CLEAVAGE, AND ZEOLITE ASSOCIATION ARE DISTINCTIVE.

Composition. A hydrous potassium calcium fluosilicate, $KCa_4(Si_4O_{10})_2F \cdot 8H_2O$. Some sodium commonly replaces potassium.

Tests. Swells and fuses easily to a white bubbly enamel. Gives water in closed tube.

Crystallization. Tetragonal; ditetragonal dipyramidal class, $\frac{4}{m}\frac{2}{m}\frac{2}{m}$. Commonly in almost square crystals with small triangular faces at the corners. Also in elongated prismatic crystals.

Occurrence. Commonly associated with zeolites, and calcite in cavities in basalt. Large pale green crystals, some 6 in. long, occur with brownish-pink stilbite in Rio Grande do Sul and Bahia, Brazil. Good crystals also occur in Paterson, New Jersey; in Nova Scotia, and near Poona, India.

PREHNITE

$Ca_2Al(AlSi_3O_{10})(OH)_2$ Orthorhombic
Hardness 6-6½   Sp. Gr. 2.9–2.95

Summary

Commonly in light green rounded masses with ridged surfaces formed by the edges of curved tabular crystals (Plate 40(*F*), facing p. 242). Found in cavities in basaltic rocks associated with calcite, datolite, pectolite, and zeolites. *Color:* usually light green, sometimes white. *Luster:* vitreous. *Streak:* white. *Cleavage:* basal; translucent; brittle. LIGHT GREEN BOTRYOIDAL AGGREGATES, AND ZEOLITE ASSOCIATIONS ARE CHARACTERISTIC. Distinguished from HEMIMORPHITE by lower specific gravity, and from SMITHSONITE by lack of effervescence in HCl.

Composition. Hydrous calcium aluminum silicate, $Ca_2Al(AlSi_3O_{10})(OH)_2$. May contain some iron.

Tests. Swells and fuses easily to a bubbly glass. Gives water in closed tube.

Crystallization. Orthorhombic; pyramidal class, 2 *m m*. Commonly in rounded aggregates of curved tabular crystals. Also stalactitic aggregates. Sometimes in aggregates of barrel-shaped crystals.

Occurrence. Prehnite occurs as a secondary mineral in cavities in basalt. Commonly associated with calcite, datolite, copper, pectolite, and zeolites. Good botryoidal masses occur in the trap rocks of West Paterson, New Jersey; in the copper deposits of Michigan; and in Fairfax and Loudon Counties, Virginia. Aggregates of barrel-shaped crystals occur in Dauphiné, France.

SERPENTINE

$Mg_3(Si_2O_5)(OH)_4$ Monoclinic
Hardness 2½–5   Sp. Gr. 2.2–2.6

Summary

Serpentine occurs in a variety of massive forms (Plate 40(*A, B*), facing p. 242). Crystals are unknown. The varieties include ANTIGORITE (compact masses); CHRYSOTILE (fibrous serpentine used for asbestos) (Plate 2(*C*), facing p. 3); LIZARDITE (extremely fine-grained matrix material that commonly contains veins

of chrysotile); and SERPENTINE MARBLE or VERDE AN-
TIQUE (rock composed of dark green serpentine and
white carbonates). *Color:* commonly, shades of green;
also yellow, red, brown, white, and black. *Luster:* waxy
to greasy, silky (fibrous varieties). *Streak:* white.
Cleavage: very poor prismatic; translucent. Smooth
greasy feel. Serpentine is a secondary mineral resulting from the alteration of such magnesium silicates as
olivine, pyroxenes, and amphiboles. RELATIVELY SOFT
COMPACT MASSES WITH A LIGHT AND DARK GREEN MOTTLED
APPEARANCE, AND A GREASY FEEL AND LUSTER ARE CHAR-
ACTERISTIC OF MASSIVE VARIETIES. Spinach green NEPH-
RITE JADE may resemble massive serpentine but is
harder (H. = 6).

Composition. Hydrous magnesium silicate, $Mg_3(Si_2O_5)(OH)_4$. Small amounts of iron and nickel may be
present.

Tests. Infusible, but decrepitates. Gives water in closed
tube. Decomposed by HCl.

Crystallization. Monoclinic. Crystals are unknown. Usually in compact masses or as fibrous asbestos.

Alteration. From olivine and enstatite. To olivine when
heated in air at about 600°C.

Occurrence. Serpentine is a common secondary mineral
of widespread occurrence formed by the hydrothermal
alteration of such magnesium silicates as olivine, pyroxenes, and amphiboles. Usually associated with
magnesite, chromite, spinel, garnet, and magnetite.
Sometimes forms large rock masses. Large deposits
of chrysotile asbestos are mined in Megantic County,
Quebec, Canada. Chrysotile also occurs in San Benito
County, California. Serpentine pseudomorphs after
olivine are found at Snarum, Norway.

Use. Chrysotile is the chief source of the asbestos
widely used in the building industry. Massive varieties
(verde antique) are used as building and ornamental
material.

PYROPHYLLITE

$Al_2(Si_4O_{10})(OH)_2$ Monoclinic
   Hardness 1–2   Sp. Gr. 2.65–2.90

Summary

Occurs as soft compact masses that resemble
massive talc; as radiating aggregates (Plate 40(*D*),
facing p. 242); and as foliated masses. Found with
mica and quartz in hydrothermal veins, and in metamorphic rocks associated with kyanite. *Color:* white,
greenish, yellowish, brownish. *Luster:* pearly to waxy.
Cleavage: perfect micaceous; yields flexible inelastic
flakes; translucent; greasy feel. VERY SOFT RADIATING
AGGREGATES WITH A PEARLY LUSTER ARE DISTINCTIVE.

Composition. Hydrous aluminum silicate, $Al_2(Si_4O_{10})(OH)_2$. May contain small amounts of iron.

Tests. Infusible. Radiating masses *exfoliate* and whiten
when heated on charcoal. Heated pyrophyllite turns
blue when moistened with cobalt nitrate and reheated
(Al test). (Talc turns violet with cobalt nitrate.)

Crystallization. Monoclinic. Usually in compact masses,
foliated masses, or radial groups.

Occurrence. Pyrophyllite is a fairly common mineral
that occurs in some shistose rocks and hydrothermal
veins. Agalmatolite is an impure compact variety of
pyrophyllite from which the Chinese made small carv-
ings. Radial pyrophyllite is found in Chesterfield,
South Carolina (Plate 40), and in North Carolina.

Use. As a filler in paint, in ceramics, and in hard rubber.
Often competitive with talc.

TALC

$Mg_3(Si_4O_{10})(OH)_2$ Monoclinic
   Hardness 1   Sp. Gr. 2.58–2.83

Summary

Commonly in white or green pearly foliated
masses (Plate 40(*E*), facing p. 242); also dark gray or
green impure compact masses (soapstone). *Color:*
white to apple green; dark in impure varieties. *Luster:*
pearly to greasy. *Streak:* white. *Cleavage:* micaceous
yielding flexible inelastic folia; translucent to opaque;
greasy feel; sectile. Talc is a secondary mineral formed
from the alteration of magnesium silicates such as

olivine, pyroxenes, or amphiboles. GREASY FEEL, FOLI-ATED STRUCTURE, AND SOFTNESS ARE CHARACTERISTIC. Resembles PYROPHYLLITE.

Composition. Hydrous magnesium silicate, $Mg_3(Si_4O_{10})(OH)_2$. May contain minor amounts of nickel or manganese.

Tests. Insoluble in acids. Difficult to fuse. Turns violet when moistened with cobalt nitrate after heating (pyrophyllite turns blue).

Crystallization. Monoclinic; prismatic class, $\frac{2}{m}$. Usually in foliated or compact masses; sometimes in radiating aggregates. Tabular pseudohexagonal crystals are rare. Impure compact varieties (soapstone, steatite, french chalk).

Alteration. From olivine, amphiboles, and pyroxenes.

Occurrence. Talc is a secondary mineral formed by the alteration of magnesium silicates. Often associated with chlorite and serpentine. Soapstone often forms large rock masses, and is quarried in the eastern United States along the Appalachian Mountains.

Use. In ceramics and cosmetics; and as a filler in paints, paper, and rubber. Slabs of soapstone are used in acid-proof laboratory table tops.

CLAY MINERALS

The clay minerals occur in aggregates of extremely small particles, and are not readily distinguishable from each other by ordinary physical properties. Individual clay minerals are best identified by such special methods as x-ray analysis, electron microscopy, and differential thermal analysis.

The most common clay minerals include kaolinite, dickite, nacrite, and halloysite (the kaolinite group); illite (hydromica), montmorillonite, vermiculite, and palygorskite. They are essentially hydrous aluminum silicates, and are usually formed from the alteration of aluminum silicates.

The clay minerals occur in soft earthy masses, which often become plastic when moistened. Massive clay usually has a smooth soapy feel and may yield an earthy odor when breathed upon. The clays are soft (hardness 2–3) and have low specific gravities (2–2.7). The color is usually white or gray, but sometimes are stained yellowish, bluish, greenish, or reddish by impurities.

KAOLINITE, $Al_2(Si_2O_5)(OH)_4$, is one of the most common clay minerals, and is usually formed by the weathering or the hydrothermal alteration of feldspars. It is widely used in the ceramic and paper industries. MONTMORILLONITE is the chief constituent of BENTONITE, an altered volcanic ash or tuff. Montmorillonite absorbs water molecules and swells when placed in water. It is used in drilling muds; as a binder in molding sands; and also as a catalyst. VERMICULITE swells into worm-like masses when rapidly heated. It is used as a lightweight heat-and-sound insulating material, and as a filler in paper, plastics, and paints.

MASSIVE CLAY IS OFTEN SOFT AND PLASTIC (WITH A SOAPY FEEL), AND HAS AN EARTHY ODOR WHEN BREATHED UPON.

Mica Group

The *elastic micas,* such as MUSCOVITE, BIOTITE, PHLOGOPITE and LEPIDOLITE are characterized by their perfect basal cleavage yielding thin flexible elastic cleavage flakes. The micaceous cleavage is the result of the mica's layered atomic structure. Mica crystals are commonly pseudohexagonal forming six-sided "mica books" (Plate 41, facing p. 243, phlogopite).

The *brittle micas,* such as MARGARITE and CHLORITOID, are harder than the elastic micas, and yield somewhat brittle cleavage flakes. All the micas can be scratched by a knife blade.

MUSCOVITE

$KAl_2(AlSi_3O_{10})(OH)_2$ Monoclinic
     Hardness $2\frac{1}{2}$–3   Sp. Gr. 2.77–2.88

Summary

Muscovite, sometimes called *common* or *white mica,* is a common rock-forming mineral. It is usually colorless or in pale shades of thin sheets (Plate 40(*C*), facing p. 242). Found usually in small flakes or foliated masses. Six-sided crystals ("mica books") are less common. *Color:* colorless, gray, brownish, amber; sometimes bright green or rose. *Luster:* vitreous to pearly. *Streak:* white. *Cleavage:* perfect basal, yielding thin flexible plates; transparent to translucent. COLORLESS OR PALE THIN FLEXIBLE CLEAVAGE FLAKES ARE DISTINCTIVE (Figure 5-6, p. 82). Pink muscovite resembles LEPIDOLITE, but is much less fusible. Pink MARGARITE cleavage flakes are harder, and are somewhat brittle. Brown PHLOGOPITE resembles some brownish muscovite, but is soluble in sulfuric acid.

Composition. Essentially a hydrous potassium aluminum silicate, but may vary considerably in composition because of atomic substitutions.

Tests. Whitens and fuses with difficulty. Insoluble in acids.

Crystallization. Monoclinic; prismatic class, $\frac{2}{m}$. Commonly in foliated masses, small flakes, or fine-grained compact masses (sericite). Sometimes in pseudohexagonal crystals, or in globular aggregates.

Alteration. Sericite commonly alters from feldspars; also from kyanite, topaz, and andalusite.

Occurrence. Muscovite is a common rock-forming mineral especially in granites, syenites, pegmatites, shists, and gneisses. Associated with quartz, orthoclase, microcline, albite, topaz, tourmaline, and beryl; also found in pegmatites.

Large pegmatite deposits occur in New Hampshire, in North Carolina, and in the Black Hills of South Dakota. Occurs also in India. Pink muscovite occurs in New Mexico, in Virginia, and in Massachusetts. Bright green muscovite (fuchsite) has been found in North Carolina and in Utah.

Use. Primary use of mica sheets is for high-temperature electrical insulation. Sericite has been used for furnace linings.

BIOTITE

K(Mg,Fe)$_3$(AlSi$_3$O$_{10}$)(OH)$_2$	Monoclinic
Hardness 2½–3	Sp. Gr. 2.7–3.3

Summary

A black mica (Plate 41(*A*), facing p. 243) commonly in small scales or foliated masses. *Color:* black, blackish-brown, greenish-black; rarely light yellow. *Luster:* vitreous. *Streak:* white. *Cleavage:* perfect basal, yielding thin flexible elastic sheets; translucent. BLACK COLOR AND MICACEOUS CLEAVAGE ARE CHARACTERISTIC.

Composition. Hydrous potassium magnesium aluminum silicate, K(Mg,Fe)$_3$(AlSi$_3$O$_{10}$)(OH)$_2$. Commonly contains small amounts of other elements. Biotite is the iron-rich member of the biotite-phlogopite series.

Tests. Whitens, fuses with difficulty, and becomes magnetic. Decomposed by boiling in sulfuric acid; yields a milky solution.

Crystallization. Monoclinic; prismatic class, $\frac{2}{m}$. Usually in small scales, scaly aggregates, or foliated masses. Pseudohexagonal or pseudo-orthorhombic crystals are uncommon.

Alteration. Alters to vermiculite, chlorite, epidote, montmorillonite. Commonly alters from amphiboles, pyroxenes, and garnet.

Occurrence. Biotite is a common rock-forming mineral, commonly associated with quartz, muscovite, orthoclase, and amphiboles. Large sheets of biotite occur in pegmatites. Small, rare, light yellow crystals occur in limestone blocks at Vesuvius.

Use. On heating, vermiculite (the hydrated alteration product of biotite or chlorite) loses water and expands. Dehydrated expanded vermiculite is used for insulation.

PHLOGOPITE

KMg$_3$(AlSi$_3$O$_{10}$)(OH)$_2$	Monoclinic
Hardness 2½–3	Sp. Gr. 2.76–2.90

Summary

A bronze mica (Plate 41(*B*), facing p. 243), commonly in six-sided tapering elongated "mica books" (Figure 8-55). Usually found in dolomite marbles. *Color:* yellowish to reddish-brown. *Luster:* pearly to

A   x 0.87

BERYL

$Be_3Al_2Si_6O_{18}$

Pale green prismatic six-sided aquamarine crystal: Minas Gerais, Brazil.

B   x 0.56

BERYL (EMERALD)

Deep green emerald crystals in mica schist: Ural Mountains, U.S.S.R.

C   x 0.80

BERYL

Light blue prismatic beryl crystal (aquamarine): Bogotá, Colombia.

D   x 0.75

BERYL

Massive pink beryl (morganite); yellow beryl crystal (heliodor); and pale green crystal (aquamarine).

E   x 0.50

BENITOITE

$BaTiSi_3O_9$

Tabular triangular benitoite crystals with uneven blue color distribution, occurring with black bladed neptunite crystals: San Benito County, California.

F   x 1.0

AXINITE

$(Ca, Mn, Fe)_3Al_2(BO_3)Si_4O_{12}(OH)$

Thin wedge-shaped crystals.

A x 0.75

Zoned tourmaline crystals showing characteristic *rounded triangular cross section:* Minas Gerais, Brazil; and San Diego County, California.

B x 0.63

Striated rose colored tourmaline crystals (rubellite): Pala, San Diego County, California.

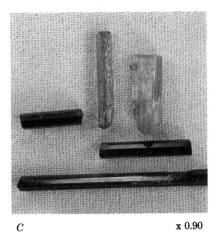

C x 0.90

Long slender prismatic crystals; Minas Gerais, Brazil.

D x 0.52

Black tourmaline crystal (schorlite): Pierrepont, New York.

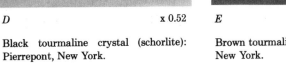

E x 0.50

Brown tourmaline crystals: Gouverneur, New York.

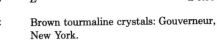

F x 0.50

Pink tourmaline (rubellite) in lepidolite: San Diego County, California.

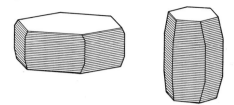

Fig. 8-55. Phlogopite crystals (six-sided mica books).

submetallic on cleavage face. *Streak:* white. *Cleavage:* perfect basal, yielding thin flexible and elastic plates; transparent to translucent. BROWN COLOR, MICACEOUS CLEAVAGE, AND ASSOCIATION WITH DOLOMITE MARBLES ARE CHARACTERISTIC.

Composition. Hydrous potassium magnesium aluminum silicate, $KMg_3(AlSi_3O_{10})(OH)_2$. May contain iron, fluorine, or manganese. If the ratio of Mg:Fe is less than 2:1 the mineral is called biotite.

Tests. Fuses with difficulty. Decomposed by boiling in concentrated sulfuric acid; yields a milky solution.

Crystallization. Monoclinic; prismatic class, $\frac{2}{m}$. Commonly in six-sided crystals, sometimes elongated and large. Also in scales and foliated masses.

Occurrence. Usually found in dolomite marbles, associated with pyroxenes, amphiboles, and serpentine. Also in such ultrabasic igneous rocks as kimberlite. Large sheets of phlogopite occur in Burgess, Ontario; and in Madagascar.

Use. Chief use is as an electrical insulator; the same as muscovite.

LEPIDOLITE

$K_2Li_2Al_3(AlSi_7O_{20})(OH,F)_4$ Monoclinic
   Hardness $2\frac{1}{2}$–4   Sp. Gr. 2.8–2.9

Summary

A lithium-bearing mica. Commonly in coarse- to fine-grained lilac scaly aggregates (Plate 41(*C*), facing p. 243). *Color:* pink, lilac, less commonly pale yellow, grayish, and white. The typical pink or lilac color of lepidolite is (as with pink muscovite) believed not to be related to the presence of lithium, but to the dominance of manganese over ferric iron ions in the absence of ferrous iron. Thus, with higher Mn/Fe ratios, the color of lepidolite deepens from shades of pale gray to pink and lilac; and, finally, to purple.[*] *Luster:* pearly. *Streak:* white. *Cleavage:* perfect basal, yielding thin flexible and elastic plates; translucent. LILAC SCALY AGGREGATES USUALLY ASSOCIATED WITH PINK TOURMALINE, SPODUMENE, AND OTHER LITHIUM-BEARING MINERALS ARE CHARACTERISTIC. Resembles PINK MUSCOVITE and the brittle mica known as MARGARITE.

Composition. A hydrous fluosilicate of lithium, potassium, and aluminum, $K_2Li_2Al_3(AlSi_7O_{20})(OH,F)_4$. May contain small amounts of Fe, Mn, Rb, Cs, or Mg.

Tests. Fuses easily to a bubbly glass, and gives a crimson flame (Li). Insoluble in acids before fusion.

Crystallization. Monoclinic; prismatic class, $\frac{2}{m}$. Usually in coarse- to fine-grained scaly aggregates. Pseudohexagonal crystals are rare.

Occurrence. Lepidolite is an uncommon mineral, usually found in lithium-bearing pegmatites; commonly associated with tourmaline (Plate 45, facing p. 275), spodumene, amblygonite, cassiterite, albite, muscovite, topaz, and quartz. In the United States, lepidolite occurs in Maine; in the Black Hills of South Dakota; in Portland, Connecticut; and in Pala, California. Also occurs in Minas Gerais, Brazil; in the Ural Mountains, U.S.S.R.; in Madagascar; and in Germany.

Use. A source of lithium.

MARGARITE

$CaAl_2(Al_2Si_2O_{10})(OH)_2$ Monoclinic
   Hardness $3\frac{1}{2}$–5   Sp. Gr. 3.0–3.1

Summary

A brittle mica. Commonly in coarse, pink, scaly aggregates formed by the alteration of corundum.

[*]E. W. Heinrich, A. A. Levinson, D. W. Levandowski, and C. H. Hewitt, "Studies in the Natural History of Micas," *University of Michigan Engineering Research Inst. Project M. 978* (final report), 1953.

Color: pink, lilac, gray, white. *Luster:* pearly on cleavage face. *Streak:* white. *Cleavage:* perfect basal, yielding thin somewhat brittle cleavage sheets; transparent to translucent. PINK MICACEOUS AGGREGATES ASSOCIATED WITH CORUNDUM ARE CHARACTERISTIC. Resembles *pink muscovite* and *lepidolite,* but its cleavage flakes are somewhat brittle, and are usually associated with corundum or emery.

Composition. Hydrous calcium aluminum silicate, $CaAl_2(Al_2Si_2O_{10})(OH)_2$. May contain some sodium or potassium.

Tests. Fuses with difficulty, and turns white. Incompletely decomposed by boiling in hydrochloric acid.

Crystallization. Monoclinic; prismatic class, $\frac{2}{m}$. Usually in coarse, scaly aggregates. Rarely in pseudohexagonal crystals.

Alteration. From corundum.

Occurrence. Usually associated with corundum or emery. In the United States, occurs with corundum in North Carolina. Associated with emery at Chester, Massachusetts, and Chester County, Pennsylvania. Two foreign localities are the emery deposits of Asia Minor, and the Greek islands.

CHLORITE GROUP

$(Mg,Fe,Al)_6(Al,Si)_4O_{10}(OH)_8$ Monoclinic
Hardness 2–2½   Sp. Gr. 2.6–3.3

Summary

The chlorite group minerals are usually *green;* they resemble the micas, but have inelastic cleavage flakes. Commonly in scaly aggregates and compact microcrystalline masses. The name chlorite comes from the Greek word for green, *chloros.* PROCHLORITE, PENNINITE, and CLINOCHLORE are common chlorite minerals. *Color:* usually shades of green; also white, yellow, pink, and red brown. *Luster:* vitreous to pearly. *Streak:* white. *Cleavage:* perfect basal, yielding flexible but inelastic cleavage folia; transparent to translucent. GREEN COLOR AND BRITTLE MICACEOUS CLEAVAGE FLAKES ARE CHARACTERISTIC.

Composition. The chlorite minerals are essentially hydrous magnesium, iron, and aluminum silicates, but vary widely in atomic substitution. Some chlorites may contain small amounts of chromium, manganese, nickel, or titanium. Chromium-bearing chlorites are often pink or violet.

Tests. Whitens and fuses with difficulty. Decomposed by boiling in sulfuric acid and yields a milky solution.

Crystallization. Monoclinic; prismatic class, $\frac{2}{m}$. Commonly in scaly aggregates, disseminated scales, foliated masses, and compact microcrystalline masses. Thin pseudohexagonal crystals are rare.

Occurrence. Chlorite is a common alteration product of such ferromagnesian minerals as pyroxene, amphibole, biotite, and garnet. Chlorite schists are composed primarily of chlorite. Associated minerals include serpentine, garnet, pyroxenes, amphiboles, chromite, magnetite, apatite, corundum, and epidote. Some chlorite is a product of hydrothermal alteration, and occurs in hydrothermal veins.

CHAIN STRUCTURES

SINGLE CHAINS
　Pyroxene Group
　　Enstatite series
　　Diopside series
　　Spodumene
　　Jadeite
　Rhodonite
　Wollastonite
　Pectolite
　Garnierite
　Chrysocolla (?)

DOUBLE CHAINS
　Amphibole Group
　　Tremolite series
　　　Tremolite
　　　Actinolite
　　　Nephrite
　　Hornblende

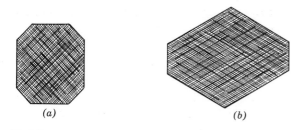

(a)   *(b)*

Fig. 8-56. *(a)* Pyroxene cross section and cleavage; *(b)* amphibole cross section and cleavage.

Pyroxene Group

The **PYROXENES** form a group of minerals that commonly occur in short four- or eight-sided prismatic crystals (Plate 42, facing p. 258) with square cross sections. The prismatic cleavage is almost at right angles (Figure 8-56). The shape and cleavage of the pyroxene crystals distinguish them from the amphibole minerals. The **AMPHIBOLES** (such as hornblende) commonly occur in six-sided elongated prismatic crystals that have a diamond-shaped cross section and cleavage.

The following pyroxenes are described:

ENSTATITE SERIES

DIOPSIDE SERIES

SPODUMENE

JADEITE

ENSTATITE SERIES

Enstatite, $Mg_2(Si_2O_6)$ Orthorhombic
Hypersthene, $(Mg,Fe)_2(Si_2O_6)$
     Hardness 5–6   Sp. Gr. 3.2–3.9

Summary

Commonly in fibrous masses. Prismatic crystals are rare. *Color:* grayish, greenish, yellowish, brown-bronze, green-black. **BRONZITE** is a variety with submetallic bronze-like luster, caused by oriented opaque inclusions. *Luster:* vitreous, silky, submetallic. *Cleavage:* good in two directions intersecting at 87° and 93°; transparent to translucent. **RESEMBLES OTHER PYROXENES. BRONZE-LIKE LUSTER OF BRONZITE VARIETY IS DISTINCTIVE. ALSO, CLEAVAGE CLOSE TO RIGHT ANGLES.**

Composition. Magnesium silicate (enstatite) and magnesium iron silicate (hypersthene). A solid solution series exists between enstatite and hypersthene, but the pure $FeSiO_3$ end member does not exist in nature. Small amounts of aluminum, manganese, and calcium may be present.

Tests. Thin edges fuse with difficulty. Hypersthene is decomposed in hot HCl, but enstatite is insoluble.

Crystallization. Orthorhombic; dipyramidal class, $\frac{2}{m}\frac{2}{m}\frac{2}{m}$. Usually in fibrous or lamellar masses. Rarely in prismatic or tabular crystals.

Occurrence. Common in basic igneous rocks like peridotites, pyroxenites, and basalts. Both enstatite and hypersthene occur in meteorites and in some metamorphosed igneous rocks. Hypersthene is darker in color, and rarer than enstatite. It is found associated with garnet in the Adirondacks region of New York.

Polymorphs. High-temperature monoclinic polymorphs clinoenstatite and clinohypersthene occur in some meteorites.

DIOPSIDE SERIES

Diopside, $CaMgSi_2O_6$ Monoclinic
Hedenbergite, $CaFeSi_2O_6$
Augite, $Ca(Mg,Fe,Al)(Al,Si)_2O_6$
     Hardness 5–6   Sp. Gr. 3.2–3.6

Summary

Commonly in prismatic crystals with square or eight-sided cross sections (Plate 42(*A*, *B*), facing p. 258). *Color:* white, light green (diopside), green (hedenbergite), black (augite). *Luster:* vitreous. *Cleavage:* imperfect prismatic, intersecting at 87° and 93°; transparent to translucent. **PRISMATIC CRYSTALS, CLEAVAGE, AND LIGHT GREEN COLOR OF DIOPSIDE ARE CHARACTERISTIC.**

Composition. Diopside, hedenbergite, and augite are members of an isomorphous series. Augite is intermediate in composition between diopside, hedenbergite, and other less common pyroxenes. It is the most common pyroxene.

Tests. Fuses with difficulty. Insoluble in acids.

Crystallization. Monoclinic; prismatic class, $\frac{2}{m}$. Commonly in prismatic crystals with a square or eight-sided cross section. Also in disseminated grains, granular, columnar, or lamellar masses.

Alteration. To serpentine, talc, chlorite, epidote, hornblende.

Occurrence. *Diopside* commonly occurs as white or pale green prismatic crystals in metamorphosed impure limestones. Crystals are sometimes large. Associated minerals include garnet, tremolite, scapolite, idocrase, and sphene. *Hedenbergite* occurs in some high-temperature ore deposits and also in contact metamorphosed limestones. *Augite* is the most common pyroxene commonly found in basic igneous rocks. Well-formed crystals are not common. Augite crystals, some up to one inch long, have been found around Mount Vesuvius. Large, light green diopside crystals and dark augite crystals occur in St. Lawrence, New York. Diopside and darker green hedenbergite crystals also occur at Nordmark, Sweden.

Use. Diopside crystals are sometimes used as gem stones.

JADEITE

$NaAlSi_2O_6$ Monoclinic
Hardness $6\frac{1}{2}$–7   Sp. Gr. 3.3–3.5

Summary

Commonly in apple green, tough, fine-grained masses, often with white or lighter green spots. The name jade is given to two minerals—the pyroxene (*jadeite*) and the less valuable amphibole (*nephrite*). For centuries, the Chinese and Japanese have greatly prized jade. The Chinese traditionally associate it with five cardinal virtues: charity, modesty, courage, justice, and wisdom. Jadeite has long been known to occur in stream-worn boulders in Upper Burma, in Tibet, and in Yünnan province in China. Carved jadeite objects have been found among the Mayan ruins of Mexico and Guatemala. For a long time the origin of jade was a mystery, but recently jadeite was found in place in Guatemala associated with serpentine. *Color:* usually light to emerald green; sometimes white, violet, pink, yellowish to reddish-brown. *Luster:* vitreous to waxy. *Streak:* white. *Cleavage:* prismatic, but rarely seen because of fine-grained massive nature; translucent to opaque. HARD, TOUGH, GREEN AGGREGATES OF COMPACT FIBERS, OFTEN WITH AREAS OF LIGHTER COLOR, ARE CHARACTERISTIC. May resemble NEPHRITE, CHRYSOPRASE (apple green chalcedony), massive green IDOCRASE (californite), SERPENTINE, or massive green GROSSULARITE GARNET.

Composition. Sodium aluminum silicate, $NaAlSi_2O_6$. Some calcium, magnesium, or iron may be present.

Tests. Fuses easily to a bubbly glass, and gives a strong yellow sodium flame.

Crystallization. Monoclinic; prismatic class, $\frac{2}{m}$. Usually in aggregates of compact fibers forming tough, compact masses. Crystals are extremely rare.

Minerals of Similar Appearance

Nephrite Jade (Plate 43, facing p. 259). Fuses with difficulty; is commonly a dark spinach green color.

Chrysoprase (apple green chalcedony, Plate 35, facing p. 211). Infusible, and has a lower specific gravity than jadeite.

Idocrase (californite, Plate 48, facing p. 306). Sometimes called California jade; is similar to jadeite in color, hardness, and specific gravity, and is thus difficult to distinguish. Small black specks are often included in massive apple green californite.

Serpentine. Softer than jadeite (can be scratched with a knife blade).

Grossularite Garnet. Sometimes called South African jade; has a specific gravity of 3.5 which is higher than most jadeite.

Occurrence. The origin of jadeite is somewhat uncertain, but it is generally believed to have been formed by the high-temperature-pressure metamorphism of igneous rocks containing such feldspathoids as nepheline. It occurs as stream-worn boulders in Upper Burma. Also in Tibet, and in Yünnan province in China. Recently, jadeite has been found in place in Burma, Japan, Guatemala, and in California. The California jadeite is white to grayish. Green jade is found near Jeffery City, Wyoming.

Use. As a gem and ornamental material.

SPODUMENE

LiAlSi$_2$O$_6$ Monoclinic
    Hardness 6½–7   Sp. Gr. 3.1–3.2

Summary

Commonly in long, flat, striated crystals with steep terminations (Plate 42, facing p. 258). Crystals are often large, and frequently have a wood-like surface alteration. Usually in pegmatites associated with tourmaline, lepidolite, beryl, quartz, and feldspar. *Color:* white, buff, grayish; transparent varieties are colorless, violet, pink, yellow, yellow-green, emerald green. *Luster:* vitreous. *Streak:* white. *Cleavage:* perfect prismatic, intersecting at 87° and 93°. *Parting:* good parallel to front pinacoid; transparent to translucent; brittle; sometimes luminesces red, orange, or yellow. CRYSTAL HABIT, PRISMATIC CLEAVAGE, AND PEGMATITE ASSOCIATIONS ARE DISTINCTIVE.

Gem Varieties

Kunzite. Pink, violet, purplish (Plate 42(*E*), facing p. 258). The cleavage surfaces of some pink kunzite from Brazil are covered with etch pits from which needle-like cavities, frequently with sharp rectangular cross sections, extend into the crystal. Under certain conditions etch solutions may dissolve holes in crystals at the centers of growth spirals (cores of screw dislocations), that are chemically more reactive than the more perfect areas of the crystals. The color of kunzite from Minas Gerais, Brazil, is often improved by heat treatment. Kunzite crystals occur in pegmatites in Pala, California, and in Madagascar.

Hiddenite. Green. Small emerald green hiddenite crystals occur in Alexander County, North Carolina.

Yellow and **yellowish-green** gem varieties of spodumene are found in pegmatites in Minas Gerais, Brazil.

Composition. Lithium aluminum silicate, LiAlSi$_2$O$_6$ (8.0% Li$_2$O, 27.4% Al$_2$O$_3$, 64.6% SiO$_2$). Some sodium may partially replace lithium. Small amounts of iron or calcium may also be present.

Tests. Fuses to a clear glass, and gives a bright crimson flame (lithium). Insoluble in acids.

Crystallization. Monoclinic; prismatic class, $\frac{2}{m}$. Commonly in large flattened vertically striated elongated crystals with steep terminations. Also commonly in cleavable masses.

Alteration. Alters to clay, albite, and muscovite.

Occurrence. Found in granite pegmatites associated with other lithium-bearing minerals like lepidolite and amblygonite; also with tourmaline, quartz, feldspar, beryl, and muscovite. Giant crystals, some over 40 ft. long, occur in the pegmatites of the Black Hills, South Dakota. Spodumene is mined at Kings Mountain, North Carolina, and in Colorado. Also found in New England pegmatites in Maine, in Connecticut, and in Massachusetts. Gem varieties occur in California, in Brazil, in Madagascar, and in North Carolina.

Use. An important ore of lithium. As a gem material.

RHODONITE

Mn(SiO$_3$)   Triclinic
    Hardness 5½–6   Sp. Gr. 3.4–3.7

Summary

Commonly in pink to rose red, rough, tabular crystals with some black manganese oxide surface alteration (Plate 41(*D*), facing p. 243, and Plate 7(*D*), facing p. 50). Also massive. *Color:* pink to rose red, gray; commonly with black alteration veinlets. *Luster:* vitreous. *Streak:* white. *Cleavage:* prismatic intersecting at 88° and 92°; transparent to translucent. ROSE TO PINK COLOR WITH BLACK ALTERATION, TABULAR CRYSTALS, AND CLEAVAGE ARE CHARACTERISTIC. Resembles pink RHODOCHROSITE, $MnCO_3$, but rhodochrosite is softer (H. $= 3\frac{1}{2}$–4), effervesces in warm HCl, and has perfect rhombohedral cleavage.

Composition. Manganese silicate, $Mn(SiO_3)$ (54.1% MnO, 45.9% SiO_2). Some zinc, calcium, or iron may be present.

Tests. Fuses to a blackish glass. In the oxidizing flame (Mn test), gives a violet color to borax and salt of phosphorus bead.

Crystallization. Triclinic; pinacoidal class, $\bar{1}$. Rough, tabular crystals are common. Also commonly in cleavable to compact masses.

Alteration. To pyrolusite, MnO_2.

Occurrence. Large crystals, some over 7 in. long, occur in calcite associated with the zinc minerals willemite, zincite, and franklinite at Franklin and Sterling Hill, New Jersey. Compact rose-colored masses with black alteration veinlets occur in the Ural Mountains, U.S.S.R., and are polished and used as decorative stone. Other notable localities include Långban, Sweden; Broken Hill, New South Wales; and Minas Gerais, Brazil.

Use. As an ornamental stone.

PECTOLITE

$Ca_2NaH(SiO_3)_3$　　　　　　　　　　　　Triclinic
　Hardness 5　　　　　　　　　　Sp. Gr. 2.7–2.8

Summary

Commonly in rounded white aggregates composed of very fine radiating needle-like crystals with a silky luster (Plate 41(*E*), facing p. 243). Brittle pectolite needles easily penetrate the skin, and therefore specimens should be handled with care. Occurs in cavities in basalt associated with zeolites, calcite, prehnite, and datolite. *Color:* white, gray, yellowish, pink. *Luster:* vitreous to silky. *Streak:* white. *Cleavage:* perfect front pinacoid and basal; translucent. Commonly luminesces orange or yellow under ultraviolet light. RADIATING NEEDLE-LIKE AGGREGATES (Plate 41) AND MINERAL ASSOCIATIONS ARE CHARACTERISTIC. Resembles fibrous zeolites, such as NATROLITE, but pectolite often occurs in finer needles, forming rounded masses.

Composition. Hydrous calcium sodium silicate, $Ca_2NaH(SiO_3)_3$. May contain some manganese, magnesium, potassium, or iron.

Tests. Fuses easily to a white enamel. Decomposed in HCl. Yields water in closed tube.

Crystallization. Triclinic; pinacoidal class, $\bar{1}$. Usually in rounded masses composed of radiating needle-like crystals. Sometimes massive.

Occurrence. Commonly occurs in cavities in basalts associated with zeolites, calcite, prehnite, and datolite. Common in trap rocks of Paterson, New Jersey.

WOLLASTONITE

$Ca(SiO_3)$　　　　　　　　　　　　　　Triclinic
　Hardness $4\frac{1}{2}$–5　　　　　　Sp. Gr. 2.8–2.9

Summary

Commonly in white fibrous aggregates of elongated crystals that resemble tremolite (Plate 41(*F*), facing p. 243). Usually occurs in contact metamorphosed limestones, frequently associated with diopside, idocrase, garnet, tremolite, epidote, and calcite. *Color:* white to colorless, gray, pinkish, yellowish, brownish. *Luster:* vitreous to silky. *Streak:* white. *Cleavage:* perfect pinacoidal intersecting at 84° and 96°; translucent; often luminesces yellow, orange, blue, green, or pink under ultraviolet light. WOLLASTONITE IS DISTINGUISHED FROM WHITE FIBROUS AGGREGATES OF TREMOLITE BY ITS CLEAVAGE AND SOLUBILITY IN HCl. Tremolite is insoluble in acid.

Composition. Calcium silicate, $CaSiO_3$. May contain iron, manganese, strontium, or magnesium.

Tests. Fuses to a white globule. Decomposed by hydrochloric acid. Commonly luminescent.

Crystallization. Triclinic; pinacoidal class, $\overline{1}$. Usually in fibrous aggregates of elongated crystals. Individual tabular crystals are rare. Also in cleavable and compact masses.

Occurrence. Found in contact metamorphosed limestones associated with diopside, tremolite, epidote, garnet, idocrase, and calcite. Good luminescent specimens of wollastonite occur at Franklin, New Jersey. Occurs in large masses at Willsboro, New York. Crystals of the rare monoclinic polymorph PARAWOLLASTONITE occur in altered limestone blocks thrown out by the volcanic eruptions at Monte Somma, Vesuvius.

Use. In ceramics and paints.

CHRYSOCOLLA

$CuSiO_3 \cdot 2H_2O$	Cryptocrystalline (Orthorhombic ?)
Hardness 2–4 (harder when impregnated with quartz)	Sp. Gr. 2–2.3

Summary

Commonly in blue-green compact masses resembling turquoise (Plate 42(F), facing p. 258). Occurs as a secondary copper mineral in the oxidized zone of copper deposits, frequently associated with malachite, azurite, cuprite, and native copper. *Color:* green, blue-green, sky blue; may be brownish or blackish when impure. *Luster:* vitreous to earthy. *Fracture:* conchoidal; sectile to brittle; translucent to opaque. BLUE-GREEN ENAMEL-LIKE BOTRYOIDAL MASSES, OFTEN WITH BLACKISH VEINLETS, ARE CHARACTERISTIC. Distinguished from TURQUOISE by inferior hardness, except when impregnated with quartz. May also resemble VARISCITE, but variscite and turquoise both give phosphate tests.

Composition. Essentially a hydrous copper silicate, $CuSiO_3 \cdot 2H_2O$, but composition varies greatly, and many specimens are impure.

Tests. Infusible. Gives tests for copper. Blackens and yields water in closed tube.

Crystallization. Cryptocrystalline. Possibly orthorhombic. Usually in compact or botryoidal masses with an enamel-like or earthy texture. Needle-like crystals have been found in Mackay, Idaho.

Alteration. Alters from such copper minerals as bornite and chalcopyrite.

Occurrence. Usually as a secondary mineral in the oxidized zone of copper veins; associated with native copper, azurite, malachite, cuprite, and limonite. Good specimens have come from Arizona, New Mexico, Chile, Africa, the U.S.S.R., and England.

Use. As a minor ore of copper. Also as an ornamental stone.

GARNIERITE

$(Ni,Mg)SiO_3 \cdot nH_2O$	Monoclinic (?)
Hardness 2–3	Sp. Gr. 2.5

Summary

Usually as green encrustations or earthy masses (Plate 42(C), facing p. 258); probably formed as an alteration product of nickel-bearing, basic igneous rocks. Found in serpentine rocks associated with chromite, olivine, talc, and chlorite. *Color:* apple green to emerald green; sometimes white. *Luster:* earthy; sometimes in pebble-like masses with a glazed surface. *Streak:* white to pale green. *Fracture:* conchoidal to earthy. GREEN COLOR, MASSIVE HABIT, AND MINERAL ASSOCIATIONS ARE CHARACTERISTIC. Pale apple green encrustations may resemble ANNABERGITE, $Ni_3(AsO_4)_2 \cdot 8H_2O$, but annabergite is fusible and occurs wth cobalt-nickel-silver ores.

Composition. Hydrous nickel magnesium silicate, $(Ni,Mg)SiO_3 \cdot nH_2O$, but composition varies widely.

Tests. Infusible, but blackens and becomes magnetic on heating. Soluble in hot HCl, but does not yield a silica gel on evaporation. Gives test for nickel with dimethylglyoxime. Gives water in closed tube.

Crystallization. Perhaps monoclinic. Crystals are unknown. Usually in earthy masses, in encrustations, or in glazed pebble-like masses.

Occurrence. Associated with serpentine and chromite in New Caledonia; in the Transvaal; in Madagascar; in the U.S.S.R.; in Cuba. In the United States, it is found in Oregon and North Carolina.

Use. An ore of nickel.

Amphibole Group

The amphiboles comprise a large group of minerals and are difficult to distinguish from each other by sight. However, it is possible to distinguish the amphiboles from the pyroxenes, which they closely resemble. The two cleavages of the amphiboles intersect at about 56° and 124°, forming a rhombic pattern that is easily distinguishable from the almost right-angular cleavage of the pyroxenes (see Figure 8-56). The amphibole minerals are fibrous or prismatic. The prismatic crystals are frequently elongated and six-sided. Hornblende is a common amphibole, often found in igneous rocks. The anthophyllite and tremolite-actinolite series are found in metamorphic rocks. The following amphiboles are described:

TREMOLITE-ACTINOLITE SERIES	Monoclinic
HORNBLENDE SERIES	Monoclinic
RIEBECKITE (crocidolite variety)	Monoclinic

TREMOLITE, ACTINOLITE

$Ca_2Mg_5Si_8O_{22}(OH)_2$;   Monoclinic
$Ca_2(Mg,Fe)_5Si_8O_{22}(OH)_2$
    Hardness 5–6   Sp. Gr. 3.0–3.35

Summary

Commonly in fibrous aggregates of long, prismatic crystals (Plate 43(*A, B, C*), facing p. 259). *Color:* white in *tremolite;* becomes green with increasing iron content in *actinolite;* pink, violet, or rose in manganiferous varieties of tremolite (hexagonite). *Luster:* vitreous to silky. *Streak:* white. *Cleavage:* perfect prismatic, intersecting at 56° and 124°; transparent to translucent. Varieties of tremolite include a matted intergrowth of asbestiform fibers called *mountain leather;* and also a tough, compact, massive form of tremolite or actinolite called *nephrite.* Nephrite is commonly spinach green in color, and is one of the true jade minerals. The other jade mineral is the pyroxene known as jadeite. FIBROUS HABIT AND CLEAVAGE ARE CHARACTERISTIC. Fibrous white tremolite strongly resembles WOLLASTONITE, but tremolite is insoluble in hot HCl.

Composition. An isomorphous series extends between tremolite and actinolite. Tremolite is the white, low-iron end of the series, and actinolite is the green iron-bearing member. May contain some manganese.

Tests. Fusible with some difficulty. Insoluble in acid.

Alteration. Tremolite alters to talc. Actinolite is an alteration product of pyroxenes, and alters to serpentine, chlorite, and epidote.

Occurrence. *Tremolite* is commonly found in metamorphosed dolomitic limestones, and in some talc schists. Associated minerals include calcite, garnet, diopside, epidote, and wollastonite. Lilac *hexagonite,* the manganiferous variety of tremolite, occurs in St. Lawrence County, New York. Green *actinolite* is common in schists. *Nephrite* jade comes from Turkestan; from New Zealand (Plate 43); and from Lander, Wyoming.

Use. The asbestiform varieties of tremolite are used, to a limited extent, as asbestos. Nephrite is a gem mineral.

HORNBLENDE

$NaCa_2(Mg,Fe,Al)_5(Si,Al)_8O_{22}(OH)_2$ Monoclinic
    Hardness 5–6   Sp. Gr. 2.9–3.4

Summary

Commonly in black or blackish-green elongated, six-sided prismatic crystals (Figure 8-57). *Color:* dark

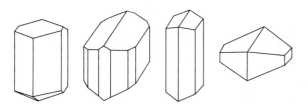

Fig. 8-57. Hornblende crystals.

green to black. *Luster:* vitreous. *Streak:* white. *Cleavage:* perfect prismatic at 56° and 124°; translucent. Hornblende is a common and widespread igneous and metamorphic rock-forming mineral. Often forms pseudomorphs after pyroxene, called *uralite.* ELONGATED BLACKISH CRYSTALS AND AMPHIBOLE CLEAVAGE ARE CHARACTERISTIC, AND MAKE THEM DISTINGUISHABLE FROM SQUARISH PYROXENE CRYSTALS. Black TORUMALINE crystals have very poor cleavage.

Composition. The hornblende series are complex silicates. Various isomorphous members of the series are *hornblende, pargasite, arfvedsonite, glaucophane,* and *riebeckite.*

Tests. Fuses with difficulty. Gives some water in closed tube.

Crystallization. Monoclinic; prismatic class, $\frac{2}{m}$. Commonly in six-sided prismatic crystals, often terminated by a low dome. Also in columnar and bladed masses.

Alteration. Alters from pyroxene into chlorite, biotite, calcite, and epidote.

Occurrence. Hornblende is common in igneous and metamorphic rocks. It is frequently formed from the alteration of pyroxene (uralite). Associated minerals include feldspars, quartz, pyroxenes, chlorite, calcite, and epidote.

CROCIDOLITE

$Na_2Fe_3Fe_2Si_8O_{22}(OH)_2$ Monoclinic

Crocidolite is a blue asbestiform variety of the amphibole known as riebeckite. It occurs in long,

silky, gray-blue fibers (Plate 43(*E*), facing p. 259) and is mined for asbestos in South Africa, Australia, and Brazil. The South African crocidolite is sometimes completely or partially replaced by quartz. The original fibrous structure of the crocidolite is retained by the quartz pseudomorph, but the iron is often oxidized to a gold-brown color. Quartz pseudomorphs after crocidolite are called TIGER'S EYE, and are used as an ornamental stone (Plate 43(*F*), facing p. 259).

RING STRUCTURES

Beryl
Benitoite
Axinite
Tourmaline
Dioptase

BERYL

$Be_3Al_2(Si_6O_{18})$ Hexagonal
     Hardness $7\frac{1}{2}$–8   Sp. Gr. 2.66–2.83

Summary

Beryl is the most important ore of beryllium; at times, may be one of the most beautiful gem minerals. Commonly in large, mottled, blue-green, or yellowish hexagonal prisms with flat (pinacoidal) terminations (Figure 8-58; Plate 44, facing p. 274; and Plate 6(*A*), facing p. 35). Crystals in pegmatites have been known to occur in giant proportions, weighing up to 200 tons. Crystals several feet long are fairly

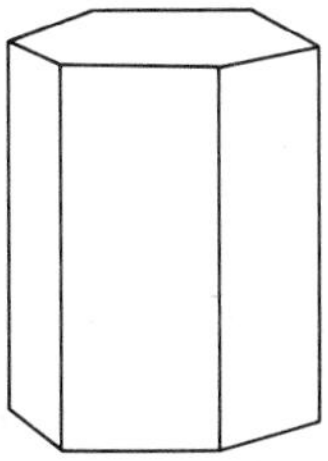

Fig. 8-58. Beryl crystal.

common. *Color:* white, green, blue-green, greenish-yellow, blue, yellow, pink. *Luster:* vitreous. *Cleavage:* imperfect basal. *Streak:* white; transparent to translucent. HARD, GREENISH, SIX-SIDED PRISMATIC CRYSTALS ARE CHARACTERISTIC. OFTEN VERTICALLY STRIATED, WITH A MOTTLED APPEARANCE. Distinguishable from hexagonal APATITE crystals by superior hardness. Apatite (H. = 5) can be scratched by a knife blade.

Transparent Gem Varieties

Emerald. The deep green color is commonly caused by traces of chromium, but chromium is not detected in all specimens. Some emeralds luminesce red or pink, especially synthetic ones.

Aquamarine. Light green, blue-green, blue. Most blue aquamarines have been heat treated. Greenish aquamarines turn blue when heated to about 400°C. Chromium is often not detected in aquamarines; the color has been attributed to the presence of small amounts of ferrous and ferric iron, and to color centers produced by structural defects.

Pink Beryl (Morganite). Frequently occurs in flat tabular crystals. The pink color has been attributed to the presence of lithium or cesium.

Golden Beryl (Heliodor). Clear yellow beryl.

Composition. Beryllium aluminum silicate, $Be_3Al_2(Si_6O_{18})$ (14.0% BeO, 19.0% Al_2O_3, 67.0% SiO_2). Usually contains some alkali ions such as Na, Li, K, and Cs. Iron, calcium, magnesium, or chromium may be present in small amounts; also some water and some gases.

Tests. Whitens and fuses with difficulty. Insoluble in acids.

Crystallization. Hexagonal; dihexagonal-dipyramidal class, $\frac{6}{m}\frac{2}{m}\frac{2}{m}$. Commonly in well-formed hexagonal prisms. Often vertically striated. Also massive and in columnar aggregates (Figure 7-12, p. 134). Pink beryl is usually massive or in flat tabular crystals.

Alteration. Commonly alters to kaolinite and muscovite. Sometimes replaced by phenakite (Be_2SiO_4). At 900°C and at higher temperatures, synthetic beryl slowly decomposes to phenakite and glass. Specimens of natural beryl have melted incongruently to liquid + phenakite at about 1475°C.[13]

Occurrence. Beryl usually occurs in pegmatites or high-temperature hydrothermal veins. Associated minerals commonly include quartz, feldspar, tourmaline, mica, cassiterite, lepidolite, spodumene, amblygonite, and columbite-tantalite.

The principal source of emeralds is Muzo, Colombia; they occur there in calcite veins in black silty limestone. In the Ural Mountains, U.S.S.R., emeralds occur with phenakite and chrysoberyl in biotite schist. Also in biotite schist at Habachtal, Salzburg Alps, Austria. In Salininha, Bahia, Brazil emerald crystals occur in decomposed limestone and soft clay.

Interesting pink and blue zoned beryl crystals, and large aquamarine crystals are found in decomposed pegmatites in Minas Gerais, Brazil. In the United States, large beryl crystals occur in pegmatites in the Black Hills of South Dakota; in San Diego County, California; in Grafton, New Hampshire; and in New Mexico. Pink beryl (morganite) is common in Madagascar and in Brazil. Peach colored morganite crystals, up to 4 × 6 inches have recently been discovered at the White Queen Mine in Pala, California.

Use. Major ore of beryllium; also as a gem mineral. (*Bertrandite* in the Spoor Mt. area of Utah is a source of beryllium.)

BENITOITE

$BaTiSi_3O_9$	Hexagonal
Hardness 6–6$\frac{1}{2}$	Sp. Gr. 3.6

Benitoite is a rare mineral found in San Benito County, California. Crystals and mineral associations are highly distinctive. THE CRYSTALS ARE FLAT TRIANGULAR, WITH AN UNEVEN SAPPHIRE BLUE AND WHITE COLOR DISTRIBUTION. They luminesce blue under ultraviolet light and are associated with black elongated NEPTUNITE crystals, $(Na,K)(Fe,Mn,Ti)(Si_2O_6)$, (Plate 44(*E*), facing p. 274). The benitoite and neptunite crystals occur in compact natrolite veins in green serpentine and schist. Benitoite crystals have a vitreous luster, poor prismatic cleavage, and are transparent to translucent. Sometimes used as a minor gem mineral.

AXINITE

$(Ca,Mn,Fe)_3Al_2BO_3(Si_4O_{12})(OH)$ Triclinic

Hardness $6\frac{1}{2}$–7   Sp. Gr. 3.26–3.36

Summary

Commonly in thin violet-brown, wedge-shaped crystals (Plate 44(F), facing p. 274). Occurs as a high-temperature mineral in cavities in granite and in contact metamorphic rocks. *Color:* violet-brown, clove brown, green, gray, rarely blue; manganese varieties are yellow, orange-red or rose. *Luster:* vitreous. *Streak:* white. *Cleavage:* one good, three poor; transparent to translucent. Manganese-rich varieties are sometimes luminescent. VIOLET-BROWN, THIN WEDGE-SHAPED STRIATED CRYSTALS ARE DISTINCTIVE. Distinguished from SPHENE crystals by greater hardness (H. sphene = 5–$5\frac{1}{2}$).

Composition. Hydrous calcium, manganese, iron, aluminum borosilicate, $(Ca,Mn,Fe)_3Al_2BO_3(Si_4O_{12})(OH)$. Some Mg and Ti may be present.

Tests. Fuses easily to a bubbly glass. Gives some water in closed tube.

Crystallization. Triclinic; pinacoidal class, $\bar{1}$. Commonly in thin, wedge-shaped crystals. Also massive.

Occurrence. Commonly found in contact-metamorphosed impure limestones. Associated minerals include calcite, pyroxene, quartz, epidote, tourmaline, zoisite, actinolite, andradite garnet, and prehnite. Outstanding crystals occur at Bourg d'Oisans, France; at Obira, Japan; and at Cornwall, England. In the United States, they occur at Riverside, California; at Luning, Nevada; and at Franklin, New Jersey. The manganese-rich Franklin crystals are red to orange-yellow in color.

Use. As a minor gem stone.

TOURMALINE

$Na(Mg,Fe,Mn,Li,Al)_3Al_6(Si_6O_{18})(BO_3)_3(OH,F)_4$

Hexagonal

Hardness 7–$7\frac{1}{2}$ Sp. Gr. 3.03–3.25

Summary

Commonly found as vertically striated elongated crystals with rounded triangular cross sections (Figure 7-22, p. 136; Plate 45, facing p. 275) in pegmatites and high-temperature veins. The color of tourmaline is extremely variable. Color zoning is common; it is usually parallel either to the length or to the base of the crystal. Watermelon tourmaline, for example, has a green exterior and a pink to rose red core. If the color banding is parallel to the base, a crystal may either be red at one end and green at the other or colorless and black at opposite ends (Plate 45). *Color:* varies with the composition. Many colored varieties have special names: the black tourmaline variety (SCHORLITE) is rich in iron; the pink to rose (RUBELLITE) is low in iron and contains divalent manganese ions, which apparently cause the pink color[2]; the green (VERDELITE) contains ferrous iron; and the brown (DRAVITE) is magnesium-rich. Tourmaline may also be blue (INDICOLITE), colorless, or yellow. *Luster:* vitreous. *Streak:* white. *Cleavage:* very indistinct; transparent to translucent; piezoelectric. CRYSTALS WITH ROUNDED TRIANGULAR CROSS SECTIONS ARE HIGHLY DISTINCTIVE. The very poor cleavage of tourmaline distinguishes it from black HORNBLENDE crystals. Green BERYL has a lower specific gravity. Brown tourmaline occurs in marble with brown ANDRADITE and with CHONDRODITE grains to which it may bear some resemblance.

Composition. Complex aluminum silicate, $Na(Mg,Fe,Mn,Li,Al)_3Al_6(Si_6O_{18})(BO_3)_3(OH,F)_4$. The composition ranges widely, forming the numerous colored varieties listed earlier. It is so complex that John Ruskin remarked in 1890 that "the chemistry of it is more like a medieval doctor's prescription than the makings of a respectable mineral."

Tests. Fusibility varies with composition. Brown, magnesium-rich varieties are the easiest to fuse. Black, iron-rich varieties fuse with difficulty, and the lithium-rich varieties (blue, green, pink) are infusible. Pyro-electric (see p. 93).

Crystallization. Hexagonal; ditrigonal-dipyramidal class, 3 *m*. Usually in crystals with three curved, convex, striated sides that are formed by combinations of prism faces which round into each other. Common forms on tourmaline crystals include the trigonal prism, the hexagonal prism, the rhombohedron, and the pedion (Plate 6, facing p. 35). In doubly termi-nated crystals, the presence of a pedion at one end and a rhombohedron at the other shows that the crys-tal is hemimorphic and lacks a center of symmetry. Tourmaline sometimes occurs in radiating groups of slender prismatic to needle-like crystals called "tour-maline suns." Also massive.

Alteration. To biotite, to muscovite, and to chlorite.

Occurrence. Tourmaline is a high-temperature mineral commonly found in pegmatites and hydrothermal veins. Sometimes found in gneisses, schists, or in con-tact-metamorphosed limestones. In pegmatites, asso-ciated minerals include quartz, muscovite, lepidolite, feldspar, topaz, beryl, and apatite. Good crystals showing interesting rose, green, white, and black color zones occur in pegmatites in Minas Gerais, Brazil; Madagascar; and Pala, San Diego County, California. The Madagascar specimens, some over seven inches long, show in cross section remarkable triangular and hexagonal color zones in concentric arrangement (see Figure 2-29, p. 22). Gem varieties of tourmaline also occur in pegmatites in Maine; Mozambique; on the island of Elba; and in the Ural Mountains. Brown, magnesium-rich tourmalines occur in St. Lawrence County, New York, in crystalline limestones. Black, fine fibrous masses of tourmaline are found at Majuba Hill, Nevada.

Use. As a gem stone, and in the manufacture of pres-sure gauges.

DIOPTASE (Copper Emerald)

$Cu_6Si_6O_{18} \cdot 6H_2O$	Hexagonal
Hardness 5	Sp. Gr. 3.3

Summary

Dioptase is an uncommon but very attractive mineral. It usually occurs as bright emerald green, short, prismatic, and rhombohedral crystals. It is found in arid regions in the oxidized zone of copper deposits. *Color:* emerald green. *Luster:* vitreous. *Cleavage:* perfect rhombohedral; transparent to translucent. Associated minerals include malachite, brochantite, chrysocolla, calcite, quartz, and limonite. DISTINGUISHED FROM OTHER GREEN COPPER MINERALS BY CRYSTALS (Figure 8-59).

Composition. Hydrous copper silicate, $Cu_6Si_6O_{18} \cdot 6H_2O$.

Tests. Infusible. Gives tests for copper. Yields water in closed tube.

Crystallization. Hexagonal; rhombohedral class, $\bar{3}$. Com-monly in short prismatic crystals terminated by rhom-bohedrons. Also rhombohedral. Less frequently in long, slender, prismatic, or needle-like crystals.

Occurrence. Occurs in the oxidized zone of copper de-posits and in limestone. Excellent crystals, some two inches in length come from Mindouli in the Niari River basin in the Congo. Good specimens are also found in Tsumeb, South-West Africa; there the crys-tals reach about an inch in length. Dioptase also occurs in Kirghiz Steppes, U.S.S.R.; in Chile; and in the United States (at the Mammoth Mine, Tiger, Pinal County, Arizona; and in Clifton, Graham County, Arizona).

Use. Minor gem mineral.

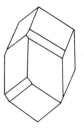

Fig. 8-59. Dioptase.

UNIT STRUCTURES

Topaz
Phenakite
Willemite
Zircon
Olivine Series
Chondrodite
Garnet Group
  Almandine
  Pyrope
  Andradite
  Grossularite
  Spessartite
  Uvarovite
Datolite

TOPAZ

$Al_2(SiO_4)(OH,F)_2$	Orthorhombic
Hardness 8	Sp. Gr. 3.49–3.57

Summary

Topaz is a high-temperature mineral commonly found in granite pegmatites in large prismatic crystals with perfect basal cleavage (Plate 46(*A*, *B*), facing p. 290, and Figure 8-60). It occurs in a wide variety of attractive colors, which, together with its great hardness, make topaz an attractive gem mineral. *Color:* colorless, white, light blue, light green, pink, light yellow, wine-yellow, yellow-brown, pinkish-brown, violet, wine red. The color of some topaz fades on exposure to heat or sunlight. Yellow topaz from Brazil turns a beautiful pink to rose color when heated from about 300°C to 450°C. Natural pink topaz is fairly rare, and occurs in Minas Gerais, Brazil, and in Sankara, the U.S.S.R. *Luster:* vitreous. *Streak:* white. *Cleavage:* perfect basal; transparent to translucent. CRYSTALS, GREAT HARDNESS, BASAL CLEAVAGE, AND RELATIVELY HIGH SPECIFIC GRAVITY ARE DISTINCTIVE. High specific gravity and crystals distinguish topaz from QUARTZ (Sp. Gr. 2.65) and BERYL (Sp. Gr. 2.66–2.83). Heat-treated smoky or amethyst quartz turns yellow
and is sometimes called *smoky topaz, citrine topaz,* or *false topaz.* The names, however, are unfortunate; they should not be confused with the mineral topaz. Similarly, yellow corundum is sometimes called *oriental topaz,* and honey yellow andradite garnet may be called *topazolite.*

Composition. Aluminum fluosilicate, $Al_2SiO_4(F,OH)_2$.

Tests. Infusible. Insoluble in acids.

Crystallization. Orthorhombic; dipyramidal class, $\frac{2}{m}\frac{2}{m}\frac{2}{m}$. Commonly in prismatic crystals, often vertically striated. Crystals are sometimes very large, some weighing hundreds of pounds. Also in fine- to coarse-grained masses.

Alteration. To sericite, fluorite, kaolinite.

Occurrence. Topaz occurs in pegmatites, in high-temperature hydrothermal veins, and in granites and cavities in rhyolite. Associated minerals include cassiterite, wolframite, quartz, tourmaline, beryl, phenakite, molybdenite, apatite, mica, and fluorite. Extremely large (600 lb.), transparent light blue, and colorless crystals have been found along with other gem varieties in Minas Gerais, Brazil. Fine, light blue, and deep wine-yellow crystals occur in Siberia. Gem topaz crystals also occur in Ceylon and Burma. The best blue crystals come from Mursinka in the Urals. Interesting brownish to peach-colored crystals that fade on exposure to sunlight occur in cavities in rhyolite in the Thomas Range of Utah. Crystals also occur in San Diego County, California; at Topsham, Maine; in Mason County, Texas; and at Pike's Peak, Colorado. Massive topaz commonly occurs with cassiterite

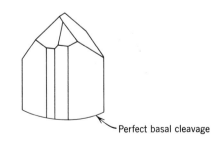
Perfect basal cleavage

Fig. 8-60. Topaz crystal.

and wolframite in Cornwall, England, and in Burnt Hill, New Brunswick, Canada.

Use. An important gem mineral.

PHENAKITE

Be$_2$SiO$_4$ Hexagonal
    Hardness 7$\frac{1}{2}$–8   Sp. Gr. 3.0

Summary

Commonly in colorless flat complex rhombohedral crystals (Plate 46(*D*), facing p. 290). Crystals are sometimes prismatic and vertically striated. Phenakite strongly resembles quartz, and is named from the Greek word for a *deceiver*. It is a high-temperature mineral found with beryl and chrysoberyl in pegmatites. *Color:* colorless or white; sometimes bluish. Bell Telephone Company has synthesized blue needle-like crystals of phenakite, pigmented blue by vanadium. *Luster:* vitreous. *Streak:* white. *Cleavage:* poor prismatic; transparent to translucent. FLAT COMPLEX CRYSTALS ARE CHARACTERISTIC (Figure 8-61). QUARTZ has a lower specific gravity, and TOPAZ has perfect basal cleavage.

Composition. Beryllium silicate, Be$_2$SiO$_4$ (45.6% BeO, 54.4% SiO$_2$). May contain minor amounts of Ca, Mg, Al, or Na.

Tests. Infusible. Insoluble in acids.

Crystallization. Hexagonal; rhombohedral class, $\bar{3}$. Usually in complex rhombohedral crystals. Sometimes prismatic. Often twinned.

Alteration. From beryl (see beryl, alteration, p. 284).

Fig. 8-61. Phenakite.

Occurrence. Phenakite is an uncommon mineral usually found in pegmatites. Associated minerals include beryl, chrysoberyl, topaz, tourmaline, fluorite, feldspar, and quartz. Large flat crystals, up to 3 in. across, occur in São Miguel de Piraçicaba, Minas Gerais, Brazil. Large prismatic crystals occur in Kragerö, Norway. Also found with emeralds and chrysoberyl in mica schist in the Ural Mountains, the U.S.S.R. In the United States, phenakite occurs in Colorado, in Maine, and in New Hampshire. It has also been reported from eastern Nevada.

Use. Source of beryllium. Minor gem mineral.

WILLEMITE

Zn$_2$SiO$_4$ Hexagonal
    Hardness 5$\frac{1}{2}$ Sp. Gr. 3.9–4.2

Summary

Commonly in yellow-green to white granular masses that frequently luminesce bright green in ultraviolet light (Plate 46(*C*), facing p. 290, and Plate 9, facing p. 66). Associated with white calcite that luminesces pink to bright red, and black franklinite at Franklin, New Jersey. *Color:* yellow-green, lemon yellow, white, reddish-brown, flesh, orange, light blue, black. *Luster:* vitreous. *Streak:* white. *Cleavage:* basal; transparent to translucent. BRIGHT GREEN LUMINESCENCE AND ASSOCIATION WITH DEEP RED ZINCITE, CALCITE, AND FRANKLINITE ARE CHARACTERISTIC OF WILLEMITE FROM FRANKLIN, NEW JERSEY. Manganese replaces some of the zinc ions in luminescent willemite.

Composition. Zinc silicate, Zn$_2$SiO$_4$ (73.0% ZnO, 27.0% SiO$_2$). Manganese may replace zinc in considerable amounts. Small amounts of iron may be present. *Troostite* is a highly manganiferous variety of willemite and is commonly brownish-red.

Tests. Frequently luminesces bright green. Infusible, but not the highly manganiferous varieties (troostite) (Plate 46(*F*), facing p. 290). Soluble in HCl. Gives test for zinc.

Crystallization. Hexagonal; rhombohedral class, $\bar{3}$. Usually massive. Crystals are either short to long six-sided prisms or rhombohedrons. Crystals are usually small, except at Franklin, New Jersey, where large prismatic troostite crystals occur in calcite (Plate 46).

Occurrence. Willemite is a major ore of zinc at Franklin-Ogdensburg, New Jersey, where it occurs in crystalline limestones with zincite, franklinite, and calcite. Large troostite crystals, some 6 in. long, have been found. Willemite also occurs in the oxidized zone of zinc bearing deposits.

Use. An ore of zinc. Minor gem mineral.

ZIRCON

ZrSiO$_4$ Tetragonal
    Hardness $7\frac{1}{2}$ Sp. Gr. 4.6–4.7

Summary

Zircon is a common accessory mineral in igneous rocks; because it is very resistant to chemical and mechanical disintegration, it is often found as a residual mineral in beach sands and gravels. Small prismatic crystals terminated by bipyramids are common (Plate 46(*E*), facing p. 290, Plate 5, facing p. 34, and Figure 8-62). *Color:* usually shades of brown, but may be colorless, yellow, green, red, orange, or blue (usually after heat treatment). Blue zircons, which are used as a gem material, are produced by heat treating brownish zircons, but in sunlight they sometimes tend to fade or revert to brown. Some brown or reddish zircons may be rendered colorless by the proper heat treatment; and orange to red gem zircons (*hyacynth*) are usually heat treated to improve the color. Green or greenish-brown zircons are usually *metamict,* which is a glassy amorphous state caused by radiation damage from included uranium and thorium ions in the zircon structure. *Luster:* adamantine. *Streak:* white. *Cleavage:* two poor; transparent to translucent. Commonly luminesces orange in ultraviolet light. Often weakly radioactive. BROWN SQUARE PRISMS TERMINATED BY BIPYRAMIDS AND FREQUENT ORANGE LUMINESCENCE ARE

CHARACTERISTIC. May resemble brown IDOCRASE crystals, but idocrase is fusible and has a lower specific gravity (3.4–3.5).

Composition. Zirconium silicate, ZrSiO$_4$. Always contains some hafnium, normally about one percent. Small amounts of uranium, thorium, and rare earths also substitute in part for zirconium. Some phosphorus may be present, probably replacing silicon.

Tests. Infusible. Insoluble in acids. Sometimes luminesces orange to orange-yellow in ultraviolet light.

Crystallization. Tetragonal; ditetragonal-dipyramidal class, $\frac{4}{m}\frac{2}{m}\frac{2}{m}$. Commonly in small prismatic crystals terminated by bipyramids. Some large crystals up to a foot long have been found. Also in irregular grains.

Alteration. Resistant to change.

Occurrence. Zircon is a common accessory mineral in igneous rocks, notably granites, syenites, nepheline-syenites, and diorites. Also in crystalline limestones, gneisses, schists; because of its great resistance to mechanical and chemical alteration, it is frequently found in heavy beach sands and gravels. Common in the beach sands of Queensland, Australia; Florida, U.S.A.; Brazil; and India. These crystals are usually small. However large zircon crystals have been found in Madagascar, and some up to a foot long have been found in Renfrew County, Ontario. Most gem zircons are found in the Mongka district of Indo-China, and in Ceylon. Rare light green transparent zircon crystals occur in Minas Gerais, Brazil. In the United States, small crystals (Plate 46) are abundant in Buncombe and Henderson Counties, North Carolina.

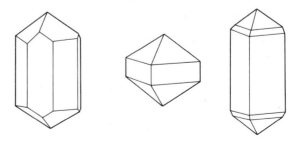

Fig. 8-62. Zircon crystals.

Use. An ore of zirconium, and a source of hafnium. Also as a gem mineral. Utilized in reactors.

OLIVINE

$(Mg,Fe)_2(SiO_4)$ Orthorhombic
        Hardness $6\frac{1}{2}$–7   Sp. Gr. 3.2–4.3

Summary

A common rock-forming mineral, usually found as grains in such dark igneous rocks as basalt, gabbro, or peridotite. Sometimes forms *dunite,* a rock composed solely of olivine (Plate 47(*A*), facing p. 291). Crystals are usually flattened. The transparent gem variety of olivine is called *peridot. Color:* usually olive green; sometimes yellowish, brownish, gray, or reddish. *Luster:* vitreous. *Streak:* white. *Cleavage:* two imperfect; transparent to translucent. OLIVE GREEN COLOR AND OCCURRENCE IN DARK IGNEOUS ROCKS ARE CHARACTERISTIC.

Composition. Magnesium iron silicate, $(Mg,Fe)_2(SiO_4)$. A complete solid solution series exists between the end members FAYALITE, Fe_2SiO_4, and FORSTERITE, Mg_2SiO_4. The name olivine is used for the intermediate members. A solid solution series exists between fayalite and TEPHROITE, Mn_2SiO_4.

Tests. Infusible; except high iron members that fuse to a black magnetic globue. Slowly soluble in HCl, and gelatinizes on evaporation.

Crystallization. Orthorhombic, dipyramidal class, $\frac{2}{m}\frac{2}{m}\frac{2}{m}$. Usually as embedded grains or granular masses. Crystals are commonly flat and consist of prisms, pinacoids, and bipyramids.

Alteration. Alters readily to serpentine, magnesite, and limonite.

Occurrence. In basic igneous rocks such as basalts and gabbros; and in ultrabasic igneous rocks such as peridotites, and dunite. Associated minerals include chromite, spinel, corundum, serpentine, magnetite, garnet, and augite. Also in contact-metamorphosed limestones and in meteorites. Peridot, the gem variety of olivine, occurs on St. John's Island in the Red Sea; there, they are large crystals up to 4 in. across. Also occur in Burma and in the lavas of Vesuvius. Some volcanic bombs composed mainly of granular olivine have been found in the Eifel district of Germany. Large crystals, some 3 in. in diameter, and which have been completely altered to serpentine, occur in Snarum, Norway.

Use. In the manufacture of refractory bricks. As a gem material.

CHONDRODITE

$Mg(OH,F)_2 \cdot 2MgSiO_4$ Monoclinic
        Hardness 6–$6\frac{1}{2}$ Sp. Gr. 3.1–3.2

Summary

Chondrodite is found in dolomitic marbles, usually as yellow to reddish-brown grains. It is the most common member of the HUMITE GROUP (norbergite, chondrodite, humite, clinohumite), whose members cannot be distinguished from each other without special tests. *Color:* yellow, red, reddish-brown. *Luster:* vitreous to resinous. *Streak:* white. *Cleavage:* poor basal; transparent to translucent. YELLOW TO YELLOWISH-BROWN EMBEDDED GRAINS IN CRYSTALLINE LIMESTONES ARE CHARACTERISTIC (Plate 47(*B*), facing p. 291). Occurs with and resembles grains of yellow-brown ANDRADITE GARNET, and brown TOURMALINE, both of which are fusible.

Composition. Magnesium fluosilicate, $Mg(OH,F)_2 \cdot 2MgSiO_4$. May contain some Ti, Al, Fe, Mn.

Tests. Infusible. Yields water in closed tube. Dissolves in boiling HCl, and gelatinizes on evaporation.

Crystallization. Monoclinic; prismatic class, $\frac{2}{m}$. Norbergite and humite are orthorhombic. Usually as embedded grains. Small crystals are complex and often rounded. Also massive.

A   x 0.65

TOPAZ

Al$_2$SiO$_4$(F, OH)$_2$

Yellow-brown striated prismatic topaz crystal: Brazil.

B   x 0.95

TOPAZ

Colorless topaz crystals showing perfect basal cleavage: Omi, Japan.

C   x 0.66

WILLEMITE

Zn$_2$SiO$_4$

Massive yellow-green willemite associated with white calcite, black franklinite, and red zincite: Franklin, New Jersey.

D   x 0.62

PHENAKITE

Be$_2$SiO$_4$

Aggregate of complex flat phenakite crystals: Minas Gerais, Brazil.

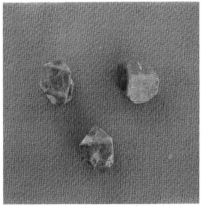

E   x 1.0

ZIRCON

ZrSiO$_4$

Small zircon crystals; square tetragonal prism and bipyramid: Buncombe County, North Carolina.

F   x 1.0

WILLEMITE (TROOSTITE)

Six-sided troostite crystal in white calcite: Franklin, New Jersey.

Plate 47. **SILICATES** ■ **UNIT STRUCTURES** OLIVINE, CHONDRODITE, GARNETS

A   x 0.90

OLIVINE
(Mg, Fe)$_2$SiO$_4$

Massive granular olivine (dunite) showing typical olive green color: Webster, North Carolina.

B   x 0.90

CHONDRODITE
Mg(OH, F)$_2$ · 2MgSiO$_4$

Brownish-yellow rounded chondrodite crystal in crystalline limestone: Franklin Furnace, New Jersey.

C   x 0.77

GARNET (ALMANDINE)
Fe$_3$Al$_2$(SiO$_4$)$_3$

Deep purple-red almandine garnet crystals (dodecahedron and trapezohedron): Fort Wrangell, Alaska.

D   x 0.66

GARNET (ALMANDINE)

Deep red almandine garnet: Gore Mountain, New York.

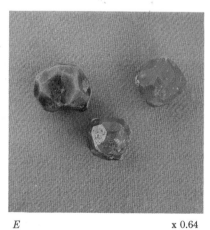

E   x 0.64

GARNET (GROSSULARITE)
Ca$_3$Al$_2$(SiO$_4$)$_3$

Light green grossularite garnet crystals: Yakutian, Russia.

F   x 0.80

GARNET (GROSSULARITE)

Pink grossularite garnet crystals with yellow chondrodite in white crystalline limestone: Morelos, Mexico.

Alteration. To serpentine, chlorite, brucite, magnesite.

Occurrence. Found in metamorphosed limestones and dolomites. Associated minerals include garnet, wollastonite, idocrase, serpentine, spinel, magnetite, pyrrhotite, calcite, olivine, diopside, and phlogopite. In the United States, chondrodite occurs in Franklin, New Jersey (Plate 47, facing p. 291) and as crystals associated with magnetite and serpentine at the Tilly Foster Mine, Brewster, New York. Notable localities include Monte Somma, Italy, and Kafveltorp, Sweden.

GARNET GROUP

Isometric

	Sp. Gr.
Pyrope, $Mg_3Al_2(SiO_4)_3$	3.58
Almandine, $Fe_3Al_2(SiO_4)_3$	4.31
Spessartite, $Mn_3Al_2(SiO_4)_3$	4.19
Grossularite, $Ca_3Al_2(SiO_4)_3$	3.59
Andradite, $Ca_3Fe_2(SiO_4)_3$	3.85
Uvarovite, $Ca_3Cr_2(SiO_4)_3$	3.90

Hardness $6\frac{1}{2}$–$7\frac{1}{2}$

Summary

Garnet is a common mineral, frequently found in well-formed dodecahedral or trapezohedral crystals (Figure 8-63 and Plates 3 and 4 [facing pp. 18 and 19], and Plate 47 [facing p. 291]). *Color:* red, green, brown, pink, yellow, white, black. *Luster:* vitreous. *Cleavage:* none, but may show parting; transparent to translucent. Commonly alters to chlorite (Plate 3). CRYSTALS AND COLOR ARE CHARACTERISTIC. (See Table 8-4.)

Tests. Fusible; but emerald green uvarovite, is almost infusible.

Composition. The compositions of garnets are rarely quite as simple as listed above, because of extensive atomic substitution.

Crystallization. Isometric; hexoctahedral class, $\frac{2}{m}\bar{3}\frac{4}{m}$. Dodecahedral and trapezohedral crystals are common for most varieties of garnet. Pyrope usually occurs in rounded grains rather than as crystals. Grossularite occurs as crystals and also in green masses called "South African Jade"; these resemble jadeite and massive green idocrase. Bright emerald green uvarovite usually occurs as small dodecahedrons. Almandine, spessartite, and andradite also form good crystals.

Occurrence. Garnet is one of the most common minerals, and therefore only a few important locations are listed for each of the garnet species.

ALMANDINE is the commonest specie of garnet, and usually occurs in metamorphic rocks, such as schist and gneiss; also in pegmatites. Large dark red to purplish-red crystals are common. Included oriented needle-like crystals of hornblende or augite produce a star effect in some almandine garnet. Almost black 4-rayed star crystals occur in India. Star crystals are

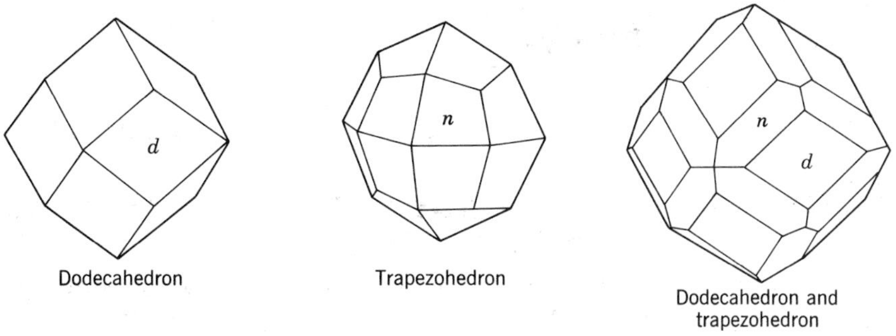

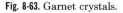

Dodecahedron   Trapezohedron   Dodecahedron and trapezohedron

Fig. 8-63. Garnet crystals.

Table 8-4

NAME	COLOR	COMMON OCCURRENCE	MINERAL ASSOCIATIONS
PYROPE	Deep vermilion	Ultra basic igneous rocks such as peridotites or kimberlites	Diamond, serpentine, olivine, pyroxene, spinel
ALMANDINE	Purplish-red	Usually in metamorphic rocks such as schists and gneisses	Andalusite, kyanite, staurolite, sillimanite
SPESSARTITE	Rich shades of brown, orange, orange-red, reddish-brown, yellowish-brown	Pegmatites and manganese-rich assemblages	Quartz, tourmaline, topaz, rhodonite, sphalerite
GROSSULARITE	Yellow, green, white, pink, cinnamon, light brown (essonite)	Metamorphosed limestones	Idocrase, diopside, wollastonite, epidote, calcite
ANDRADITE	Brown, reddish-brown, greenish-brown, black (melanite), apple green (demantoid), honey-yellow (topazolite)	Contact metamorphosed limestones; syenites; demantoid and topazolite varieties occur mainly in serpentine and chlorite schists.	
UVAROVITE	Deep emerald green	Chromium deposits	Chromite, serpentine

also found in Benewah County, Idaho. Good quality gem almandine crystals are found in Tanganyika; Ceylon; and Madagascar. Deep lustrous purplish-red crystals occur in Fort Wrangell, Alaska (Plate 47, facing p. 291). Large almandine crystals are mined for use as abrasives at Gore Mountain, North Creek, New York.

PYROPE OR PYROPE-ALMANDINE usually occurs in ultrabasic igneous rocks, such as peridotites and serpentine derived from altered peridotites. Sometimes in schists and gneisses. Pyrope rarely occurs as crystals. Deep red gem pyrope grains occur in the Kimberley diamond mines of South Africa, and are sometimes referred to as "Kimberley rubies" or "Cape rubies." Rounded grains also occur in peridote in the diamond-bearing pipes of Arkansas. Pyrope is also found in Kentucky, Utah, Arizona, and New Mexico.

SPESSARTITE crystals are usually a rich shade of brown or orange (orange-brown, orange-yellow, orange-red, reddish-brown). Good crystals of spessartite occur in cavities in granite pegmatites. Spessartite is also found as masses associated with pink rhodonite in manganese-rich metamorphic rocks. Large orange-brown crystals occur with black tourmaline, albite and quartz in the pegmatites of the Ramona district, San Diego County, California. Orange-brown crystals, some four inches in diameter, occur in white platy albite (cleavelandite) in the pegmatites of Amelia County, Virginia. Large, orange gem crystals occur in Minas Gerais, Brazil; also in Madagascar.

GROSSULARITE, the lighest colored of all the garnets, commonly occurs in contact-metamorphosed impure limestones associated with idocrase, wollastonite, diopside, and epidote. Light pink, or white dodecahedral crystals, some over five inches across, occur in Morelos, Mexico (Plate 47, facing p. 291). Good olive green crystals are found in Siberia. Light yellow to brownish-orange crystals are called *hesson-*

ites or *essonites* and are common in the gem gravels of Ceylon. Massive green grossularite associated with black chromite occurs in the Transvaal, South Africa, and is commonly referred to as "South African jade" or "Transvaal jade." Minute black crystals, often imbedded in the grossularite, help distinguish it from green jadeite. Massive pink grossularite also occurs in the Transvaal, but is much less common than the green variety.

ANDRADITE occurs in contact metamorphosed limestones. The varieties *demantoid* (yellow-green to apple green) and *melanite* (black) occur in serpentine. Bright apple green small *demantoid* crystals are found in serpentine asbestos near the Bobrovka River in the Ural Mountains. *Melanite* occurs in altered serpentine in San Benito County, California, and *schorlomite,* a black titanium-rich andradite is found in Magnet Cove, Arkansas. Large brown, yellowish-brown, and black andradite crystals occur in the zinc deposits of Franklin, New Jersey.

UVAROVITE is a relatively rare garnet, and is usually intimately associated with chromite. Small bright emerald green uvarovite crystals occur in narrow crevices in chromite in the Ural Mountains, U.S.S.R. The crystals are usually small ($\frac{1}{8}$ inch across), and are commonly associated with the violet chlorite (kämmererite). Unusually large uvarovite crystals, up to an inch across, have been found at Outokumpu, Finland. The uvarovite crystals occur with associated chromium minerals in a copper deposit in metamorphosed limestone.

DATOLITE

CaB(SiO₄)(OH)   Monoclinic
     Hardness 5–5½   Sp. Gr. 2.9–3.0

Summary

Datolite is found chiefly as a secondary mineral in cavities in basalt associated with prehnite, calcite, and zeolites. Commonly found as complex, almost equidimensional glassy crystals with a faint greenish tinge. Also occurs in dull white, opaque, porcelain-like masses associated with copper. *Color:* colorless, pale greenish. Opaque datolite is commonly white, but is often stained pink, yellow, or brown, and is often mottled. *Luster:* vitreous (crystals), dull (fine-grained masses). *Streak:* white. *Cleavage:* none. *Fracture:* conchoidal to uneven; transparent to opaque. Sometimes luminesces blue in ultraviolet light. CLEAR GLASSY COMPLEX CRYSTALS (Figure 8-64) WITH A PALE GREENISH TINGE ASSOCIATED WITH ZEOLITES ARE DISTINCTIVE. Massive porcelaneous datolite resembles WHITE CHALCEDONY, but is softer, fusible, and often contains inclusions of copper.

Composition. Calcium borosilicate, CaB(SiO₄)(OH). Usually very little or no variation in composition. Minor amounts of Al, Fe, Mn, Mg, Na, and K are sometimes present.

Tests. Swells and fuses very easily to a clear glass, and gives a green boron flame. Yields water in closed tube.

Crystallization. Monoclinic; prismatic class, $\frac{2}{m}$. Complex, almost equidimensional crystals are common. Also in compact, fine-grained masses.

Occurrence. Usually as a secondary mineral in cavities in basic lavas such as basalt. Associated minerals include prehnite, calcite, apopyllite, stilbite, heulandite, natrolite, and analcite. Also occurs in the Michigan copper deposits around Lake Superior, where crystals and dull porcelain-like masses occur associated with copper. Datolite is common in cavities in the dark trap rocks of Paterson, New Jersey. Good crystals also occur in the Harz Mountains, Germany; in the Italian Alps; and in Arendal, Norway.

Use. As a minor gem material.

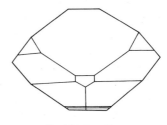

Fig. 8-64. Datolite.

COMBINED PAIRS AND UNITS

Epidote Series
    Epidote
    Clinozoisite
    Piemontite
    Idocrase (Vesuvianite)

PAIR STRUCTURE

Hemimorphite

EPIDOTE SERIES

                                  Monoclinic

Epidote, $Ca_2(Al,Fe)_3(SiO_4)_3(OH)$
Clinozoisite, $Ca_2Al_3(SiO_4)_3(OH)$
Piemontite, $Ca_2(Al,Fe,Mn)_3(SiO_4)_3(OH)$
    Hardness $6-6\frac{1}{2}$ Sp. Gr. 3.2–3.5

Summary

Commonly in elongated striated prismatic crystals or in pistachio-green masses (Plate 48 (*A*), facing p. 306). *Color:* In EPIDOTE—yellow-green, brownish-green, or green-black. In CLINOZOISITE—gray, light yellow, or light green; and in PIEMONTITE—violet-red, red-brown, or black. *Luster:* vitreous, pearly on cleavage. *Cleavage:* perfect basal; transparent to translucent. PISTACHIO-GREEN COLOR AND ONE PERFECT CLEAVAGE ARE CHARACTERISTIC OF EPIDOTE. Transparent epidote crystals are strongly pleochroic, showing green and brown colors as the crystal is viewed in different directions.

Composition. Epidote is a solid solution series consisting of *clinozoisite*, $Ca_2Al_3(SiO_4)_3(OH)$; *epidote,* $Ca_2(Al,Fe)_3(SiO_4)_3(OH)$; and *piemontite,* $Ca_2(Al,Fe,Mn)_3(SiO_4)_3(OH)$.

Tests. Swells and fuses to a bubbly glass.

Crystallization. Monoclinic; prismatic class, $\frac{2}{m}$. Commonly in elongated striated crystals; also massive. Clinozoisite is dimorphous with orthorhombic zoisite.

Alteration. From the hydrothermal alteration of plagioclase (saussuritization). Also from olivine, hornblende, scapolite, augite, biotite, and garnet.

Occurrence. Epidote (Figure 8-65) is usually formed by the metamorphism of calcium-rich igneous and sedimentary rocks. Commonly associated with chlorite, albite, and actinolite in gneisses and schists. Also in contact-metamorphosed limestones at times with scheelite; in pegmatites; and in basalt cavities with prehnite and zeolites. Larger, greenish-black, elongated crystals have come from Knappenwand, Austria, and well-formed, short prismatic crystals have come from Prince of Wales Island, Alaska (Plate 48(*D*), facing p. 306). Commonly occurs with garnet in California. In the United States, crystals up to a foot long have been recorded from Adams County, Idaho.

Use. Minor gem stone.

ZOISITE

$Ca_2Al_3(SiO_4)_3(OH)$ Orthorhombic
    Hardness 6   Sp. Gr. 3.15–3.36

Summary

Usually in aggregates of long, gray striated crystals or columnar aggregates (Plate 48(*C*), facing p. 306), in calcium-rich mica schists. *Color:* usually gray or greenish-brown; sometimes bright pink (*thulite*). *Luster:* vitreous, pearly on cleavage faces. *Cleavage:* perfect side pinacoid; translucent. DISTINGUISHABLE FROM SOME OF THE AMPHIBOLES BY SINGLE PERFECT CLEAVAGE WITH PEARLY LUSTER ON THE CLEAVAGE FACES.

Fig. 8-65. Epidote crystal.

Composition. Hydrous calcium aluminum silicate, $Ca_2Al_3(SiO_4)_3(OH)$. Small amounts of manganese are present in thulite variety. Chromium-bearing zoisites are rare.

Tests. Swells and fuses easily to a bubbly mass. Thulite sometimes luminesces yellow in long-wave ultraviolet light.

Crystallization. Orthorhombic; dipyramidal class, $\frac{2}{m}\frac{2}{m}\frac{2}{m}$. Commonly in elongated prismatic crystals, and columnar aggregates. Also in masses and in aggregates of fine, interlocking, needle-like crystals. Zoisite is dimorphous with monoclinic *clinozoisite*.

Alteration. To anorthite at high temperatures. The hydrothermal alteration of plagioclase feldspars yields an alteration product called *saussurite,* composed of zoisite, epidote, clinozoisite, albite, and calcite.

Occurrence. Commonly found in schists. Also as a hydrothermal alteration product of calcium-rich plagioclase. Thulite is sometimes found in quartz veins or pegmatites. Attractive pink thulite specimens are found in Mitchell County, North Carolina, and at Tröndelag, Norway (Plate 48, facing p. 306). Deep red, tabular ruby crystals are embedded in a rare chromium-bearing green zoisite in Tanganyika.

IDOCRASE (Vesuvianite)

$Ca_{10}(Mg,Fe)_2Al_4(Si_2O_7)_2(SiO_4)_5(OH)_4$ Tetragonal
Hardness $6\frac{1}{2}$ Sp. Gr. 3.3–3.5

Summary

Commonly in prismatic crystals or striated columnar masses. The green, massive variety (californite) (Plate 48(*F*), facing p. 306) resembles jadeite. Commonly found in contact-metamorphosed limestones associated with diopside, grossularite garnet, brown tourmaline, sphene, chondrodite, wollastonite, and epidote. *Color:* usually brown or green; sometimes blue, red, or yellow. *Luster:* vitreous. *Streak:* white. *Cleavage:* poor prismatic; transparent to translucent.

PRISMATIC BROWN OR GREEN CRYSTALS WITH SQUARE CROSS

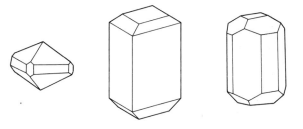

Fig. 8-66. Idocrase crystals.

SECTIONS ARE CHARACTERISTIC (Plate 5(*B*), facing p. 34). May resemble ZIRCON crystals, but zircon is heavier (Sp. Gr. 4.6–4.7), infusible, commonly luminesces orange, and is often found in igneous rocks.

Composition. Hydrous calcium, magnesium, iron, aluminum silicate. The composition is variable. May contain some B, F, Mn, Ti, Na, K, and Li.

Tests. Swells and fuses easily to a bubbly glass.

Crystallization. Tetragonal; ditetragonal-dipyramidal class, $\frac{4}{m}\frac{2}{m}\frac{2}{m}$. Crystals are prismatic or pyramidal (Figure 8-66). Also commonly in striated columnar aggregates. Sometimes massive.

Occurrence. Commonly found in contact-metamorphosed limestones. Also in chlorite schists, gneisses, and serpentine. Well-formed, clear green crystals have been found in Ala, Italy; and at Eden, Vermont. The massive, jade-like green variety known as "californite" occurs in Tulare, Siskiyou, and Fresno Counties in California. Bipyramidal crystals in light blue calcite also occur in California.

Use. A minor gem stone.

HEMIMORPHITE

$Zn_4Si_2O_7(OH)_2 \cdot H_2O$ Orthorhombic
Hardness $4\frac{1}{2}$–5   Sp. Gr. 3.4–3.5

Summary

Commonly in rounded aggregates with crystallized surfaces that resemble "segmented worms"

(Plate 48(E), facing p. 306). Found in the oxidized zone of zinc deposits. *Color:* white; sometimes stained blue, green, yellowish, or brown by copper or iron. *Luster:* vitreous. *Streak:* white. *Cleavage:* prismatic; transparent to translucent. Sometimes luminesces orange in long-wave ultraviolet light. ROUNDED GROUPS WITH RIDGED SURFACES, FORMED BY THE EDGES OF MINUTE DIVERGENT PLATE-LIKE CRYSTALS ARE CHARACTERISTIC. Resembles botryoidal SMITHSONITE and PREHNITE. Distinguishable from SMITHSONITE ($ZnCO_3$) by lack of effervescence in HCl and is heavier than yellow-green PREHNITE (Sp. Gr. 2.8–2.9).

Composition. Hydrous zinc silicate, $Zn_4Si_2O_7(OH)_2 \cdot H_2O$ (67.5% ZnO, 25.0% SiO_2, 7.5% H_2O). May contain small amounts of Al, Fe, or Pb.

Tests. Fuses with difficulty. Gives test for zinc.

Crystallization. Orthorhombic; pyramidal class, $m\ m\ 2$. Usually in rounded masses with crystallized surfaces. Also in smooth botryoidal aggregates and earthy masses. Seldom in large tabular crystals with opposite ends differing from each other (hemimorphic).

Alteration. From other zinc minerals such as sphalerite.

Occurrence. Hemimorphite is a secondary mineral derived from the alteration of zinc ores in the oxidized zone of zinc deposits. Commonly associated with smithsonite, sphalerite, anglesite, galena, and cerussite. Worm-like aggregates have been found at Sterling Hill, New Jersey (Plate 48), and good hemimorphic crystals from Mapimi, Mexico, and from Algeria. Good specimens have also been found in Leadville, Colorado, and in Elkhorn, Montana.

Use. Ore of zinc.

SUBSATURATES

Staurolite
Andalusite
Kyanite
Sillimanite
Sphene

UNCERTAIN STRUCTURE

Dumortierite

STAUROLITE

$(Fe^{2+},Mg)_2(Al,Fe^{3+})_9O_6(SiO_4)_4(O,OH)_2$ Monoclinic
(pseudo-orthorhombic)
Hardness 7–7½   Sp. Gr. 3.7–3.8

Summary

Commonly in brown cruciform twins with a rough surface (Plate 8(D), facing p. 51; and Plate 49(F), facing p. 307). Hence, the name is derived from the Greek words *stauros* and *lithos* meaning stone cross. Found in metamorphic rocks such as schists and gneisses; associated with kyanite, muscovite, and almandine garnet. *Color:* dark brown. *Luster:* vitreous to dull. *Streak:* white. *Cleavage:* moderate pinacoidal; translucent. BROWN, CROSS-TWINNED, OR FLAT ELONGATED CRYSTALS IN MICA SCHISTS ARE DISTINCTIVE.

Composition. Iron magnesium aluminum silicate. Some rare varieties may contain small amounts of zinc, cobalt, or nickel.

Tests. Infusible and insoluble.

Crystallization. Monoclinic; prismatic class, $\frac{2}{m}$ (pseudo-orthorhombic). Commonly in cruciform twins with the two individuals crossing at almost 90° or 60°. Also in elongated tabular crystals (Figure 8-67).

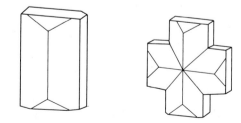

Fig. 8-67. Staurolite.

Alteration. To kaolinite. Under higher grades of metamorphism, staurolite may be replaced by kyanite and almandine garnet, or by sillimanite and almandine.

Occurrence. Usually occurs in mica schists. Also in gneisses and slates. Well-formed cross twins occur in Fannin County, Georgia, and in Fairfax County, Virginia, where they are called "fairy stones" and are sold as good-luck charms. Elongated, lustrous red-brown single crystals of staurolite occur with blue kyanite in Monte Campione, Switzerland.

Use. Minor gem stone.

ANDALUSITE

$Al_2(SiO_4)O$ Orthorhombic
Hardness $6\frac{1}{2}$–$7\frac{1}{2}$ Sp. Gr. 3.1–3.2

Summary

Commonly in dull, rough, prismatic crystals with almost square cross sections (Plate 49(*C*), facing p. 307). CHIASTOLITE is a variety of andalusite; in cross section, shows a black cruciform design that is formed as a result of enclosed carbonaceous material which was pushed aside into definite areas as the crystal grew in metamorphosed shales (Plate 49(*B*), facing p. 307). Alters readily to sericite. *Color:* gray, brownish-pink, white, rose-red, red-brown, green, yellow, violet. *Luster:* vitreous, dull when altered. *Streak:* white. *Cleavage:* good prismatic; strong pleochroism (green and red-brown); transparent to translucent. ROUGH PRISMATIC CRYSTALS WITH MUSCOVITE (SERICITE) SURFACE ALTERATION, AND ALMOST SQUARE CROSS SECTIONS ARE CHARACTERISTIC. CHIASTOLITE VARIETY IS DISTINCTIVE.

Composition. Aluminum silicate, $Al_2(SiO_4)O$. May contain small amounts of iron or manganese.

Tests. Infusible and insoluble.

Crystallization. Orthorhombic; dipyramidal class, $\frac{2}{m}\frac{2}{m}\frac{2}{m}$. Usually in elongated rough crystals. Prisms are sometimes rounded (cigar-shaped). Also massive.

Alteration. Alters readily to sericite. Inverts to polymorphs sillimanite or kyanite by a rise in temperature or pressure. At elevated temperatures, dissociates to mullite and quartz.

Occurrence. In mica schists which have usually been formed from shales by contact metamorphism. Associated minerals include garnet, muscovite, kyanite, sillimanite, cordierite, and tourmaline. Chiastolite (Plate 49, facing p. 307) is common in California, and at Sterling and Lancaster, Massachusetts. Gem varieties of andalusite occur in Ceylon, and also as waterworn crystals in Minas Gerais, Brazil. The cut stone shows both green and red colors as a result of the differential absorption of light waves in different directions (pleochroism). In Mono County, California, andalusite has been mined for refractory material.

Use. Refractory material and as a minor gem stone.

KYANITE

$Al_2(SiO_4)O$ Triclinic
Hardness 5 (parallel to   Sp. Gr. 3.53–3.65
length of crystal)
7 (across length)

Summary

Usually in bladed crystals with patchy blue or greenish color distribution (Plate 49(*A*), facing p. 307). Found in schists and gneisses associated with staurolite, garnet, rutile, lazulite, and corundum. *Color:* blue, green, white (patchy distribution); sometimes pink, yellow, or black. *Luster:* vitreous to pearly. *Streak:* white. *Cleavage:* perfect pinacoidal; transparent to translucent. BLADED CRYSTALS, COLOR, AND VARIABLE HARDNESS ARE DISTINCTIVE.

Composition. Aluminum silicate, $Al_2(SiO_4)O$. May contain a little ferric iron.

Tests. Infusible and insoluble.

Crystallization. Triclinic; pinacoidal class, $\bar{1}$. Usually in long bladed crystals. Also in bladed aggregates.

Alteration. To muscovite, pyrophyllite, or kaolinite. May invert to the polymorphs, andalusite or sillimanite by a change in temperature-pressure conditions. At about 1300°C and atmospheric pressure, converts to the refractory material called mullite.

Occurrence. Usually found in metamorphic rocks such as schists and gneisses. Large crystals, some 6 in. long, occur in Minas Gerais, Brazil. Large crystals occur in St. Gotthard, Switzerland (Plate 49(*A*)). Long blade-like blue-green polysynthetically twinned crystals up to a foot in length and an inch wide occur in the Machakos District of Kenya, East Africa. Kyanite is mined as a refractory mineral in Kenya and in India. In the United States, it is mined in North Carolina, in Georgia, and in Virginia.

Use. In heat-resistant ceramics, and as a minor gem mineral.

SILLIMANITE

$Al_2(SiO_4)O$ Orthorhombic
   Hardness $6\frac{1}{2}-7\frac{1}{2}$ Sp. Gr. 3.23–3.27

Summary

Usually occurs in aggregates of slender prismatic crystals, or fibrous masses in high temperature metamorphic rocks. Andalusite and kyanite are polymorphs. *Color:* white, colorless, sometimes yellow, brown, green, gray, or bluish. *Luster:* vitreous to silky. *Streak:* white. *Cleavage:* perfect pinacoidal; transparent to translucent. FIBROUS AGGREGATES ASSOCIATED WITH ANDALUSITE AND CORUNDUM IN GNEISSES AND SCHISTS ARE CHARACTERISTIC. May resemble fibrous TREMOLITE or WOLLASTONITE, but is infusible.

Composition. Aluminum silicate, $Al_2(SiO_4)O$. May contain a little iron.

Tests. Infusible and insoluble in acids.

Crystallization. Orthorhombic; dipyramidal class, $\frac{2}{m}\frac{2}{m}\frac{2}{m}$. Usually in fibrous masses. Also aggregates

of slender prismatic crystals. Sometimes in tough compact masses.

Alteration. Muscovite, pyrophyllite, kaolinite. May convert to kyanite under stress or high-pressure conditions. Above 1545°C, sillimanite converts to mullite $(3Al_2O_3 \cdot 2SiO_2)$ plus a liquid; on further heating above 1810°C mullite breaks down to corundum (Al_2O_3) plus a liquid.[*]

Occurrence. Usually found in gneisses and schists. Associated minerals include corundum, almandine, zircon, cordierite, andalusite, muscovite, and quartz. Fibrous masses are common at Worcester, Massachusetts, and at Norwich, Connecticut.

Use. In heat-resistant ceramics.

SPHENE

$CaTi(SiO_4)O$ Monoclinic
   Hardness $5-5\frac{1}{2}$ Sp. Gr. 3.45–3.55

Summary

The name sphene is derived from the Greek word for a wedge, *sphenos,* in allusion to sphene's typically flat wedge-shaped crystals (Figure 8-68) (Plate 7(*B*), facing p. 50, and Plate 49(*D*), facing p. 307). *Color:* usually yellow-green or brown; also colorless or black. *Luster:* adamantine to resinous. *Cleavage:* good prismatic; transparent to translucent. COLOR, LUSTER, AND CRYSTALS WITH WEDGE-SHAPED CROSS SECTIONS ARE CHARACTERISTIC. Distinguishable from wedge-shaped AXINITE crystals by inferior hardness. (H. axinite = $6\frac{1}{2}-7$.)

Composition. Calcium titanium silicate, $CaTi(SiO_4)O$. May contain some cerium, yttrium, manganese, thorium, aluminum, strontium, barium, or iron.

Tests. Fuses with intumescence to a glass. Decomposed by H_2SO_4. Gives test for titanium.

Crystallization. Monoclinic; prismatic class, $\frac{2}{m}$. Commonly in wedge-shaped crystals or in twinned crystals.

---

[*]W. A. Deer, R. A. Howie, and J. Zussman, *Rock Forming Minerals,* Vol. 1, John Wiley and Sons, New York, 1962, p. 124.

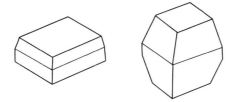

Fig. 8-68. Sphene crystals.

Sometimes as prismatic crystals, disseminated grains, or lamellar masses.

Alteration. Alters to anatase; or, less frequently, rutile.

Occurrence. Sphene is a common accessory mineral in igneous rocks. Also common in metamorphic rocks such as gneisses and schists; and in contact-metamorphosed limestones. Sphene is mined as an ore of titanium in the Kola Peninsula, the U.S.S.R. Good gem crystals occur in Zillertal, Tyrol; in St. Gotthard, Switzerland; and in Mexico. Large, dark brown crystals occur in crystalline limestones in Renfrew, Ontario, Canada. In the United States, crystals are common in St. Lawrence County, New York, and in Riverside, California.

Use. Minor ore of titanium; as gem stone.

DUMORTIERITE

$(Al,Fe)_7BSi_3O_{18}$ Orthorhombic
Hardness 7   Sp. Gr. 3.3–3.4

Summary

Usually occurs as violet or blue fibrous masses (Plate 49(*E*), facing p. 307). Found in gneisses and schists, less frequently in pegmatites. *Color:* violet, blue, pink. *Luster:* vitreous. *Streak:* white. *Cleavage:* good pinacoidal; translucent. **COLOR AND FIBROUS HABIT ARE DISTINCTIVE.** Sometimes luminesces purple under ultraviolet light.

Composition. Aluminum iron borosilicate, $(Al,Fe)_7BSi_3O_{18}$.

Tests. Infusible.

Crystallization. Orthorhombic; dipyramidal class, $\frac{2}{m}\frac{2}{m}\frac{2}{m}$ Usually in fine fibrous masses. Small prismatic crystals are rare.

Alteration. To mica. At high temperatures, transforms to refractory the material *mullite*.

Occurrence. Usually in metamorphic rocks such as gneisses and schists. Sometimes in pegmatites. Dumortierite is mined at Oreana, Nevada, for use in the manufacture of spark plug porcelain. Also common in Dehesa, San Diego County, California; Minas Gerais, Brazil; Lyons, France; and Madagascar. Bright blue needles of dumortierite disseminated in quartz may resemble massive blue sodalite.

REFERENCES

1. A. J. Cohen, "Anisotropic Color Centers in α-Quartz" (Part 1, Smoky Quartz), *J. Chem. Phy.,* Vol. 25, 1956, p. 908.
   ———, "Color Centers in the α-Quartz Called Amethyst," *Am. Mineralogist,* Vol. 41, 1956, p. 874.
2. W. A. Deer, R. A. Howie, and J. Zussman, *Rock Forming Minerals,* John Wiley and Sons, New York, 1963.
3. Dunn, *Econ. Geo.,* Vol. 48, pp. 58, 729.
4. C. Frondel, *Dana's System of Mineralogy,* 7th ed., Vol. 3 (Silica Minerals), John Wiley and Sons, New York, 1962.
5. C. Frondel and C. Hurlbut, "Determination of the Atomic Weight of Silicon by Physical Measurements on Quartz," *J. Chem. Phy.,* Vol. 23, 1955, p. 1215.
6. E. F. Holden, "The Cause of Color in Smoky Quartz and Amethyst, *Am. Mineralogist,* Vol. 10, 1925, p. 203.
   ———, "The Color of Three Varieties of Quartz," *Am. Mineralogist,* Vol. 8, 1923, p. 117.
7. O. P. Komarov, "On a New Find of a Gigantic Quartz Crystal," *Mém. Soc. Russe Min.,* Vol. 80, 1951, p. 153.
8. Parker, *Am. Mineralogist,* Vol. 47, 1961, p. 1201.
9. K. Przibram, *Irradiation Colors and Luminescence,* Pergamon Press, London, 1956.
10. *Smithsonian Inst. Ann. Rep. for 1956, 1957,* p. 329.
11. Spencer and Percival, *Econ. Geo.,* Vol. 47, 1952, p. 365.
12. L. I. Tsinober and L. G. Chentsova, *Kristallografiya,* Vol. 4, 1960, p. 593.
13. A. Van Valkenburg and C. E. Weir, "Beryl Studies $(3BeO \cdot Al_2O_3 \cdot 6SiO_2)$," *Bull. Geo. Soc. Am.,* Vol. 68, 1957, p. 1808 (abstract).

Appendix 1.

Meteorites—"Rocks" from Outer Space

Meteorites are fragments that have fallen on the earth from outer space, and range in size from dust particles to masses weighing many tons. Materials represented are nickel-iron alloys (iron meteorites), silicates (stony meteorites), nickel-iron and silicates (stony-iron meteorites), minor amounts of carbonaceous matter, and glass. At present, meteorites constitute our most tangible source of extraterrestrial material, and represent fragmentary bodies from the solar system.

Age measurements have been made on a considerable number of meteorite groups by the lead isotope, the potassium-argon, and the rubidium-strontium methods. These give concordant results, and indicate that the crystallization of the different meteorite types took place some $4\frac{1}{2}$ billion years ago.

With few exceptions, the minerals found in meteorites also occur in terrestrial rocks. Likewise, almost all known elements have been detected spectroscopically; no new elements have been found that do not exist on the earth. The most abundant minerals are olivine, pyroxenes, and nickel-iron alloys. Meteorites are usually divided into three major groups:

1. **STONY METEORITES** or **AEROLITES** (silicate minerals usually with some nickel-iron metal)
   (*a*) Chondrites—Usually containing small rounded bodies called chondrules.

(*b*) Achondrites—Without chondrules, and poor in nickel-iron.
2. **IRON METEORITES** or **SIDERITES** (nickel-iron with small amounts of accessory minerals)
3. **STONY-IRON METEORITES** or **SIDEROLITES** (approximately 50% nickel-iron and 50% silicate minerals)

Carbonaceous Chondrites

The carbonaceous chondrites account for about 4 percent of observed meteorite falls. They do not resemble other meteorites, and are small (the largest measure only about 8 or 9 in. across), sooty black, crumbly objects that are quickly destroyed by weathering. The most primitive (Type 1) carbonaceous chondrite contains no chondrules, few recognizable minerals, and considerable amounts of volatile components (up to about 20 percent water, 7 percent sulfur, 5 percent carbon and hydrocarbons).

The iron in most carbonaceous chondrites is completely oxidized, in contrast to chondrites that contain metal particles of nickel-iron.

According to one theory, the chemical composition of the original dust cloud from which some authorities believe the earth, other planets, and meteorites formed is closely approximated by Type 1 carbonaceous chondrite.[2,3,4,7] Also, other types of meteorites could be derived from large quantities of this aggregated material by processes of heating, reduction, metamorphism, melting, and differentiation. According to a second theory (H. C. Urey) the evolutionary process has gone in the opposite direction, and the carbonaceous chondrites are secondary objects formed by alteration.

The presence of "organic" compounds in carbonaceous chondrites has also aroused much interest. At present, it is still a puzzling question whether these organic compounds are the remains of extraterrestrial life, or whether they are nonbiological features.

Chondrites

The chondrites make up about 51 percent of known meteorites. They usually contain small rounded bodies or chondrules about 1 mm. in diameter, commonly composed of olivine and pyroxene. These chondrules are embedded in a matrix that is usually composed of a fragmentary mixture of the same minerals. Small grains of nickel-iron are usually distributed through the matrix and sometimes in the chondrules. The texture of the chondrites is quite un-

Fig. A-1. Stony meteorite, chondrite (Miller, Arkansas). A thin fusion crust with ablation furrows covers the surface (courtesy of the American Museum—Hayden Planetarium).

like any terrestrial rock. An average nickel-iron content of about 12 percent further differentiates them from terrestrial rocks. The surface of these meteorites consists of a thin layer of glass underlaid by an impregnation zone in which glass has penetrated between the granular constituents (Figure A-1).

Achondrites

Achondrites (without chondrules) are less common than the chondrites, and account for only about 4.2 percent of recognized meteorites. They are structurally similar to terrestrial rocks but frequently contain small amounts of nickel-iron (about 1 percent). Achondrites range in structure from fine-grained masses to coarse crystal aggregates that may resemble certain pegmatites.

Iron Meteorites (Siderites)

Iron meteorites, the second largest group, are composed mainly of nickel-iron alloys, usually containing a small amount of cobalt. When an iron meteorite enters the atmosphere, the atmospheric friction causes weak portions to burn, resulting in the formation of surface pits often referred to as "thumb marks" (Figure A-2).

As friction reduces velocity, the glowing surface cools to form a black FUSION CRUST composed of magnetite (Fe_3O_4) underlaid by a narrow alteration layer of fine granular iron. The magnetite crust is usually less than one millimeter thick. Because of the short exposure to heat on entry into the earth's atmosphere, the extremely cold interior of a fair sized meteorite absorbs little heat, and the internal structure usually remains relatively unchanged. Small meteorites, however, are often so disintegrated by their passage through the atmosphere that they fall to the earth as dust.

Some nickel-iron meteorites (known as OCTAHEDRITES) when sectioned, polished, and etched with acid show a lattice of intersecting crystals known as a WIDMANSTÄTTEN STRUCTURE (Figure A-3). This lattice-like pattern is formed by the intergrowth of two nickel-iron alloys; *nickel-poor* KAMACITE (about 5.5

percent nickel, body-centered cubic structure), and *nickel-rich* TAENITE (ranging from 27 to 65 percent nickel, face-centered cubic structure). The Widmanstätten pattern results from the unmixing of an original high-temperature solid solution (see exsolution, p. 55). As the solid nickel-iron alloy is slowly cooled, kamacite is precipitated preferentially along the octahedral planes of the original homogeneous high-temperature taenite phase. The kamasite bands are bordered by thin taenite lamallae. The presence of a coarse Widmanstätten pattern suggests that precipitation occurred under conditions of VERY SLOW COOLING, controlled by diffusion. PLESSITE, an intimate mixture of kamacite and taenite, often fills the angular spaces between the kamacite and taenite lamallae.

The octahedrites commonly contain about 6 to 14 percent of nickel, and there is a general relationship between the amount of nickel, the presence of the

Fig. A-2. Iron meteorite showing surface pits (Mexico). (Courtesy of the American Museum—Hayden Planetarium.)

Widmanstätten structure, and the coarseness of the bands. Generally, the pattern is FINE in octahedrites of high nickel content; but other factors (not all of which are yet understood) such as the influence of phosphorous or cobalt impurities may play an important role in the formation of the Widmanstätten structure.

Nickel-iron meteorites with less than 6 percent of nickel, known as HEXAHEDRITES, do not show a Widmanstätten structure for they are essentially of only one component, kamacite. Nickel-rich ATAXITES also show no Widmanstätten structure. At roughly above 27 percent nickel, all are taenite.

The most prevalent inclusions in nickel-iron meteorites are TROILITE, an iron sulfide (FeS); SCHREIBERSITE, a phosphide of iron, nickel, and cobalt $(Fe,Ni,Co)_3P$; COHENITE, a carbide of iron nickel and cobalt $(Fe,Ni,Co)_3C$; and GRAPHITE, carbon (C). Carbon in the form of diamond is rare. Among the common constituents of iron meteorites, only the mineral schreibersite has not been found in terrestrial rocks.

Iron meteorites can usually be identified by their fused pitted surfaces coated with black to brownish-black iron oxide, and by the presence of nickel. Meteoric iron is *always* alloyed with nickel.

Stony-iron Meteorites

The stone-irons are not common and are composed primarily of nickel-iron metal and silicate minerals in approximately equal amounts. The silicate is usually olivine, but may sometimes be pyroxene and plagioclase.

The PALLASITES are olivine stone-iron meteorites (Figure A-4). They are composed of more or less rounded olivine crystals, commonly about $\frac{1}{2}$–1 cm. across, which are surrounded by nickel-iron that frequently show a Widmanstätten structure. Small inclusions of troilite, FeS, and schreibersite, $(Fe,Ni,Co)_3P$, are often included in the nickel-iron.

It has been proposed that the pallasites were formed by the intrusion of a liquid-iron melt (containing considerable amounts of iron sulfide) into an accumulation of olivine crystals (i.e. an olivine achondrite), and that the metal phase of the pallasites represents samples of the parent melt from which the

Fig. A-3. Section of an iron meteorite showing fine Widmanstätten structure (Australia). (Courtesy of the American Museum—Hayden Planetarium.)

various types of iron meteorites differentiated (Lovering[2]).

The **BENCUBBIN** stone-iron meteorite from Australia contains an interesting mixture of both achondrite and chondrite fragments surrounded by nickel-iron metal.

Tektites

Tektites are small bottle green to blackish siliceous glass bodies that resemble obsidian (volcanic glass), but differ in chemical composition from all known terrestrial glasses. Common shapes of tektites are lenses, buttons, dumbells, pear shapes, and spindels. Although the origin of tektites is still disputed, their surfaces may be pitted, grooved, and molded in a manner that suggests a prolonged whirling motion. Tektites occur in the Nullarbor Plain of Southern Australia; in Bohemia, Moldavia, Borneo, Java, Tasmania; and in Texas, the U.S.A.

Size of Meteorites

The largest known meterite is the **HOBA IRON**, which measures $9 \times 9 \times 3\frac{1}{2}$ ft. and is estimated to weigh about 60 tons. This giant meteorite lies where

Fig. A-4. Polished surface of a pallasite composed of rounded olivine crystals; they appear dark but are enclosed in a light nickel-iron groundmass. The specimen is 10 cm. across (Ollague, Bolivia). (Courtesy of the American Museum—Hayden Planetarium.)

gases on impact; because of their great magnitude the earth's atmosphere is relatively ineffective in reducing their high velocities. These meteorites are known only by the craters they produce on the earth's surface and by the surviving fragments of the exploded meteorite.

A large crater believed to be produced by the impact of a meteorite is Meteor Crater near Winslow, Arizona. It measures $\frac{3}{4}$ of a mile across and is about 700 ft. deep. Thousands of meteorites have been found within $2\frac{1}{2}$ miles of the crater, whereas fused silica glass (lechatelierite) and high-density silica (coesite) occur within the crater. Meteor Crater was produced in prehistoric times, probably about 20,000 to 50,000 years ago, but meteoric craters have also been formed in recent years. In 1947, a huge meteorite produced a cluster of over 100 craters at Ussuri, Siberia, some of them being 40 ft. deep and 90 ft. wide.

it was found in 1920, near Grootfontein, South-West Africa. The second largest meteorite is the $36\frac{1}{2}$-ton AHNIGHITO IRON that was successfully transported from Cape York, Greenland, by Admiral Peary in 1897, and is now in the Hayden Planetarium, New York City.

No stony meteorite of comparable size has ever been found. The largest stony meteorite, found in 1948, is the NORTON (KANSAS) ACHONDRITE, described as weighing 2000 lb. This is the largest meteorite of any type recovered from an observed fireball. On August 2, 1946, a similar type of coarse-grained achondrite, weighing 155 lb., landed in a swimming pool at the Gage Ranch in Brewster County, Texas. Twenty-four persons were within a few hundred feet of the fall, and one of them actually saw the meteorite in flight.

CRATER-FORMING METEORITES. The largest meteorites that strike the earth are shattered or reduced to

REFERENCES

1. W. M. Latimer, "Astrochemical Problems in the Formation of the Earth," *Science 112,* 1950, p. 101.
2. J. F. Lovering, "The Evolution of Meteorites," *Researches on Meteorites,* John Wiley and Sons, New York, 1962.
3. B. Mason, *Meteorites,* John Wiley and Sons, New York, 1962.
4. B. Mason, "Organic Matter from Space," *Scientific American,* March, 1963.
5. H. H. Nininger, *Out of the Sky,* Dover Publications, New York, 1952.
6. S. H. Perry, "The Metallography of Meteoric Iron," *Bulletin 184,* Smithsonian Institution, United States National Museum, Government Printing Office, Washington, D.C., 1944.
7. A. E. Ringwood, "Present Status of the Chondritic Earth Model," *Researches on Meteorites,* John Wiley and Sons, New York, 1962.
8. H. C. Urey, *The Planets,* Yale University Press, New Haven, 1952.
9. H. C. Urey, "Primary and Secondary Objects," *J. Geophys. Res.,* Vol. 64, 1959.

INDEX

Boldface type is used for minerals that are described in detail; page numbers in **bold-face** type refer to the principal mineral descriptions.

A   x 0.66

EPIDOTE
$Ca_2(Al, Fe)_3(SiO_4)_3(OH)$

Massive epidote showing characteristic pistachio green color: Yancy County, North Carolina.

B   x 0.50

ZOISITE (THULITE)

Massive pink thulite: Leksirken, Tröndelag, Norway.

C   x 0.50

ZOISITE
$Ca_2Al_3(SiO_4)_3(OH)$

Columnar aggregate of grayish zoisite.

D   x 0.70

EPIDOTE

Short prismatic black epidote crystals: Prince of Wales Island, Alaska.

E   x 0.70

HEMIMORPHITE
$Zn_4Si_2O_7(OH)_2 \cdot H_2O$

Aggregates of white hemimorphite crystals which resemble "segmented worms": Sterling Hill, New Jersey.

F   x 0.65

IDOCRASE
$Ca_{10}(Mg, Fe)_2Al_4(Si_2O_7)_2(SiO_4)_5(OH)_4$

Massive green variety of idocrase known as *californite*. Resembles jade, and massive green grossularite garnet: California.

A   x 0.65

KYANITE
$Al_2(SiO_4)O$

Bladed kyanite crystals with uneven blue and whitish color distribution: St. Gotthard, Switzerland.

B   x 0.8

ANDALUSITE (CHIASTOLITE)
$Al_2(SiO_4)O$

Chiastolite variety of andalusite showing black maltese cross as a result of selectively distributed carbonaceous material: Modoc County, California.

C   x 0.82

ANDALUSITE

Rough elongated prismatic crystals with an almost square cross-section, coated with muscovite: Tyrol.

D   x 0.70

SPHENE
$CaTi(SiO_4)O$

Typical flat wedge-shaped crystals.

E   x 0.70

DUMORTIERITE
$(Al, Fe)_7BSi_3O_{18}$

Massive dumortierite with characteristic fibrous habit: Oreana, Nevada.

F   x 0.60

STAUROLITE
$(Fe^{2+}, Mg)_2(Al, Fe^{3+})_9O_6(SiO_4)_4(O, OH)_2$

Cruciform twin with rough surface: Charleston, New Hampshire.